Nanoscale Interface
for Organic Electronics

Nanoscale Interface
for Organic Electronics

Editors

Mitsumasa Iwamoto
Tokyo Institute of Technology, Japan

Young-Soo Kwon
Dong-A University, Korea

Takhee Lee
Gwangju Institute of Science and Technology, Korea

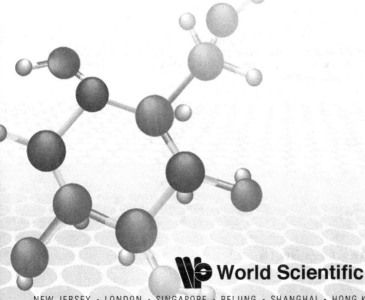

World Scientific

NEW JERSEY · LONDON · SINGAPORE · BEIJING · SHANGHAI · HONG KONG · TAIPEI · CHENNAI

Published by

World Scientific Publishing Co. Pte. Ltd.
5 Toh Tuck Link, Singapore 596224
USA office: 27 Warren Street, Suite 401-402, Hackensack, NJ 07601
UK office: 57 Shelton Street, Covent Garden, London WC2H 9HE

British Library Cataloguing-in-Publication Data
A catalogue record for this book is available from the British Library.

NANOSCALE INTERFACE FOR ORGANIC ELECTRONICS

Copyright © 2011 by World Scientific Publishing Co. Pte. Ltd.

All rights reserved. This book, or parts thereof, may not be reproduced in any form or by any means, electronic or mechanical, including photocopying, recording or any information storage and retrieval system now known or to be invented, without written permission from the Publisher.

For photocopying of material in this volume, please pay a copying fee through the Copyright Clearance Center, Inc., 222 Rosewood Drive, Danvers, MA 01923, USA. In this case permission to photocopy is not required from the publisher.

ISBN-13 978-981-4322-48-5
ISBN-10 981-4322-48-2

Printed in Singapore by B & Jo Enterprise Pte Ltd

PREFACE

Since 2001, groups of Japanese and Korean researchers have been discussing on the development of optoelectronic devices based on organic materials in view points of nanoscale materials and devices engineering. For this purpose, an annual conference called International Discussion and Conference on Nano Interface Controlled Electronic Devices (IDC-NICE) has been held in Korea and Japan alternately and the leading scientists have gathered in this conference to discuss research findings and suggest diverse practical methods to take advantage of research results. The first workshop was held at Gyeongju, Korea in 2001, and most recently the 9th conference was held at Chiba University, Chiba, Japan in 2009. The 10th conference will be held in Jeju, Korea in 2010. This book was aimed for celebrating the 10th anniversary of this conference and summarizing the achievements so far.

The conference particularly focuses on organic materials and devices as a future device technology. Since the invention of organic semiconductors, these materials have progressed into a variety device applications including organic switching elements, diodes, field-effect transistors, liquid crystal display elements, artificial muscles, etc. Towards more practical applications and technological advances in this field, understanding and designing the functional organic materials, and the control of their functions in the organic-organic and organic-inorganic junctions as the actual device platforms are required. Often, the interface at the junctions plays a critical role in the device functionalities and performances.

This book treats the important issues of interface control in organic devices in a wide range of applications that cover from electronics, displays, and sensors to biorelated devices. This book is composed of three parts:

Part 1. Nanoscale interface
Part 2. Molecular electronics
Part 3. Polymer electronics.

In Part 1, the charge injection, the effect of space-charge, and the contact resistances are discussed at the nanoscale interface in organic transistors which are used as a driving circuit element in the organic flexible displays. In Part 2, the charge conduction mechanisms in molecular-based electronics are mainly discussed. Scanning tunneling microscopic tool was used to investigate the charge conduction in the molecular memory, sensors as well as solid-state type molecular electronic devices. In Part 3, synthesis and deposition techniques to prepare polymer materials and devices are discussed for more variety of devices applications, such as organic optoelectronic devices and ID tags. Bioelectronic devices using biomolecules are also discussed.

Tremendous progress was seen in the last decade in the organic devices not only for the academic understanding of molecular scale electronics, but also for the realistic device application. The chapters of this book were written by the leading experts specializing in the organic electronic elements of the future devices. This book covers a variety of research topics, but conceptional approach for understanding and controlling the nanoscale interface is common over the fields. This book will be a valuable resource for researchers working in the fields of nanotechnology, physics, materials science, electrical engineering, chemistry, and biology, and for individuals who are interested in the tremendously developing organic electronics.

We are grateful for all the contributed authors and valuable advices from a number of people. In particular, we appreciate Prof. Hoon-Kyu Shin in Pohang University in Science and Technology, Korea who put tremendous efforts in the editorial work for this book.

Mitsumasa Iwamoto
Young-Soo Kwon
Takhee Lee

Preface vii

Mitsumasa Iwamoto graduated from Tokyo Institute of Technology and obtained B.E. in 1975. He obtained M.E. and D.E. from Tokyo Institute of Technology in 1977 and 1981, respectively. Now, he is a professor of Tokyo Institute of Technology, His current interests focus on organic and molecular materials electronics, and on the electrical, optical, physical and dielectric properties of thin films, monolayers, liquid crystals and molecules, and soft matter physics for electronics. He has published more than 450 papers including review articles in international journals, and published more than 10 books as authors, co-authors, and editors, e.g., he wrote a book entitled "The Physical Properties of Organic Monolayers", World Scientific, Singapore, pp. 1-200 (2001). He has chaired international conference many times and also made a contribution as guest editors of international Journals such as Thin Solid Films, and Japanese Journal of Applied physics more than 10 times.

Young-Soo Kwon received his B.S. degree in Electrical Engineering at Yeungnam University, M.S. degree in Electrical Engineering at Kyungpook National University, Korea in 1973 and 1976, respectively, and he received his Ph.D. degree in Physical Electronics at Tokyo Institute of Technology, Japan in 1988. He was a Visiting Professor in Department of Chemistry and Bioengineering at University of Houston, USA from 2007 until 2008. Currently he is a professor in Department of Electrical Engineering and Nano Engineering, Dong-A University, Busan, Korea. His research interests are in the areas of molecular electronics, organic electronic devices, ultra-thin films, OLED, nano-biotechnology and its applications. He has edited a book titled "Measuring Technology of Electrical Properties in Molecule Level" and written 4 book chapters, 12 review articles, 42 patents and about 350 international journal articles. He has experience as an invited speaker more than 12 times in international conferences and guest editors of international journals such as "Journal of Nanoscience and Nanotechnology".

**Takhee Lee** received his B.S. and M.S. degrees in physics at Seoul National University, Seoul, Korea in 1992 and 1994, respectively, and he received his Ph.D. degree in physics at Purdue University, West Lafayette, USA in 2000. He was a postdoctorer in the Department of Electrical Engineering at Yale University, New Haven, USA from 2000 until 2004. In 2004 he joined the Department of Materials Science and Engineering, Gwangju Institute of Science and Technology, Gwangju, Korea, currently he is an Associate Professor. His research interests are characterization of the electrical and structural properties of nanostructures involving single molecules, nanoparticles, nanowires, and their assembly of these nanobuilding blocks into electronic devices. He has edited a book titled "Molecular Nanoelectronics" with Mark Reed (Yale University) and written 9 book chapters, 5 review articles, and about 100 journal articles.

CONTENTS

Preface   v
    Mitsumasa Iwamoto, Young-Soo Kwon and Takhee Lee

Part 1: Nanoscale Interface

1. Introduction to Nanoscale Interface   3
   Mitsumasa Iwamoto, Young-Soo Kwon and Takhee Lee

2. Analysis of Contact Resistance and Space-Charge Effects in Organic Field-Effect Transistors   9
   Martin Weis, Takaaki Manaka and Mitsumasa Iwamoto

3. Interface Control of Vertical-Type Organic Transistors   27
   Yasuyuki Watanabe and Kazuhiro Kudo

4. Electrochemical Properties of Self-Assembled Viologen Derivative and Its Application to Hydrogen Peroxide Detecting Sensor   49
   Dong-Yun Lee, Hyen-Wook Kang, Sang-Hyun Park and Young-Soo Kwon

5. Zinc (II), Iridium (III) and Tin (IV) Complexes for Nanoscale OLED Devices   69
   Trinh Dac Hoanh and Burm-Jong Lee

6. Structure Optimization for High Efficiency White Organic Light-Emitting Diodes   93
   Ji Hoon Seo, Ji Hyun Seo and Young Kwan Kim

Part 2: Molecular Electronics

7. Statistical Analysis of Electronic Transport Properties of Alkanethiol Molecular Junctions — 121
   Tae-Wook Kim, Gunuk Wang, Hyunwook Song and Takhee Lee

8. A Hysteric Current/Voltage Response of Redox-Active Ruthenium Complex Molecules in Self-Assembled Monolayers — 151
   Kyoungja Seo, Junghyun Lee, Gyeong Sook Bang and Hyoyoung Lee

9. Characteristics of Charge Transport and Electric Conduction in Viologen Self-Assembled Monolayers — 175
   Nam-Suk Lee, Dong-Yun Lee, Dong-Jin Qian, Young-Soo Kwon and Hoon-Kyu Shin

10. Time-Averaged Deuterium NMR Studies of the Dynamic Properties for a Low Molar Mass Nematic — 193
    Akihiko Sugimura and Geoffrey R. Luckhust

11. Training and Fatigue of Conducting Polymer Artificial Muscles — 223
    Keiichi Kaneto

Part 3: Polymer Electronics

12. Surface Plasmon Excitations and Emission Lights in Nanostructured Organic Films — 243
    Keizo Kato

13. Morphology Control of Nanostructured Conjugated Polymer Films — 273
    Mitsuyoshi Onoda and Kazuya Tada

14. Way of Roll-To-Roll Printed 13.56 MHz Operated RFID Tags   297
    Minhun Jung, Jinsoo Noh, Hwangyou Oh, Hwiwon Kang, Dongsun Yeom, Donghwan Kim and Gyoujin Cho

15. Physical Vapor Deposition of Polymer Thin Films and Its Application to Organic Devices   319
    Hiroaki Usui

16. Nanoscale Bioelectronic Device Consisting of Biomolecules   347
    Jeong-Woo Choi, Taek Lee and Junhong Min

Part 1
Nanoscale Interface

CHAPTER 1

INTRODUCTION TO NANOSCALE INTERFACE

Mitsumasa Iwamoto[1], Young-Soo Kwon[2] and Takhee Lee[3]

[1]*Department of Physical Electronics, Tokyo Institute of Technology, 2-12-1 O-okayama, Maguro-ku, Tokyo 152-8552, Japan*
[2]*Department of Electrical Engineering and NTRC, Dong-A University, Busan 604-714, Korea*
[3]*Department of Materials Science and Engineering, Gwangju Institute of Science and Technology, Gwangju 500-712, Korea*

In this chapter, an introduction to nanoscale interface is described in terms of organic electronics as well as molecular electronics. In contrast to scale-down technology, so-called "top-down" technology, "bottom-up" technology is a key in this field. Using this technique, artificial molecular systems comprised of organic molecules are designed, and utilized to fruition novel functions at nanoscale. As a result, understanding of alignment of molecules, interaction between molecules and substrates and so forth are getting important in nano-device and organic electronic devices. In this chapter we briefly summarize the feature of nanoscale organic interface from viewpoint of organic and molecular device physics.

1. Introduction

In this book many chapters deal with nanoscale interface in terms of organic electronic devices, including molecular switching elements, molecular diodes, organic field effect transistors, liquid crystal display elements, artificial muscles, biosensors, etc. In this chapter, as a guide to the nanoscale organic interface, we briefly discuss features of nanoscale interface from viewpoint of device physics.

Since the discovery of organic conducting materials, organic devices have drawn our attention in electronics, and much effort has been intensively devoted to improve organic device performance. We could see related stories in the development of organic electro-luminescent devices, organic transistors, and liquid crystal displays, organic solar-cells, organic batteries, etc. Among them are synthesis of new organic molecules, preparation method of organic thin films, fabrication techniques for improving carrier mobility and enhancement of carrier injection, assembling techniques for realizing novel functional devices, etc. Along with these efforts, new technologies based on scanning tunneling microscopy families, new techniques to characterize organic molecules and mono- and multi-layers have been proposed, and have encouraged us to study function and role of organic nanoscale interface. However, owing to ambiguities and complexities at the organic-metal and organic-organic interfaces, device operation mechanism is still not fully understood. As our final targets are to fruition organic devices and molecular devices, it is essential to make clear organic device physics, on paying attention to characteristic feature of organic molecules and organic films. For this purpose, we here focus on the characteristic feature of molecules in actual space, which is ascribed to molecular configuration, alignment, etc. The other is the characteristics of organic materials from viewpoints of energetic space. In the following we briefly discuss these two points.

2. Order Parameter and Organic Device Application

For simplicity let us consider organic film system comprised of rod-shaped molecules with a permanent dipole along molecular-long axis. Among them, floating monolayers on a water surface as well as organic monolayers on solid substrate are typical examples. Other examples are molecular systems comprised of liquid crystalline phase such as nematic and smectic phase. Also we can see ferroelectric molecular system comprised of aligned molecules with permanent dipoles.

The state of these organic film systems is expressed by using at least two kinds of order parameters, i.e., planar order parameter and orientational order parameter.[1,2] The former expresses the molecular configuration describing the positional distribution of the heads of the rod-molecules, and the latter expresses the orientational distribution of the rod-like molecules. Interestingly, physico-chemical properties of the molecular systems are dependent on these parameters. The electrical, optical, and mechanical properties of the systems are described using these parameters. Hence we have a chance to describe the specific feature of organic material systems by using these parameters.

As the first step, we see how molecular system can be classified in terms of orientational parameters. The orientational order parameter S_n, (n = 1, 2, 3) is defined using the thermodynamic average of the Legendre polynomials, $P_n(\cos\theta)$ (n = 1, 2, and 3), of the orientational angle θ of molecules.[1,2] In Fig. 1, the angle θ is defined as an angle from the z-direction. Figure 1 shows a typical example of molecular systems. Figure 1(a) shows the isotropic bulk materials, where rod molecules are randomly distributed, thereby $S_1 \equiv <P_1(\cos\theta)> = <\cos\theta> = 0$, $S_2 \equiv <P_2(\cos\theta)> = <(3\cos^2\theta - 1)/2> = 0$, and $S_3 \equiv <P_3(\cos\theta)> = <(5\cos^3\theta - 3\cos\theta)/2> = 0$. Figure 1(b) shows a case of nematic phase of bulk liquid crystals, where the orientational direction of molecules is restricted and molecules are totally distributed up and down directions, but there is no distinction between up and down directions. Thereby, $S_1 = 0$, $S_3 = 0$, but $S_2 \neq 0$. Figure 1(c) shows the monolayer composed of rod molecules, where all molecules randomly orient upper direction. Thereby S_1, S_2, $S_3 \neq 0$. From Fig. 1, it is found that the physical properties of nematic liquid crystals can be described using S_2. Actually we use this property, $S_2 \neq 0$, in the application of liquid crystal displays. Similarly, the physical properties of monolayers can be described using non-zero order parameters S_1, S_2, and S_3. Obviously, among these non-zero three parameters, S_1 and S_3 are the specific parameters that are connected to symmetry breaking. In other words, the specific physico-chemical properties of monolayers can be expressed using non-zero S_1 and S_3.

From the sketch of Fig. 1, we see that to generate spontaneous polarization in molecular systems, alignment of rod molecules, i.e, $S_1 \neq 0$ is key. This means that preparation of sophisticated molecular system, highly orientated at the nano-interface is needed, and development of nanoscale interface fabrication method is truly expected.

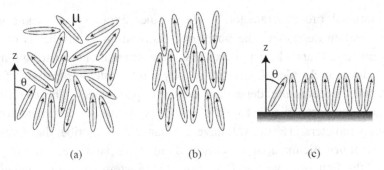

Fig. 1. Molecular system and orientational order parameter (a) $S_1 = 0$, $S_2 = 0$, $S_3 = 0$, (b) $S_1 = 0$, $S_2 \neq 0$, $S_3 = 0$, (c) $S_1 \neq 0$, $S_2 \neq 0$, $S_3 \neq 0$.

In liquid crystals, we use alignment layer to regulate the orientational distribution of molecules, on paying attention to spatial distribution of molecules. That is, anchoring surface energy profile is purposefully designed by means of rubbing or non-rubbing methods. Homeotropic, homogeneous alignment and others are available nowadays in liquid crystal technologies. In the field of organic device application, we use dipolar layer to reduce barrier height etc., on paying attention to energy diagrams of device systems, as we describe in Sec. 3. Alignment method based on self-assembling technique, Langmuir-Blodgett method and so forth, manipulation technique of molecules using scanning microprobe are used for the preparation of dipolar layer to create surface layer with $S_1 \neq 0$.

From the discussion above, it is easy to see that nanoscale interface control is the key to improve device performance as well as to realize novel function of molecules. In the next section we briefly look at the feature of organic molecular system from viewpoints of energy.

3. Nanoscale Interface for Organic Electronic Devices

The interface dipole layer greatly influences the energy band alignment, such a way that the contact barrier between organic and metal electrode can become larger or smaller, based on the direction of the interface dipoles.[3] Figure 2 schematically describes the energy band diagrams for organic-metal junctions in terms of the interface dipole. When an organic material makes a contact to a metal, the Fermi level of the metal is generally located within the highest occupied molecular orbital (HOMO)-lowest unoccupied molecular orbital (LUMO) gap. And in many cases, the Fermi level is located closer to the HOMO of organic material (like p-type semiconductor), as shown in Fig. 2(a).

Then, the hole-injection barrier (Φ_h) can be increased or decreased, based on the direction of the interface dipoles, as shown in Figs. 2(b) and 2(c). This effect significantly influences the electrical properties of junctions in terms of the efficiency of charge injection from metal to organic materials, and may impact the device performance in many of organic electronic devices.

The amount of the decrease or increase in the charge injection barrier is governed by the magnitude of the interface dipoles. Therefore, controlling the interface dipole can enable to tune the energy band alignment. For example, by inserting a proper molecular layer between metal and organic materials, one can actually control the interface; as a result one can tune the charge injection efficiency. These ideas are also utilized in molecular devices, where electron states of molecules in organic devices and molecular devices are intentionally regulated by means of carrier doping, external stimuli such as electric field, heating, light, etc., on the basis of quantum chemistry.

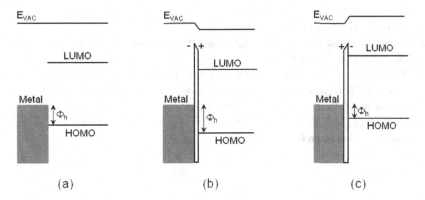

Fig. 2. Energy band diagrams of organic-inorganic (metal) junctions.

The interface is more critical in the device properties in molecular electronic devices as a structure of metal-molecule-metal junctions.[4] The contact resistance is significant in molecular scale electronic devices. Therefore, not only the functional molecule itself but also the interface governed by the molecular end-group is equally important element. Particularly, designing appropriate anchoring end-groups is required on metal electrodes in order to enhance the charge injection at the metal-molecule contacts.

4. Outlook

As has been described above, nano-interface control is a key in the field of nanoscale organic electronics, as well as in molecular electronics, where we need to make clear the behind physics of nanoscale interface from viewpoints of actual space as well as energy space, in terms of nanoscale fabrication technology. As a guide to this field, we comprise this book with following three parts: Part 1. nanoscale interface, Part 2. molecular electronics, and Part 3. polymer electronics. The reader will find that this book covers a variety of research topics, but conceptional approach is common over the field, and this field is actually interdisciplinary, and growing as a new field in science and technology.

References

1. M. Iwamoto and C. X. Wu, *The Physical Properties of Organic Monolayers* (World Scientific, Singapore, 2001).
2. G. R. Luckhurst and G. W. Gray, *The Molecular Physics of Liquid Crystals* (Academic Press, London, 1979).
3. W. R. Salaneck, K. Seki, A. Kahn and J. J. Pireaux, *Conjugated Polymers and Molecular Interfaces: Science and Technology for Photonic and Optoelectronic Applications* (Marcel Dekker, New York, 2002).
4. M. A. Reed and T. Lee, Eds., *Molecular Nanoelectronics* (American Scientific Publishers, Stevenson Ranch, 2003).

CHAPTER 2

ANALYSIS OF CONTACT RESISTANCE AND SPACE-CHARGE EFFECTS IN ORGANIC FIELD-EFFECT TRANSISTORS

Martin Weis, Takaaki Manaka and Mitsumasa Iwamoto*

*Department of Physical Electronics, Tokyo Institute of Technology,
2-12-1 O-okayama, Maguro-ku, Tokyo 152-8552, Japan
E-mail: iwamoto@pe.titech.ac.jp

We present a brief review on the limitation of charge injection to organic field-effect transistor due to interface metal-organic phenomenon, the contact resistance. Experimental methods for estimation of the contact resistance are summarized and their analysis is discussed. Voltage dependence of the contact resistance is carefully studied in respect to the definition of contact resistance and its time dependence during the device charging is illustrated. The physical background provides design of contact resistance.

1. Introduction

Semiconductor devices using organic materials,[1] such as thin film transistors,[2] and light emitting diodes,[3] have attracted a lot of research interests. With the development of organic materials with high mobilities, recent trend in the research on the organic field-effect transistor (OFET) has been concentrated on the applied research mostly focused on increase of carrier mobility and many experimental approaches have been exerted. Among them are development of modified surface gate insulator, use of single crystal semiconductor layer, etc.[4,5] Along with these researches, the importance of the basic research such as injection, accumulation and transfer mechanisms is being recognized to improve the OFET performance. Even though there is significant study, the device physics of OFET is not yet clear in comparison with that of inorganic FET structures. Various theoretical studies have been carried out to clarify the device physics of OFETs,[6-8] but attention was focused on transport phenomenon only. However, recently was research aimed on OFET devices where it has been shown that carriers injected from a source electrode dominate OFET operation. Therefore deep understanding of injection processes is crucial for further application of OFET.

However, improvement of the carrier transport in the OFETs revealed another one bottleneck of OFETs: the charge injection. This barrier is expressed by the contact resistance (R_c) and in OFETs it is a serious problem for practical applications.[9,10] The contact resistance R_c has many origins, such as the non-uniformity of organic semiconductors, the presence of dipole layers at the interface, electrode resistance, and the interfacial energy states.[11-13] At least two requirements — low-voltage operation and high-frequency performance — are responsible for reducing the values of the contact resistance R_c.[14] Therefore, to optimize the device performance, advanced techniques for preparing OFETs must be developed. To drive OFETs more efficiently, modeling of OFETs,[15] while accounting for R_c, where both the effect of barrier height for carrier injection[16] and charge accessing time from the electrode to organic material plays a crucial role, must be carried out. Note that the device contact effect influences the device contact operation conditions, such as the potential energy distribution in the channel of OFET.[17] The carrier mechanism is influenced by R_c at the metal-organic material interface when the energy difference between the Fermi level of metal electrode and the highest occupied molecular orbital (HOMO) of organic semiconductor dominates over the hole injection. In addition, here is a force from the applications side. Commercial use of organic devices requires decrease of the OFET channel length as much as possible, since the cut-off frequency as a main parameter for high-frequency response is proportional to $1/L^2$ (L: channel length). Hence so-called 0.7 scale-down rule is continuously employed for the improvement of device performance. However, for a sub-micrometer channels the contact resistance it is expected to be a major contribution to the device resistance.[18] Hence, we cannot simply employ the scale-down rule, and contact resistance is an important parameter for design of nanoscale

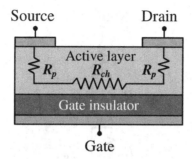

Fig. 1. A schematic view of the organic field-effect transistor (OFET) with channel resistance R_{ch} and parasitic resistance R_p.

devices based on organic semiconductors. This is the most crucial problem in nanoscale devices.

2. Contact Resistance Definition

The charge transport through the OFET device is usually discussed by three processes: (i) carrier injection from the source electrode, (ii) transport through the channel and (iii) extraction from the channel to the drain electrode. This can be roughly described by three independent resistors, as is illustrated in Fig. 1. The resistance associated with carrier injection and accumulation bellow the electrode is denoted as a parasitic resistance R_p, while resistance depicting carrier transport across the channel region is denoted as a channel resistance R_{ch}. Intuitively, the channel resistance is proportional to the channel length[19]:

$$R_{ch} = \frac{dV_{ds}}{dI_{ds}} = \frac{L}{WC_g\mu(V_{gs} - V_{th})}, \qquad (1)$$

where the drain-source current I_{ds} is ruled by the drain-source and gate-source voltages, V_{ds} and V_{gs}, and is proportional to the gate insulator capacitance per unit of area C_g, mobility μ and channel width and length W and L, respectively. This relation valid in the linear region (i.e. $V_{gs} - V_{th} \gg V_{ds}$) gives us possibility to separate parasitic resistance by electrical measurements of transistors with various channel lengths.

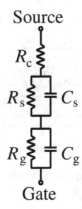

Fig. 2. The Maxwell-Wagner model of OFET as a double layer capacitor.

The parasitic resistance describes carrier injection and transport through the organic semiconductor film to the interface with the gate insulator. For simplicity, we neglect ohmic wire resistance (lead resistance). Hence, the parasitic resistance can be separated to the contact resistance R_c and semiconductor resistance R_s expressing injection and transport[20]

$$R_p = R_c + R_s. \qquad (2)$$

Here the contact resistance is usually significantly higher than resistance of organic semiconductor thin film. Thus the parasitic resistance is usually unified with the contact resistance. Description by the equivalent electric circuit can be useful and the Maxwell-Wagner model[21] is found as a powerful tool for description of charging of transistor.[22,23] There, the OFET is modeled as metal-insulator-semiconductor (MIS) structure with a contact resistance (see Fig. 2).

Using this equivalent circuit the surface charge Q_s accumulated on the organic semiconductor–gate insulator interface is in steady state approximately calculated as

$$Q_s = LWC_g \left(1 - \frac{\tau_s}{\tau_g}\right)\left(V'_{gs} - \frac{1}{2}V_{ds}\right), \qquad (3)$$

where τ_s and τ_g stand for the Maxwell relaxation times of semiconductor and insulator, respectively. Here we must point out reduced gate voltage through the potential drop on the resistor R_c, $V'_{gs} = R_g/(R_c + R_s + R_g)V_{gs}$. This charge is conveyed along the OFET channel to the drain electrode by the average electric field $E' = V'_{ds}/L$ formed between the source and drain electrodes, where $V'_{ds} = V_{ds} - R_c I'_{ds}$. Note here idea of ohmic contact for the contact resistance, i.e. potential drop across the resistance is $\Delta V = R_c I'_{ds}$, which is valid only if the channel resistance is larger than parasitic resistance. Hence, as a result through the OFET flows current

$$I'_{ds} = \frac{Q'_s}{L}\mu E' = \frac{Q'_s}{L}\mu \frac{V_{gs} - R_c I'_{ds}}{L}. \tag{4}$$

In other words, potential drop across the R_c resistor influences carrier accumulation as well as carrier transport. By substitution of Eq. (3) to Eq. (4) we can obtain

$$I'_{ds} = \frac{Q'_s}{L}\mu_{eff}\frac{V_{ds}}{L} \tag{5}$$

with effective mobility

$$\mu_{eff} = \left(\frac{\tau_{tr}}{\tau_{ch} + \tau_{tr}}\right)\mu, \tag{6}$$

where transit time τ_{tr} reaches form

$$\tau_{tr} = \frac{L^2}{\mu(V'_{gs} - V_{th} - \frac{1}{2}V_{ds})}, \tag{7}$$

and time required for charging of organic semiconductor–gate insulator below the source electrode time τ_{ch} is expressed as follows

$$\tau_{ch} = R_c C_{gs}. \tag{8}$$

Here, the $C_{gs} = C_g WL$ represents capacitance of source-gate electrode system. In detail, the charging time directly corresponds to the contact resistance and thus can be used for R_c evaluation in the experiment.

However, the question what means the contact resistance still remains. Although the contact resistance is common used term to reach deep understanding of this phenomenon we need go to the physical background. The contact resistance is the element of the equivalent electric circuit only and stands for a potential drop due to metal/organic interface. Note that this potential drop does not require to be just on the metal/organic interface only, but it is distributed through the organic semiconductor in film thickness direction. In detail, after application of the gate-source voltage V_{gs} it is established potential drop $\Delta V_0 = V_{gs} C_g/(C_s + C_g)$ across the organic semiconductor film. This creates an electric field $E_s = \Delta V/d_s$ (d_s: thickness of semiconductor film) as a motive force for carrier injection. If there is present no injection barrier the charge is successfully accumulated on the semiconductor–gate insulator interface and its space-charge field fully compensates applied electric field. As a result, no potential drop across the semiconductor film is observed. However, if the transport of the charge is comparable with charge injection and accumulation (i.e. $\tau_{tr} \approx \tau_{ch}$), there is not enough charge accumulated on the insulator surface and potential drop still remain also in the steady state. Although the slow injection originates from the carrier injection barrier $\Delta\Phi$ (for hole (electron) injection it is difference between the Fermi energy of metal and HOMO (or LUMO) level of organic semiconductor), there can be distinguished various microscopic mechanisms[17]: (a) thermionic (Schottky) injection, (b) field emission (Fowler-Nordheim tunneling) or (c) defect-assisted injection. This leads to the voltage dependence of contact resistance, that is, contact resistance is defined nothing but by carrier injection mechanism.

3. Effect of Electric Field on Contact Resistance

Let us discuss here the thermionic injection, which is common observed in metal/organic interface. Hence, the current density j on the interface is described[24] by relation

$$j = AT^2 \exp\left(-\frac{\Delta\Phi - \beta\sqrt{E}}{k_B T}\right), \qquad (9)$$

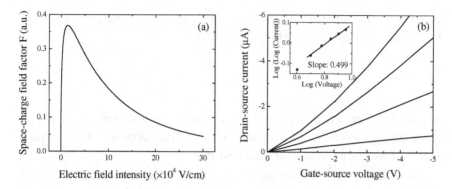

Fig. 3. (a) The space-charge field factor F as a function of local electric field intensity and (b) illustration of field effect on carrier injection for output characteristics of pentacene OFET for various gate-source voltages (from -10 to -40 V) in low electric field region. Inset depicts $\log_{10}$-$\log_{10}$ scale for illustration of the Schottky injection.

where k_B and A are the Boltzmann and Richardson constants, and β is Schottky parameter ($=\sqrt{e^3/4\pi\varepsilon_0\varepsilon_r}$, where e is elementary charge, and $\varepsilon_0\varepsilon_r$ is dielectric constant of organic semiconductor). Hence, the resistivity ρ as a reciprocal value of conductivity σ is in linear approximation defined as

$$\rho = \frac{1}{\sigma} = \left(\frac{\partial}{\partial E} j\right)^{-1} = \left(\frac{k_B T}{\beta}\right)^2 \frac{x}{\exp(x)} \frac{2}{AT^2} \exp\left(\frac{\Delta\Phi}{k_B T}\right), \quad (10)$$

with $x = \beta\sqrt{E}/k_B T$ as an electric field parameter. The fraction $x/\exp(x)$ depicts the local electric field dependence of the contact resistance; it is denoted as a space-charge field factor F and its electric field dependence is shown in Fig. 3(a). This electric field represents local field which consists of applied (external) field as well as space-charge (internal) field. Also note that in Eq. (10) the interface conductivity is independent on mobility. This is in contradict with standard approach used for bulk conductivity ($\sigma = en\mu$, where n is carrier density).

For derivation of contact resistance is organic semiconductor treated as a dielectric material with low intrinsic carrier density and an excess injected charge accumulated on semiconductor–gate insulator interface only. Since the electric field is assumed uniformly distributed across the organic semiconductor film from electrode to accumulated charge layer, the contact resistance is evaluated as follows

$$R_c = \rho \frac{d}{S}, \qquad (11)$$

where S is effective injection area and d is potential drop length representing distance on which potential drop appears (in our case identical with semiconductor thickness d_s).

In addition we must mention impact of contact resistance on carrier injection in low electric field region. In ideal case the drain-source current is only a linear function of drain-source voltage for $V_{gs} - V_{th} \gg V_{ds}$. However, in the experiment it can be observed deviation from ohmic contact as is illustrated in Fig. 3(b). Detail analysis reveals thermionic injection mechanism represented by the $I \propto \exp(V^\alpha)$ relation with power exponent $\alpha \approx 0.5$ (i.e. square root) of voltage, which is in accordance to Eq. (9).

4. Measurements of Contact Resistance

4.1. *Transmission Line Method (TLM)*

As is discussed above the total device resistance is the sum of the contact resistance and the channel resistance, which is linearly dependent on the channel length L. Hence, one of the simplest ways to evaluate contact resistance is based on extrapolation of total resistance to the zero channel length.[18,25,26] In detail, by measurement of the electrode system with various channel lengths for the same semiconductor film it is possible to construct total resistance versus channel length plot. There, the contact resistance is estimated as the intercept for zero channel length, see Fig. 4.

Even though simplicity if this measurement is a great advantage, it has few important drawbacks. First, we are forced to fabricate several devices with identical properties in order to obtain total resistance versus channel length plot. Second, contact resistance is affected by the charge accumulated in the channel region, the TLM experiment should be done with constant drain-source field, i.e. for precise measurements it is necessary to scale drain-source voltage to channel length. Furthermore, this method cannot distinguish contribution of the source and drain electrodes.

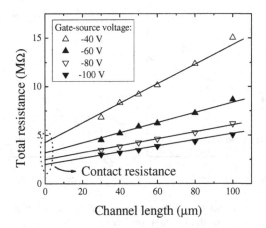

Fig. 4. Transmission line method of top contact pentacene film with silver electrodes for different gate-source voltages. Solid lines represent linear fit.

4.2. Other Electrical Measurements

Beside common used TLM setup also alternative electrical measurements of the contact resistance were proposed: the gated four-probe measurement[27-29] and the time-of-flight method.[30] The gated four-probe measurement introduces two additional narrow electrodes into the OFET channel, which are connected to high input impedance electrometer to avoid drain-source current loss through these electrodes. Extracted two additional potentials V_1 and V_2 from the channel region give us in approximation of linear voltage dependence across the channel possibility to estimate potential drops on the source and drain electrodes as is illustrated in Fig. 5(a). Although this method has advantage of using single device only, complicated electrode system can be used only for longer channel lengths. In addition, for correct estimation of potential drops on the source and drain electrodes it requires linear potential profile across the channel, which is in steady state not always satisfied.[17]

On the other hand, the time-of-flight method is a technique focused on estimation of the time required for charge transport from the source to drain electrode. As was discussed earlier the total transport time consists of time required for carrier injection and accumulation bellow the source electrode and carrier transport across the channel. In accordance to

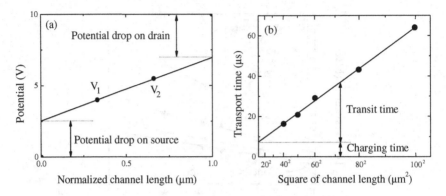

Fig. 5. Example of the result for (a) the gated four-probe measurement (for $V_{ds} = 10$ V) and (b) the time-of-flight method, where the charging time corresponds to the contact resistance.

Eq. (7) the transit time is proportional to square of the channel length L^2. Therefore, the linear extrapolation of the transport time to the zero channel length (see Fig. 5(b)) results to the charging time, which is related to the contact resistance as is described by Eq. (8). In addition, it is interesting to point out that the transient signal (voltage pulse) for estimation of the contact resistance is used in experiment. In the transient state the channel is not fully charged and still no space-charge field is created. As a result, without space-charge field only ohmic transport is established and the potential drop across the channel follows Eq. (1), i.e. the linear potential across the channel is well satisfied. This is proved by time-resolved experiment and simulation results[31] and can be understand as a charging of semi-infinite interface.

4.3. *Kelvin Probe Force Microscopy (KFM)*

In contrast to previous methods based on linear approximation of the potential across the channel region, the KFM gives chance to directly measure surface potential profile. In this technique the atomic force microscopy (AFM) with conductive tip is used for scanning of surface morphology as well as surface potential.[17,32] Since the potential drop on the electrode edge is measured directly, KFM is a strong tool for estimation of contact resistance. Study of surface potential by Bürgi et al. showed that contact resistance is mostly situated on the injection electrode, while on the drain electrode is only resistance of semiconductor film observed.

4.4. *Time-Resolved Microscopic Second-Harmonic Generation (TRM-SHG)*

Recently was proposed an alternative method for contact resistance estimation based on time-resolved measurement of electric field profile in OFET. In accordance to the nonlinear optics the optical second-harmonic generation (SHG) is activated by induced polarization of the organic semiconductor, an effective dipole originating in the distortion of electron distribution by the local electric field.[33] This is known as an electric-field-induced SHG (EFISHG) and is useful for probing the electric field profile in OFET. Here, the intensity of SH light is proportional to square of the local electric field. In addition, as an improvement of this method was developed time-resolved microscopic SHG (TRM-SHG) measurement for electric field evaluation with high spatial and temporal resolution.[34] In contrast to KFM technique which is applicable for the steady state only, the TRM-SHG measurement enables us to evaluate also the transient state.[35] Further this method enables us to probe individually carrier injection process as well as carrier transport process by study of electric field at the electrode and in the channel region, respectively. Hence, interfacial phenomena resulting contact resistance is well investigated.

Our previous discussion showed that the source-gate electrode system requires charging time dependent on the contact resistance, see Eq. (8). Therefore, the potential drop on the source electrode can be expressed as

$$\Delta V = (\Delta V_0 - \Delta V_\infty)\exp\left(-\frac{t}{\tau_{ch}}\right) + \Delta V_\infty. \qquad (12)$$

Here, the ΔV_0 ($= V_{gs}C_g/(C_s + C_g)$) and ΔV_∞ stands for initial and steady state potential drop, respectively, as depicted in Fig. 6. Hence, we have two independent ways for the contact resistance evaluation: (i) the

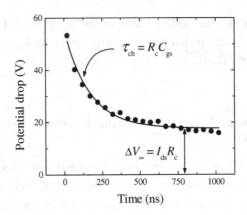

Fig. 6. Time-evolution of potential drop on silver source electrode in pentacene OFET evaluated by the TRM-SHG technique (V_{gs} = -100 V). Solid line illustrates exponential fit.

transient signal represented by the relaxation time τ_{ch} of the electric field[23] and (ii) the steady state potential drop. Surprisingly, a great discrepancy of few orders is observed for the contact resistance estimated from the transient and steady state signal.[35] Difference between these values reflects effect of charges accumulated in the OFET. The transient contact resistance illustrates injection properties in absence of excess charges in OFET, and depends mainly on the difference between the electrode metal work function and HOMO level (in case of holes). On the other hand, the contact resistance in steady state depicts carrier injection energy barrier together with the space-charge field of accumulated charges. The applied electric field is mostly compensated by the field of excess charges and thus contact resistivity increases as can be seen in Fig. 3(a). As a result, the contact resistance is time-dependent as continues charging of OFET. This conclusion cannot be obtained by common steady state measurements like the TLM or KFM techniques.

5. Design of Contact Resistance

5.1. *Influence of Device Setup*

As we discussed above, the contact resistance is dependent on the carrier injection barrier $\Delta\Phi$ as well as local electric field on the metal/organic interface. In contrast to the common assumption of the constant contact resistance as a material parameter this understanding gives us powerful tool for contact resistance engineering. In this section is a brief summary of common approaches to modify contact resistance. However, we do not show list of R_c measured values for difficulty in comparison. One must be careful for different experimental conditions, e.g. applied voltage. Nevertheless, basic concept of contact resistance modification is still valid.

As we showed previously the contact resistance for specific organic semiconductor depends on the Fermi energy of electrode metal. Hence, fine tuning of the interface energetics plays a key role for smooth carrier injection. Rough adjustment of the injection barrier can be done by choice of the electrode metal with work function similar to HOMO level (for hole injection).[36] However, also the effective work function of metal can be changed. It was shown that the exposure to the UV and ozone environment is helpful for the surface cleaning[37] and can remove contamination lowering the work function.[38] Moreover, this treatment creates metal oxides on the electrode surface and thus increases the work function. Here should be noted that for the UV/ozone surface cleaning the decrease[37] as well as increase[39] of injection barrier was reported. Unfortunately, this simple change of the injection barrier can be applied for the bottom contact OFET only, where the electrode is evaporated prior to the organic semiconductor film deposition.

Another common approach to modify injection properties is introducing a dipolar layer. Usually a self-assembled monolayer (SAM) is created on metal/organic interface for bottom contact OFET[40,41] or on the gate insulator surface for top contact OFET.[42,43] In case of SAM grown on the electrode surface it is changed injection barrier by the interfacial dipole.[41] In other words, electric field of the dipole can decrease the injection barrier in means of Eq. (9). In similar way, the SAM grown on the gate insulator surface induces field in organic semiconductor film and metal/organic interface.[43] However, surface modification of the gate insulator has a great effect on the crystallinity of organic semiconductor.[42,44] Increase of the grain boundaries (i.e. decrease of the grain size) is related to the rise of trapped carriers due to interface traps and grain boundary resistance.[45] Trapped (immobile) carriers are additional source of the electric field which compensates applied electric field. As a result, the contact resistance increases together with amount of traps. Similar effect was also reported for traps on metal/organic interface for top contact OFET.[11]

This studies force us to include to our consideration also the field all excess charges in the device. Especially high density of injected carriers accumulated on the organic semiconductor–gate insulator surface can influence local field on the metal/organic interface. In more details, electric field E_i induced on injection electrode (source) is proportional to the potential drop V_i created by charge Q_i situated on the gate insulator surface as follows

$$E_i \propto \frac{Q_i}{C_{si}+C_{di}+C_{gi}}\frac{1}{d} \approx \frac{Q_i}{C_{si}d(1+C_{gi}/C_{si})}. \qquad (13)$$

Here the capacitances C_{si}, C_{di}, C_{gi} are related to the induced electric field on the source, drain and gate electrode, respectively and d is the thickness of the organic semiconductor. Therefore detail evaluation of electric field effect requires charge integration across the channel as well as device geometry. However, for a simplified problem when are taken in account charges bellow the source electrode only the induced electric field decreases with increase of the film thickness d. Figure 7 depicts

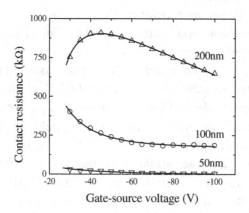

Fig. 7. Voltage dependence of contact resistance for top contact pentacene OFET (gold electrodes) for three different pentacene film thicknesses: 50 nm (the down triangles), 100 nm (the circles), and 200 nm (the up triangles). Solid lines represent fit by Eq. (10).

example of the voltage dependence of the contact resistance for three different film thicknesses, solid lines represent fit by Eq. (10). It is obvious that increase of film thickness increases the contact resistance. Moreover, in case of slower injection than charge transport from the source to the drain electrode accumulated charge is not fully established and the potential drop increases, i.e. contact resistance rises. Note that in bottom contact OFET the charge cannot be accumulated bellow the electrode and thus its effect is minimized. Hence, just weak electric field influence on the contact resistance for this device geometry was reported.[46] Here it should be noted that the space-charge plays an important role in the charge transport, as was recently shown.[47,48]

5.2. Where is the Limit for Contact Resistance?

Previous discussion can induce impression that by careful choice of electrode metal, device geometry, and design of internal fields we can suppress contact resistance to the zero value. Here, we show that although this is our final goal, there is present limitation for contact resistance.

The common used TLM experiment is based on idea of the transmission line with distributed parameters. Hence, the channel resistance scales with channel length. However, this theory also predicts reflection of the propagating signal in case if the resistance on the end of the transporting channel is different from the characteristic impedance, $Z = V/I$, which depends on OFET channel properties. In detail, the current I representing charge flow illustrates the increment of the charge dQ on the infinitesimal small capacitance dC by voltage V:

$$I = \frac{dQ}{dt} = \frac{VdC}{dt} = C'v. \qquad (14)$$

Here the increase of capacitance is related to the propagation as $dC = C'v$, where C' is capacitance per unit of length ($= C_g W$) and v is the group velocity, which can be obtained from Eq. (7) as $v \approx \mu(V_{gs} - V_{th})/L$. Hence, by substitution of these relation we obtain characteristic impedance Z_0 the channel

$$Z_0 = \frac{1}{C'v}. \qquad (15)$$

Interesting is to note that the characteristic impedance is a function of the channel length ($Z_0 \propto L$) or voltage ($Z_0 \propto 1/(V_{gs} - V_{th})$). This termination resistance is representing minimal contact resistance present in the OFET device. In other words, if there is not impedance (resistance) matching the backward-travelling wave induces additional losses. For typical values of the experimental variables in OFET (L/W = 50 μm/3 mm, μ = 1 cm^2/V.s, ($V_{gs} - V_{th}$) = -10 V and C_g = 5.4 nF/cm^2) we obtain characteristic impedance of Z_0 ~ 50 kΩ. One can wonder how it is possible to use linear extrapolation in TLM evaluation of contact resistance if the minimal contact resistance is a function of the channel length. Reason is in still high contact resistance, which value is usually greatly larger than theoretical minimum.

6. Conclusions and Outlook

In this brief summary we focused on the charge injection problem of organic devices. The contact resistance as an element of equivalent electrical circuit representing potential drop due to insufficient injection was defined and relaxation time of organic field-effect transistors were proposed on the basis of the Maxwell-Wagner model. The microscopic description of the metal/organic interface was also employed to explain voltage dependence of the contact resistance as well as non-ohmic behavior for low voltage region. Various novel experimental techniques applicable for the contact resistance evaluation were listed and their capabilities were discussed.

Methods for contact resistance engineering were summarized: choice of electrode metal, UV/ozone or SAM treatment, and device geometry design. We propose that the careful design of internal electric field is powerful tool to modify the contact resistance. The contact resistance is not more discussed as a material parameter with constant value, but as a field-dependent property of nanoscale interfaces.

References

1. H.E. Katz, *J. Mater. Chem.* **7**, 369 (1997).
2. H. Sirringhaus, *Adv. Mater.* **17**, 2411 (2005).

3. C.W. Tang, S.A. VanSlyke, *Appl. Phys. Lett.* **51**, 913 (1987).
4. C. Kim, A. Facchetti, T.J. Marks, *Science* **318**, 76 (2007).
5. D. Knipp, R.A. Street, A. Völkel, J. Ho, *J. Appl. Phys.* **93**, 347 (2003).
6. R. Tecklenburg, G. Paasch, S. Scheinert, *Adv. Mater. Opt. Electron.* **8**, 285 (1998).
7. J. Zaumseil, K.W. Baldwin, J.A. Rogers, *J. Appl. Phys.* **93**, 6117 (2003).
8. G. Horowitz, *Adv. Mater.* **10**, 365 (1998).
9. S.D. Wang, T. Minari, T. Miyadera, K. Tsukagoshi, *Appl. Phys. Lett.* **91**, 203508 (2007).
10. G.B. Blanchet, C.R. Fincher, M. Lefenfeld, J.A. Rogers, *Appl. Phys. Lett.* **84**, 296 (2004).
11. S.D. Wang, T. Minari, T. Miyadera, Y. Aoyagi, K. Tsukagoshi, *Appl. Phys. Lett.* **92**, 63305 (2008).
12. T. Miyadera, M. Nakayama, S. Ikeda, F. Saiki, *Curr. Appl. Phys.* **7**, 87 (2007).
13. Y. Minari, T. Nemoto, S. Isoda, *J. Appl. Phys.* **96**, 769 (2004).
14. V. Wagner, P. Wöbkennerg, A. Hopp, J. Seekamp, *Appl. Phys. Lett.* **89**, 243515, (2004).
15. D. Natali, L. Fumagalli, M. Sampietro, *J. Appl. Phys.* **101**, 14501 (2007).
16. N. Koch, A. Kahn, J. Ghijsen, J.-J. Pireaux, J. Schwartz, *Appl. Phys. Lett.* **82**, 70 (2003).
17. W.R. Silveira, J.A. Marohn, *Phys. Rev. Lett.* **93**, 116104 (2004).
18. H. Klauk, G. Schmid, W. Radlik, W. Weber, L. Zhou, C.D. Sheraw, J.A. Nichols, T.N. Jackson, *Solid-State Electron.* **47**, 297 (2003).
19. G. Horrowitz, *Adv. Mater.* **10**, 365 (1998).
20. M. Weis, M. Nakao, J. Lin, T. Manaka, M. Iwamoto, *Thin Solid Films* **518**, 795 (2009).
21. T. Manaka, E. Lim, R. Tamura, M. Iwamoto, *Thin Solid Films* **499**, 386 (2006).
22. R. Tamura, E. Lim, T. Manaka, M. Iwamoto, *J. Appl. Phys.* **100**, 114515 (2006).
23. E. Lim, T. Manaka, M. Iwamoto, *J. Appl. Phys.* **104**, 54511 (2008).
24. S.M. Sze, *Physics of Semiconducting Devices*, 2nd ed., Wiley, New York, 1981.
25. D.J. Gundlach, L. Zhou, J.A. Nichols, T.N. Jackson, P.V. Necliudov, M.S. Shur, *J. Appl. Phys.* **100**, 24509 (2006).
26. J. Zaumseil, K.W. Baldwin, J.A. Rogers, *J. Appl. Phys.* **93**, 6117 (2003).
27. R.J. Chesterfield, J.C. McKeen, C.R. Newman, C.D. Frisbie, P.C. Ewbank, K.R. Mann, L.L. Miller, *J. Appl. Phys.* **95**, 6396 (2004).
28. P.V. Pesavento, R.J. Chesterfield, C.R. Newman, C.D. Frisbie, *J. Appl. Phys.* **96**, 7312 (2004).
29. P.V. Pesavento, K.P. Puntambekar, C.D. Frisbie, J.C. McKeen, P.P. Ruden, *J. Appl. Phys.* **99**, 94504 (2006).
30. M. Weis, J. Lin, D. Taguchi, T. Manaka, M. Iwamoto, *J. Phys. Chem. C* **113**, 18459 (2009).
31. T. Manaka, F. Liu, M. Weis, M. Iwamoto, *Phys. Rev. B* **78**, 121302(R) (2008).
32. K.P. Puntambekar, P.V. Pesavento, C.D. Frisbie, *Appl. Phys. Lett.* **83**, 5539 (2003).

33. R.W. Terhume, P.D. Maker, C.M. Savage, *Phys. Rev. Lett.* **8**, 404 (1962).
34. T. Manaka, E. Lim, R. Tamura, M. Iwamoto, *Nat. Photonics* **1**, 581 (2007).
35. M. Nakao, T. Manaka, M. Weis, E. Lim, M. Iwamoto, *J. Appl. Phys.* **106**, 14511 (2009).
36. L. Diao, C.D. Frisbie, D.D. Schroepfer, P.P. Roden, *J. Appl. Phys.* **101**, 14510 (2007).
37. J.R. Vig, *J. Vac. Sci. Technol. A* **3**, 1027 (1985).
38. Y. Suzue, T. Manaka, M. Iwamoto, *Jpn. J. Appl. Phys.* **44** (2005).
39. A. Wan, J. Hwang, F. Amy, A. Kahn, *Org. Electron.* **6**, 47 (2005).
40. I.H. Campbell, J.D. Kress, R.L. Martin, D.L. Smith, N.N. Barashkov, J.P. Ferraris, *Appl. Phys. Lett.* **71**, 3528 (1997).
41. I.H. Campbell, S. Rubin, T.A. Zawodzinski, J.D. Kress, R.L. Martin, D.L. Smith, N.N. Barashkov, J.P. Ferraris, *Phys. Rev. B* **54**, 14321 (1996).
42. K.P. Pernstich, S. Haas, D. Oberhoff, C. Goldmann, D.J. Gundlach, B. Batlogg, A.N. Rashid, G. Shitter, *J. Appl. Phys.* **96**, 6431 (2004).
43. Y. Jang, J.H. Cho, D.H. Kim, Y.D. Park, M. Hwang, K. Cho, *Appl. Phys. Lett.* **90**, 132104 (2007).
44. C. Kim, A. Facchetti, T.J. Marks, *Science* **318**, 76 (2007).
45. T.W. Kelley, C.D. Frisbie, *J. Phys. Chem. B* **105**, 4538 (2001).
46. P.V. Necliudov, M.S. Shur, D.J. Gundlach, T.N. Jackson, *Solid-State Electron.* **47**, 259 (2003).
47. M. Weis, J. Lin, D. Taguchi, T. Manaka, M. Iwamoto, *Jpn. J. Appl. Phys.* **49**, 071603 (2010).
48. M. Weis, D. Taguchi, T. Manaka, M. Iwamoto, *Jpn. J. Appl. Phys.* **49**, 04DK15 (2010).

CHAPTER 3

INTERFACE CONTROL OF VERTICAL-TYPE ORGANIC TRANSISTORS

Yasuyuki Watanabe[1] and Kazuhiro Kudo[2,*]

[1]*Faculty of Systems Engineering, Tokyo University of Science, Suwa, 5000-1 Toyohira, Chino, Nagano 391-0282, Japan*
[2]*Faculty of Engineering, Chiba University, 1-33 Yayoi-cho, Inage-ku, Chiba 263-8522, Japan*
E-mail: kudo@faculty.chiba-u.jp

Organic field-effect transistors (OFETs) have been studied extensively for flexible displays, organic sensors and radio frequency identification tags. However, conventional OFETs have remaining issues such as higher driving voltage, lower current operation and lower frequency response due to lower carrier mobility of organic semiconductors than their inorganic counter parts. Vertical-type organic transistors are strong candidates for improvement in the performance of the OFETs. This article describes an overview of the vertical-type organic transistors. We will focus on the progress in our recent experimental studies on organic static induction transistors (OSITs) as one type of the vertical-type organic transistors. OSITs are proposed for application in flexible sheet displays, logic devices, and organic light-emitting transistors (OLETs). In particular, this article describes our recent efforts to improve the performance of the OSITs by controlling the interface states between source electrode and organic semiconductor (pentacene). These results showed that both the high on/off ratio more than 10^3 and the high-current value of larger than 40 µA were achieved. In addition, vertical-type OLET and logic devices were fabricated using the OSITs for the realization of flexible sheet displays. These results exhibit that the OSITs attract much attention for application in organic flexible displays.

1. Introduction

Recently high performance electric and optoelectronic devices based on organic semiconductors have been demonstrated, such as organic light-emitting diodes (OLEDs), organic field-effect transistors (OFETs), and organic photovoltaics. These organic devices show promise for low-cost, large-area and flexible devices. In particular, display panels based on OLEDs are expected for mobile electronic devices and excellent stability and high efficiency OLEDs have been reported. On the other hand, rapid progress of OFETs has been made in recent years.[1] Furthermore, all-organic display devices are expected by combining the OLEDs with organic transistors,[2-4] because organic transistors driving OLEDs are necessary to achieve flexible and large scale active-matrix-displays. To be practical, however, the performance of organic transistors are required to obtain high-current density with a drive voltage as low as a few volts and have sufficient reliability. Conventional OFETs have a low-speed, a low-power, and a relatively high driving voltage mainly due to the low-mobility and low-carrier density of the organic semiconductors. It is known that vertical-type organic transistors such as organic static induction transistors (OSITs) are suitable for a driving element of displays because of the high-speed and high-power of operation.[5-7] The excellent characteristics arise from the vertical structure with a very short channel length between the source and drain electrodes. From this point of view, the vertical-type organic light-emitting transistors (OLETs), which combined with the OSITs and OLEDs, are promising for flexible sheet displays.[4-7]

Our recent efforts to improve the performance of OSITs by optimizing the device structure and organic materials are reported. This paper describes the interface control of OSITs for the improvement in the electrical characteristics, and the application for organic inverters and vertical OLETs based on the OSITs for flexible sheet displays. To improve the performance of pentacene-based OSITs, the effect of the interface control on the characteristics of OSITs were investigated. An ultra-thin copper phthalocyanine (CuPc) layer was inserted between indium-tin-oxide (ITO) as a source electrode and pentacene thin-film. The results showed that both the high on/off ratio more than 10^3 and the high-current value of larger than 40 µA of the OSITs were achieved, which demonstrate that it is a key factor for high performance OSITs to optimize the electronic states such as the hole injection barrier and band bending at the interface between the organic semiconductor layer and the source electrode. The improved OSITs were applied to fabricate the organic inverters which can operate at low-voltage. From the measurement results of the inverter transfer characteristics, it was found that the operational voltage of the organic inverters based on the OSITs was from -2 to +2 V. In addition, the vertical-type OLETs, which combined with the OSITs based on pentacene/CuPc and conventional OLEDs based on Alq3 and NPD, were fabricated. The results show that the current is controlled by relatively small gate voltage of -1 V and typical OSITs characteristics are obtained for the OLETs. The luminance also varies corresponding to the *I-V* characteristics.

2. Vertical-Type Organic Transistors

Based on the carrier transport direction, organic transistors can be classified into two groups; lateral-type organic transistors, and vertical-type organic transistors. Conventional OFETs is well-known as lateral-type organic transistors, where the carrier flow is parallel to the substrate. On the other hand, many kinds of vertical-type organic transistors have also been reported, where the carrier flow is perpendicular to the substrate. Figure 1 shows cross-sectional schematic illustrations of a various vertical-type organic transistors such as (a) polymer grid triode (PGT),[8] (b) organic static induction transistor (OSIT),[5] (c) vertical-channel polymer field-effect transistor,[9] (d) vertical organic field-effect transistor (VOFET),[10] (e) metal-base organic transistor (MBOT),[11] (f) OSIT using colloidal lithography,[12] (g) dual self-aligned vertical multichannel organic transistor (DSA-VMCOT),[13] (h) three-dimensional organic field-effect transistor (3D-OFET),[14] (i) step-edge vertical-type channel organic field-effect transistor (SVC OFET).[15] Their operational mechanisms are based on vertical-type inorganic transistors such as metal-base transistor (MBT),[16] SIT[17] and permeable-base transistor (PBT)[18] and so on. Here, we have focused on the vertical-type organic transistors that operate as a SIT, that is, the OSITs (Fig. 1(b)).

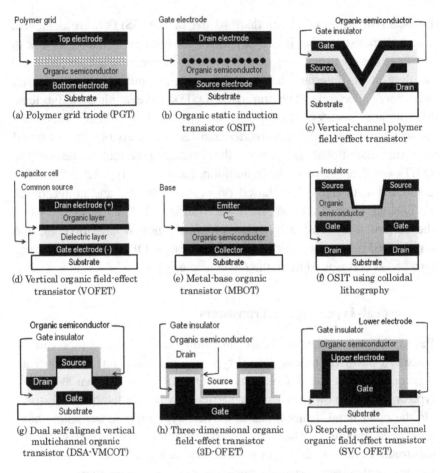

Fig. 1. Various structures of vertical-type organic transistors.

The advantages of vertical-type OSITs over lateral-type OFETs are the realization of low-voltage operation and ease of fabrication on flexible substrates. In lateral-type OFETs, the current flows along the channel formed in the lateral direction. In general, the channel length between the source and drain electrodes is limited by the size of the evaporation mask (10-20 μm). On the other hand, the current flows in the vertical direction through the organic thin-film in a vertical-type OSIT (Fig. 1(b)). The channel length corresponds to the thickness of the organic semiconductors (0.1-0.2 μm). Therefore, the channel length of the OSIT is shorter than that of the OFET, which leads to lower operational voltage. Furthermore, the effect of surface roughness on the characteristics of OFETs may be significant, because the channel is formed along the interface of the gate-insulator and the semiconductor film. In contrast, the characteristics of OSITs may be less affected by surface roughness, because the channel is formed in the vertical direction through the thin-film. The lateral-type OFET was proposed as a prototype for the thin-film transistor (TFT), where a highly doped silicon (Si) substrate functions as the gate electrode.[19] However, in order to fabricate flexible OFETs, not only should non-flexible Si wafers be replaced with flexible substrates such as plastic, but the non-flexible SiO_2 gate-insulator should also be replaced with a flexible material. In contrast, vertical-type OSITs consist of organic semiconductors and metal electrodes. It should be noted that no gate-insulator is required for fabrication of the OSITs; therefore, only the non-flexible glass substrate requires replacement by a flexible substrate. This means that OSITs can be fabricated on a flexible substrate in the same manner as those on glass substrates. Considering this, a simple method such as a vacuum evaporation technique could be applied to the fabrication of OSITs on either glass or flexible plastic substrates. To confirm the advantages of vertical-type FETs, OSITs using pentacene or CuPc evaporated films were fabricated on glass substrates and the basic characteristics were evaluated,[20,21] as detailed in Sec. 3. Furthermore, OSITs based on pentacene evaporated film were fabricated on flexible substrates[22] and the electrical characteristics under bending were measured.[23] These results show that OSITs have potential for use as flexible organic transistors with low operational voltages.

In application, optoelectronic elements using organic materials show promise for low-cost, large-area, light-weight and flexible devices. In particular, liquid crystals and OLEDs are expected to be used as display components of mobile electronic devices, and OLEDs have been developed with excellent stability and high efficiency.[24,25] On the other hand, organic transistors have been greatly improved in recent years, and all-organic displays are expected to be developed using organic transistors.[26-29] However, for practicality, it is necessary to operate with a drive voltage as low as a few volts and with sufficient reliability. The vertical transistors, particularly OSITs, are suitable for flexible displays and the basic characteristics of OSITs as driving elements for organic display devices have already been reported.[4] In addition, OLETs combined with OSITs and OLEDs,[7] and organic logic devices using OSITs have also been reported.[21]

3. Interface Control of Vertical-Type Organic Transistor

In OLEDs, it has been reported that the electron carrier injection characteristics were improved by a thin insulating layer such as LiF^{30} or $Al_2O_3^{31}$ inserted between the cathode and organic layers. In addition, the hole carrier injection characteristics were also improved by a thin copper phthalocyanine (CuPc) inserted between the anode (typically indium-tin-oxide (ITO) on glass) and organic layers.[32-34] These phenomena demonstrate importance of the effect of the inserted layer between the electrodes and the organic layer on the carrier injection in the organic devices which is a key factor for designing organic devices. On the other hand, in order to realize the high performance OFETs, many works have been carried out to obtain the high field-effect mobility[35,36] and to understanding the operation mechanism of OFETs.[37,38] Nevertheless, it has been known to be difficult to operate OFETs at lower voltage because of the low-carrier density and the low-carrier mobility of organic semiconductors compared with inorganic semiconductors such as silicon. Many works have been carried out to obtain the high field-effect mobility by surface treatment on gate dielectric material.

OSITs have been studied for an attractive device in respect to the realization of the high-speed and high-power operation at the lower driving voltage.[4-6,22,23,39-42] The cross-sectional illustration and electrical circuits of OSITs are shown in Fig. 2(a). The excellent characteristics result from the vertical structure with very short channel length which corresponds to the thickness of the organic semiconductors. Recently, we have fabricated the OSITs using pentacene thin-film on ITO formed on the flexible substrate as shown in Fig. 2(b) and have reported their basic electrical property[22] under the bending conditions.[23] The flexible OSITs showed stable electrical characteristics under bending conditions. However, several problems have remained in the OSITs with low on/off ratio and low current density for use in the electronic circuits. The schematic energy diagram of the OSITs was illustrated in Fig. 2(c). In the OSITs, hole carriers were injected through the interface between ITO source electrode and pentacene film and transported to Au drain electrode. Therefore, we focus on the interface electronic states in relation to the hole injection for improving the characteristics of the OSITs.

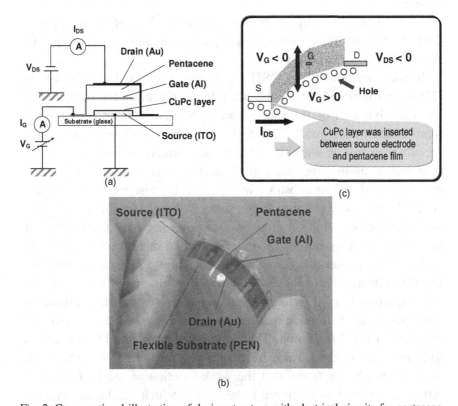

Fig. 2. Cross-sectional illustration of device structure with electrical circuits for pentacene OSITs with ultra-thin CuPc layer (a), photograph of flexible OSITs (b), and energy band diagram for pentacene-based OSITs (c). CuPc was deposited on ITO to improve the on/off ratio of OSITs.

In our recent research on the OSITs, an ultra-thin CuPc layer was inserted between the ITO source electrode and the pentacene in order to improve the carrier injection from the ITO source electrode. The CuPc layer is well-known to the key materials to improve hole injection into OLEDs.[32,33] As a result, the characteristics of OSITs were dramatically improved.[41] In addition, the dependence of the CuPc thickness on the static characteristics of the OSITs was investigated, which indicates that the optimized CuPc thickness was from 1 to 3 nm.[42] There are many reports on the research with regards to the CuPc/ITO systems in the OLEDs. However, the pentacene/CuPc/ITO system has not been reported.

In this report, we report that the effects of the electronic states of CuPc at the interface between the pentacene and the ITO on the characteristics of the OSITs.

In this experiment, the OSITs were also fabricated directly on an ITO/glass substrate without the ultra-thin CuPc layer for comparison. The effective area of the source and drain electrodes of the OSIT is approximately 1.76 mm^2. The devices are fabricated by conventional vacuum evaporation with the substrate temperature maintained at room temperature during the vacuum deposition. Prior to the deposition process, the ITO/glass substrate was cleaned because the work function of ITO is extremely sensitive to the surface cleaning,[43] which affects the characteristics of OSITs.[44] Here, we describe not only the method of the surface cleaning but also the work function of ITO obtained by the results of ultraviolet photoemission spectroscopy (UPS) measurements. The ITO substrate was subjected to ultrasonic agitation in water, acetone and isopropanol at 70°C for 10 min, respectively, and then expose to isopropanol vapor. In addition, the substrate was exposed to UV-ozone treatment at 50°C for 5 min. The work functions of the ITO substrates with and without the surface treatment were measured by UPS. From these results, it was found that the work function of the ITO with the surface treatment was estimated to be 5.3 eV, which was higher than the ITO work function of 4.3 eV without the surface treatment.

In this study, the high work function ITO with the surface treatment was used for fabricating the OSITs. The OSITs based on pentacene thin-film with an ultra-thin inserting CuPc layer was fabricated as follows.

The different process from our past research is that the CuPc layer with a thickness of 1 nm is deposited on the ITO source electrode formed on the glass substrate. On the other hand, the deposition conditions of pentacene, Al gate electrode, Au drain electrode are the same as our past research, which are described in detail in Refs. 22, 23, 41 and 42.

Here, we compare the static characteristics for the pentacene OSITs between with and without the CuPc layer as shown in Fig. 3. In both samples, source-drain current (I_{DS}) at a constant drain-source voltage (V_{DS}) decreased with increasing gate voltage (V_G) and typical OSITs characteristics were exhibited: the slope of the I_{DS}-V_{DS} curves increased with increasing V_{DS} without current saturation. In this experiment, V_{DS} was changed continuously from 0 to -3 V while V_G was changed from -1 to +5 V in +0.2 V steps. I_{DS} was controlled by the V_G. I_{DS} was found to increase as V_G decreased from 0 to -1 V, and decrease as V_G increased from 0 to +5 V. In the OSITs without the CuPc layer, the on/off ratio of I_{DS} at V_{DS} = -3 V under applied V_G from -1 V (on state) to +5 V (off state) was 2. On the other hand, in the OSITs with CuPc layer, the on/off ratio was 1080, which was two orders higher than that of the OSITs without CuPc layer. In addition, the current value of the OSIT with CuPc layer was also higher by three orders of magnitudes than those of the OSIT without an CuPc layer. The obtained results also indicated that the

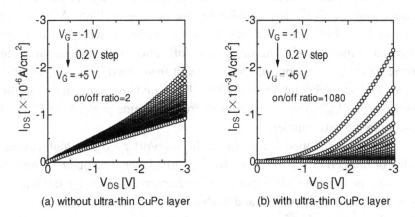

(a) without ultra-thin CuPc layer   (b) with ultra-thin CuPc layer

Fig. 3. Static characteristics of pentacene OSITs without ultra-thin CuPc layer (a) and with ultra-thin CuPc layer (b). High-current on/off ratio of OSITs was achieved by inserting ultra-thin CuPc layer between ITO source electrode and pentacene film.

high on/off ratio arise from the increase in I_{DS} at the on state with no variation in I_{DS} at the off state.

To investigate the effect of CuPc layer between the ITO and the pentacene on the characteristics of the OSITs, the electronic structure of pentacene/ITO and pentacene/CuPc/ITO was evaluated UPS. The UPS measurement was carried out to clarify the influence of the CuPc layer on the electronic states at the interface of the pentacene/CuPc/ITO. Figures 4(a) and 4(b) show the UPS spectra of pentacene/ITO and pentacene/CuPc/ITO systems as a function of the incremental deposition of the pentacene overlayer in the secondary region and the HOMO region. In secondary-cutoff region, the work function of ITO was estimated approximately 5.3 eV. In the HOMO band region, the hole injection barrier height of pentacene (1 nm, 10 nm)/ITO was approximately 0.35 eV and the peak of the HOMO feature was 0.85 eV as shown in Fig. 3(a). In this case, the hole injection barrier height is defined as the energy difference between the low binding energy edge of the HOMO feature and the Fermi level (E_F^{sub}).

On the other hand, the hole injection barrier height of CuPc (1 nm)/ITO was approximately 0.15 eV and that of pentacene/CuPc (1 nm)/ITO was decreased from 0.35 to 0.15 eV with decreasing the pentacene film thickness from 10 nm to 1 nm which originates a HOMO band tailing located between the low binding energy edge of the HOMO feature and the E_F^{sub}. The HOMO feature peaking at 0.85 eV was also seen in Fig. 4(b). From the results of the static characteristics of the OSITs and the UPS spectra, the high on/off ratio was achieved when the increases in I_{DS} at the on state which result from the decrease in the hole injection barrier height by the CuPc layer and the prevention of the increasing in I_{DS} at the off state result from the same peak of the HOMO feature in the two samples.

We have developed that both the high on/off ratio and high-current value of the OSITs was achieved by the CuPc layer. The obtained results demonstrate that the CuPc layer plays an important role in the fabrication of OSITs with high-current and high-current on/off ratio by decreasing the hole injection barrier at the interface between pentacene and ITO. The controllability of HOMO and LUMO level of pentacene films is strongly depends on the crystal phase and grain orientation. In order to

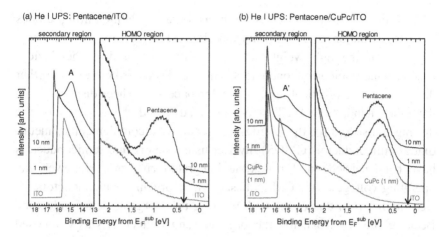

Fig. 4. He I UPS spectra of pentacene/ITO (a) and pentacene/CuPc/ITO (b) system as a function of deposition amount of pentacene overlayer in secondary region and HOMO region. All of the spectra are measured at 295 K with a -5 V bias applied to the sample to observe the vacuum level. The EB scale refers to E_F^{sub}.

improve the performance of the OSITs, it is a key factor to optimize the electronic structure such as the hole injection barrier and band bending at the interface between the organic semiconductor layer and the source electrode.

4. Organic Inverter Based on Vertical-Type Organic Transistors

Vertical-type OSITs have advantage with regard to the lower operational voltages compared to the conventional lateral-type OFETs. For the above reason, OSITs were applied to fabricate the organic inverters which can operate at low-voltages. In this study, we have fabricated two types of organic inverters with OSITs based on pentacene films. One is the horizontally aligned structure, the other is vertical combined structure. In particular, the organic inverter with vertically combined structure has advantage of the effective device area compared to that with horizontally aligned structure. From the measurement results of the inverter transfer characteristics, it was found that the operational voltage of the organic inverters based on the OSITs was from -2 to +2 V.

Organic electronic devices such as organic light-emitting diodes (OLEDs)[45] and organic thin-film transistors (OTFTs)[19] are very attractive in terms of the light-weight and flexibility compared to the devices based on inorganic semiconductors. The OLEDs contribute to the further development of practical application for paper-like sheet displays.[46] On the other hand, the OTFTs have been advanced the research of the driving transistors for the OLEDs.[2,47] In addition, complementary metal-oxide-semiconductor (CMOS) inverter using the organic semiconductors have been carried out for flexible integrated circuits.[48] However, the operating voltage of the CMOS based on the organic semiconductors[49-59] were higher than that of the CMOS using the inorganic semiconductors. Because the CMOS using the organic semiconductors was composed of the OTFTs with lateral-type structure which operate at higher voltage arise from low-carrier density and low-mobility in organic semiconductors.

To overcome the disadvantages of the OTFTs mentioned above, we have researched on the OSITs with vertical-type structure.[4-7,20-23] Recently, other research group also reported on the OSITs.[40] Besides the OSITs, various vertical-type organic transistors[8-16] have been proposed to fabricate high performance organic thin-film transistors. The SIT based on inorganic semiconductor was proposed by Nishizawa, and showed high-speed and high-power operations and their excellent characteristics mainly arise from the vertical structure which has a very short length between the source, drain and gate electrodes.[60] In our past research, the SITs have been applied to fabricate the organic transistors and were also excellent characteristics with low driving voltage and high-speed operation due to the short channel length corresponds to the thickness of the organic semiconductor layer. We have succeeded to fabricate the OSIT on the flexible substrate[22,23] and to improve in on/off ratio of the OSITs with ultra-thin CuPc layer.[41,42]

In this study, organic inverters based on the OSITs were fabricated and their characteristics were investigated. Two types of device structures for the organic inverters were proposed: one is the conventional horizontally aligned structure, the other is the vertically combined structure. We report their fabrication process and the basic characteristics of inverters using the OSITs. Figures 5(a) and 5(b) show

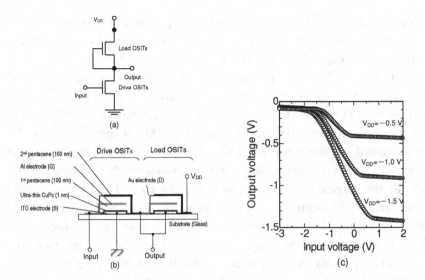

Fig. 5. Inverter circuit schematic (a), cross-sectional illustration of the organic inverter based on OSITs with horizontally aligned structure (b) and transfer characteristics of organic inverter based on OSITs with horizontally aligned structure (c).

the inverter circuit schematic and the cross-sectional illustration of the organic inverter based on OSITs acted as load transistor and drive transistor, respectively.

Before the measurement results of OSITs characteristics, we explain the operational mechanism of OSITs. I_{DS} under applied V_{DS} condition was controlled by the applied V_{GS}. When the applied V_{GS} was negative, the hole injected from the source electrode was enhanced by increasing the electric field at the interface between organic semiconductor and source electrode. When the applied V_{GS} was positive, the hole carrier flow between the source and drain electrodes was restricted by the formation of depletion layer around Al gate electrode and by decreasing the electric field at the interface between organic semiconductor and source electrode. In both cases, the gate leak current was much smaller than I_{DS} due to the formation of the depletion layer around Al gate electrode.

The fabrication process of the OSITs based on the pentacene thin-film was as follows. The ITO/glass substrate was subjected to ultrasonic agitation in water, acetone and isopropanol at 70°C for 10 min, respectively, and then expose to isopropanol vapor. In addition, the substrate was exposed to UV-ozone treatment at 50°C for 5 min. First, an ultra-thin CuPc with a thickness of 1 nm is deposited on the cleaned ITO/glass substrate to improve in on/off ratio of the OSITs.[41,42] The ultra-thin CuPc layers are evaporated under a vacuum of 2×10^{-4} Pa and the evaporation rate is 0.01 nm/s. Second, a 100 nm pentacene thin-film is deposited on the ultra-thin CuPc layer. Third, a slit-type Al gate electrode with a thickness of 25 nm is formed on the pentacene film. Fourth, the Al gate electrode is covered with a second 100 nm pentacene film. Finally, the drain Au with a thickness of 30 nm is fabricated on the pentacene film. The pentacene films are evaporated under a vacuum of 2×10^{-4} Pa and the evaporation rate is 0.1 nm/s. Further points of the fabrication process are described in detail in Refs. 46 and 47. The effective area of the source and drain electrodes of the OSIT is approximately 1.76 mm^2.

Electrical measurements were performed using a semiconductor parameter analyzer (4156C, Agilent) in air at room temperature. Measurements were carried out in the dark in order to obtain OSIT characteristics without the photovoltaic effect. Prior to investigate the voltage transfer characteristics of the inverters integrated two OSITs, the static characteristics such as the source-drain current (I_{DS}) as a function of the source-drain voltage (V_{DS}) of two OSITs were measured.

The organic inverters with horizontally aligned structure was composed of two OSITs which interconnecting between the electrodes as shown in Fig. 5(b). The inverter transfer characteristics were measured under a variety of bias conditions. The supply voltage was varied in the range from -0.5 to -1.5 V and the transfer characteristics measured in each case as shown in Fig. 5(c). For low values of V_{in}, the drive OSITs is on and the load OSITs is off. As the input voltage is increased the drive OSITs is gradually turned off. A further increase in input voltage causes the drive OSITs to be turned off and the load OSITs to be turned on. The voltage at the output is the supply voltage. From the results, it was found that the operating voltage of the inverter based on OSITs was lower (-2 to +2 V) than that of the inverters using the OFETs.[49-59] The slight hysteresis at the switch-on and switch-off voltage was observed.

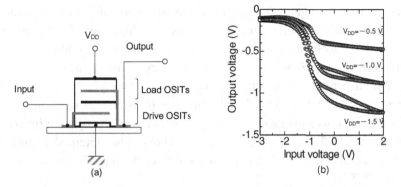

Fig. 6. Cross-sectional illustration (a) and transfer characteristics (b) of the organic inverter based on OSITs with vertically combined structure.

Organic inverters which are vertically combined structure with two OSITs were also fabricated and the device characteristics are investigated. In this case, the OSITs acted as load transistor were fabricated on the OSITs acted as drive transistor as shown schematically in Fig. 6(a). This vertically combined structures have several advantages as a simple fabrication process compared to that with horizontally aligned structure. In addition, the effective electrode area of the organic inverter with vertical stacked structure was smaller than that with horizontally aligned structure as mentioned above. The inverter transfer characteristics were measured under the same condition with the horizontally aligned organic inverters, which was shown in Fig. 6(b).

Compared with the horizontally aligned organic inverters, the input and output levels do not match at the output voltage level. In addition, the vertically aligned organic inverters have a hysteresis around the switching voltage range. These characteristics were seemed to be due to the low on/off ratio of the drive OSITs and the load OSITs. However, it is difficult to determine the reason of the levels mismatch and the hysteresis because the organic inverter with vertically combined structure cannot be divided in two OSITs acted as load transistor and drive transistor to measure the static characteristics of the OSITs. For the above reason, the horizontally aligned organic inverters composed of two OSITs with low on/off ratio was fabricated to investigate the reason of the levels mismatch and the hysteresis. The results showed a similar inverter transfer characteristics to that of the vertically aligned organic inverters. These results demonstrate that the high on/off ratio of the OSITs acted as load transistor and drive were needed to realization high performance organic inverter with vertically combined structure.

We have reported the characteristics of the two types of the organic inverters based on the OSITs. The inverter transfer characteristics were observed at lower operational voltage compared to the organic inverters based on OFETs as reported several authors. The obtained results as mentioned above demonstrate the feasibility of the organic inverter based on the OSITs.

5. Vertical-Type Organic Light-Emitting Transistor

OLETs which are vertically combined with the OSITs and OLEDs are fabricated and the device characteristics are investigated. One of the device structure of the OLETs is shown schematically in Fig. 7(a). The OLET consists of the OSITs based on pentacene with CuPc layer and conventional OLED based on α-NPD and a simple structure similar to the OLEDs.[7,61] All layers were fabricated on ITO-coated glass substrate using the vacuum evaporation technique at approximately 10^{-4} Pa. The substrate temperature was maintained at room temperature during evaporation.

The OLET has a slit-type Al gate electrode in the organic semiconductor layers of OSITs. The slit-type gate electrode was formed using a shadow evaporation mask, as described in Secs. 2 and 3. Both the line and space dimensions of the shadow mask are 20 µm. In the OSIT, typical thicknesses for both the first and second pentacene layers are 100 nm. On the other hand, in the OLED, typical thicknesses for both the α-NPD layer used as an hole transport layer and tris-(8-hydroxyquinoline) aluminum (Alq3) layer as an emitting layer are approximately 50 nm. The effective electrode area of the OLED using OSIT and the OLED is approximately 1.76 mm^2. For fabrication of the OLETs, the controllability of the slit-type Al gate electrode is important, as with the fabrication of OSITs. The estimated gate gap, W, was checked using optical microscopy, and W was less than 1 µm. Figure 7(b)

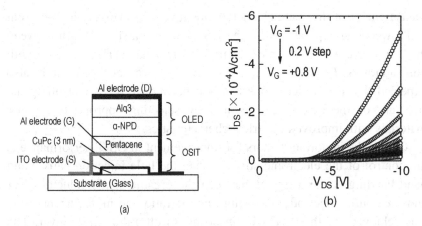

Fig. 7. Cross-sectional structures (a) and I-V characteristics (b) of OLET.

shows the I-V characteristics of the OLET as a function of gate voltage V_G. The I-V curve indicates that the Al-slit-gate forms a Schottky barrier to pentacene layer, and pentacene behaves as a p-type semiconductor. The I_{DS} at a constant V_{DS} decreases with increasing V_G. The current I_{DS} is controlled by relatively small V_G (from -1 to +0.8 V) and typical OSIT characteristics are also obtained for the OLETs. We confirmed that the luminance also varies corresponding to the I-V characteristics. These results are reasonable, because the potential barrier formed by the Schottky gate restricts the current flow from the source to the drain electrodes, and the emitted light is also obstructed by the Al gate electrode (slit gate part in the device structure). Future developments of OLETs operating with higher power and speed are expected from optimization of the device structure, such as the thickness of the organic layers and the dimensions of the gate electrode. In particular, the modulation depth and operational frequency of the OLETs are strongly dependent on the edge features of the gate electrode.

6. Summary

We have researched on vertical-type organic transistors that operate as SITs and recently succeeded to improve the device performance.

Measurement of the static characteristics revealed relatively high-current and low-voltage operations for the OSITs due to a short length between the source, drain, and gate electrodes. Typical SIT operation, with non-saturating I_{DS} versus V_{DS} characteristics, was observed. It is also expected that the device properties will be improved by controlling the structure of the gate electrode, and by selecting appropriate organic materials with improved electrical characteristics.

One of the important factors for improvement in OSIT performance is the control of the electronic structure, i.e. the hole injection barrier and band bending at the interface between the organic semiconductor layer and the source electrode. Both high on/off ratio and high-current value was achieved for the OSITs by inserting an ultra-thin CuPc layer. The experimental results demonstrate that the effect of the ultra-thin CuPc layer on the characteristics of OSITs is not only that the hole injection barrier at the interface between pentacene and ITO is decreased, but also that the pentacene film orientation may be controlled.

For the realization of flexible displays based on OSITs, flexible transistors, logic devices, and light-emitting transistors based on OSITs were fabricated. In our past experimental results showed that the flexible OSITs exhibited stable electrical characteristics at compressive and tensile strains up to a bending radius of 5 mm. Organic inverters using two OSITs displayed low operational voltage compared to those based on conventional OFETs. The device structures and performance of OLETs combined with OSITs and OLEDs were also investigated. Relatively high luminance modulation with relatively low gate voltage (±1 V) was observed in the OLET by optimizing the gate electrode and layer thicknesses. The results obtained here demonstrate that OSITs are a suitable element for flexible sheet displays.

References

1. C. D. Dimitrakopoulos and P. R. L. Malenfant, Adv. Mater. **14**, 99-117 (2002).
2. A. Dodabalapur, Z. Bao, A. Makhija, J. G. Laquindanum, V. R. Raju, Y. Feng, H. E. Katz and J. Rogers, Appl. Phys. Lett. **73**, 142-144 (1998).
3. H. Sirringhaus, N. Tessler and R. H. Friend, Science **280**, 1741-1743 (1998).
4. K. Kudo, D. X. Wang, M. Iizuka, S. Kuniyoshi and K. Tanaka, Synth. Met. **111-112**, 11-14 (2000).

5. K. Kudo, D. X. Wang, M. Iizuka, S. Kuniyoshi and K. Tanaka, Thin Solid Films **331**, 51-54 (1998).
6. K. Kudo, M. Iizuka, S. Kuniyoshi and K. Tanaka, Thin Solid Films **393**, 362-367 (2001).
7. K. Kudo, Curr. Appl. Phys. **5**, 337-340 (2005).
8. Y. Yang and A. J. Heeger, Nature **372**, 344-346 (1994).
9. N. Stutzmann, R. H. Friend and H. Sirringhaus, Science **299**, 1881-1884 (2003).
10. L. Ma and Y. Yang, Appl. Phys. Lett. **85**, 5084-5086 (2004).
11. S. Fujimoto, K. Nakayama and M. Yokoyama, Appl. Phys. Lett. **87**, 133503 (2005).
12. K. Fujimoto, T. Hiroi, K. Kudo and M. Nakamura, Adv. Mater. **19**, 525-530 (2007).
13. H. Naruse, S. Naka and H. Okada, Appl. Phys. Exp. **1**, 011801 (2008).
14. M. Uno, I. Doi, K. Takimiya and J. Takeya, Appl. Phys. Lett. **94**, 103307 (2009).
15. T. Takano, H. Yamauchi, M. Iizuka, M. Nakamura and K. Kudo, Appl. Phys. Exp. **2**, 071501 (2009).
16. S. M. Sze and H. K. Gummel, Solid State Electron. **9**, 751-769 (1966).
17. J. Nishizawa, T. Terasaki and J. Shibata, IEEE Trans. Electron. Dev. **22**, 185-197 (1975).
18. C. O. Bozler, G. O. Alley, R. A. Murphy, D. C. Flanders and W. T. Lindley, IEEE Tech. Dig. Int. Electron Dev. Meet. 384-387 (1979).
19. K. Kudo, M. Yamashina and T. Moriizumi, Jpn. J. Appl. Phys. **23**, 130-130 (1984).
20. K. Kudo, M. Iizuka, S. Kuniyoshi and K. Tanaka, Thin Solid Films **393**, 362-367 (2001).
21. Y. Watanabe, H. Iechi and K. Kudo. Thin Solid Films **516**, 2729-2732 (2008).
22. Y. Watanabe and K. Kudo, Appl. Phys. Lett. **87**, 223505 (2005).
23. Y. Watanabe, H. Iechi and K. Kudo, Appl. Phys. Lett. **89**, 233509 (2006).
24. S. A. Van Slyke, C. H. Chen and C. W. Tang, Appl. Phys. Lett. **69**, 2160-2162 (1996).
25. G. Parthasarathy, P. E. Burrows, V. Khalfin, V. G. Kozlov and S. R. Forrest, Appl. Phys. Lett. **72**, 2138-2140 (1998).
26. H. E. A. Huitema, G. H. Gelinck, J. B. P. H. van der Putter, K. E. Kuijk, C. M. Hart, E. Cantatore, P. T. Herwig, A. J. J. M. van Breemen and D. M. de Leeuw, Nature **414**, **599** (2001).
27. C. D. Sheraw, L. Zhou, J. R. Huang, D. J. Gundlach and T. N. Jackson, Appl. Phys. Lett. **80**, 1088-1090 (2002).
28. J. A. Rogers, Z. Bao, M. Meier, A. Dodabalapur, O. J. A. Schueller and G. M. Whitesides, Synth. Met. **115**, 5-11 (2000).
29. H. Sirringhaus, T. Kawase, R. H. Friend, T. Shimoda, M. Inbasekaran, W. Wu and E. P. Woo, Science **290**, 2123-2126 (2000).
30. L. S. Hung, C. W. Tang and M. G. Mason, Appl. Phys. Lett. **70**, 152-154 (1997).
31. F. Li, H. Tang, J. Anderegg and J. Shinara, Appl. Phys. Lett. **70**, 1233-1235 (1997).
32. I. G. Hill and A. Kahn, J. Appl. Phys. **86**, 2116-2122 (1999).
33. S. M. Tadayyon, H. M. Grandin, K. Griffiths, P. R. Norton, H. Aziz and Z. D. Popovic, Organic Electron. **5**, 157-166 (2004).

34. S. F. Nelson, Y.-Y. Lin, D. J. Gundlach and T. N. Jackson, Appl. Phys. Lett. **72**, 1854-1856 (1998).
35. H. Klauk, M. Halik, U. Zschieschang, G. Schmid, W. Radlik and W. Weber, J. Appl. Phys. **92**, 5259-5263 (2002).
36. O. D. Jurchescu, J. Baas and T. T. M. Palstra, Appl. Phys. Lett. **84**, 3061-3063 (2004).
37. M. Nakamura, N. Goto, N. Ohashi, M. Sakai and K. Kudo, Appl. Phys. Lett. **86**, 122112 (2005).
38. T. Manaka, E. Lim, R. Tamura and M. Iwamoto, Appl. Phys. Lett. **87**, 222107 (2005).
39. D. X. Wang, Y. Tanaka, M. Iizuka, S. Kuniyoshi, K. Kudo and K. Tanaka, Jpn. J. Appl. Phys. **38**, 256-259 (1999).
40. S. Zorba and Y. Gao, Appl. Phys. Lett. **86**, 193508 (2005).
41. Y. Watanabe, H. Iechi and K. Kudo, Jpn. J. Appl. Phys. **45**, 3698-3703 (2006).
42. Y. Watanabe, H. Iechi and K. Kudo, Jpn. J. Appl. Phys. **46**, 2717-2721 (2007).
43. K. Sugiyama, H. Ishii, Y. Ouchi and K. Seki, J. Appl. Phys. **87**, 295-298 (2000).
44. Y. Watanabe, H. Iechi and K. Kudo, Trans. Mater. Res. Soc. Jpn. **31**, 593-596 (2006).
45. C. W. Tang and S. A. VanSlyke, Appl. Phys. Lett. **51**, 913-915 (1987).
46. L. Zhou, A. Wanga, S. Wu, J. Sun, S. Park and T. N. Jackson, Appl. Phys. Lett. **88**, 083502 (2006).
47. M. Kitamura, T. Imada and Y. Arakawa, Appl. Phys. Lett. **83**, 3410-3412 (2003).
48. H. Klauk, M. Halik, U. Zschieschang, F. Eder, G. Schmid and C. Dehm, Appl. Phys. Lett. **82**, 4175-4177 (2003).
49. A. R. Brown, A. Pomp, M. Hart and D. M. de Leeuw, Science **270**, 972-974 (1995).
50. A. Dodabalapur, J. Baumbach, K. Baldwin and H. E. Katz, Appl. Phys. Lett. **68**, 2246-2248 (1996).
51. A. Dodabalapur, J. Laquindanam, H. E. Katz and Z. Bao, Appl. Phys. Lett. **69**, 4227-4229 (1996).
52. H. Klauk, D. J. Gundlach and T. N. Jackson, IEEE Electron. Dev. Lett. **20**, 289-291 (1999).
53. Y. Y. Lin, A. Dodabalapur, R. Sarpeshkar, Z. Bao, W. Li, K. Baldwin and V. R. Raju, Appl. Phys. Lett. **74**, 2714-2726 (1999).
54. G. H. Gelinck, T. C. T. Geuns and D. M. de Leeuw, Appl. Phys. Lett. **77**, 1487-1489 (2000).
55. M. G. Kane, J. Campi, M. S. Hammond, F. P. Cuomo, B. Greening, C. D. Sheraw, J. A. Nichols, D. J. Gundlach, J. R. Huang, C. C. Kuo, L. Jia, H. Klauk and T. N. Jackson, IEEE Electron. Dev. Lett. **21**, 534-536 (2000).
56. B. K. Crone, A. Dodabalapur, R. Sarpeshkar, R. Sarpeshkar, R. W. Filas, Y. Y. Lin, Z. Bao, J. H. O'Neill, W. Li and H. E. Katz, J. Appl. Phys. **89**, 5125-5127 (2001).
57. Y. Inoue, Y. Sakamoto, T. Suzuki, M. Kobayashi, Y. Gao and S. Tokito, Jpn. J. Appl. Phys. **44**, 3663-3668 (2005).

58. M. Ahles, R. Schmechel and H. V. Seggern, Appl. Phys. Lett. **87**, 113505 (2005).
59. D. J. Gundlach, K. P. Pernstich, G. Wilckens, M. Gruter, S. Hees and B. Batlog, J. Appl. Phys. **98**, 064502 (2005).
60. J. Nishizawa, T. Terasaki and J. Shibata, IEEE Trans. Electron. Dev. **22**, 185-197 (1975).
61. K. Kudo, S. Tanaka, M. Iizuka and M. Nakamura, Thin Solid Films **438-439**, 330-333 (2003).

CHAPTER 4

ELECTROCHEMICAL PROPERTIES OF SELF-ASSEMBLED VIOLOGEN DERIVATIVE AND ITS APPLICATION TO HYDROGEN PEROXIDE DETECTING SENSOR

Dong-Yun Lee, Hyen-Wook Kang, Sang-Hyun Park and Young-Soo Kwon*

*Department of Electrical Engineering and NTRC, Dong-A University,
Busan 604-714, Korea
E-mail: yskwon@dau.ac.kr

Viologen derivatives, which have been widely investigated their well-behaved electrochemistry including electron transfer mediation, are attractive materials because of their chemical stability, their relatively simple behavior of redox reaction and their possible practical applications due to their electrochemical properties. The self-assembled monolayers (SAMs) were prepared surfaces by use of a viologen. The electrochemical property of the self-assembled viologen derivatives has been investigated with a quartz crystal microbalance (QCM), which has been known as a nanogram order mass detector. After then we attempted to apply this viologen derivative for biosensor design. A gold electrode modified by viologen-thiol modification is used to design a biosensor in corporate with hemoglobin (Hb). By incorporating with SAMs viologen and Hb, viologen can act as an electron transfer mediator between Hb and gold electrode. Experimental conditions influencing the biosensor performance such as pH and/or potential are optimized and assessed. This sensor offered an excellent electrochemical response for H_2O_2 concentration below μmol level with high sensitivity, selectivity and short time response.

1. Introduction

Electrochemical properties of a redox molecule incorporated in a phospholipid monolayer, bilayer, or membrane have attracted much attention in the past several decades.[1] Controlled modification of an electrode surface with a monolayer film of organic molecular gas has attracted much attention because of its relevance to the construction of molecular devices. The modified electrode is also useful for the understanding of the electron transfer mechanism at electrode/electrolyte interfaces at a molecular level. An oriented monolayer assembly can be created by both the Langmuir-Blodgett and the self-assembling methods using various alkane derivatives and substrates.[2] The self-assembled monolayers of alkane derivatives with sulfur containing head groups on gold substrates have been widely examined recently, since the binding between S atoms and the Au surface is strong and the S-anchored monolayers thus formed have a well-oriented structure.[3,4] To understand the structure of the assembled monolayer and electron transfer mechanism, it is essential to conduct quantitative in situ monitoring of mass transport both in the assembly process and in the electron transfer process so that we can quantify the number of adsorbed molecules and the movement of the ion and solvent accompanying an electron transfer process. A recently developed electrochemical quartz crystal microbalance (EQCM) offers one of the best ways to obtain this quantitative information.[5]

Viologen derivatives are actually desirable electron mediators.[6] They have been widely studied for their chemical stability, their relatively simple behavior in redox reactions, and their possible practical applications due to their electrochemical properties.[7] The electrochemical behavior of the alkyl viologen as an electron transfer mediator and a redox-active linker on a gold substrate, confined on the self-assembled monolayer (SAM) coated on Au electrodes in the presence of different supporting electrolytes. The investigation of a viologen derivative, VC_8SH, has been carried out by cyclic voltammetry which indicated that anions adsorbed and desorbed during redox reaction. Viologen exists in three main oxidation states, namely, $V^{2+} \leftrightarrow V^{\cdot+} \leftrightarrow V^0$. These redox reactions, especially the first one ($V^{2+} \leftrightarrow V^{\cdot+}$), are highly reversible and can be cycled many times without significant side reactions.[8] This well-known electrochemical behavior has used in various studies, such as an electron transfer mediator.[9]

The QCM technique is based on the tendency of a piezoelectric crystal to change its resonant frequency when additional mass adsorption or desorption on the crystal electrodes takes place.[10] The analytical technology has been improved or newly developed as to the incremental requirement for electrochemical phenomena at the interface. The QCM combined with electrochemical technique, namely EQCM, is a powerful tool in investigating the interfacial processes that occur on surface and thin-films.[11] The EQCM method is used in situ experiments to measure mass changes at electrode surfaces after electrochemical deposition of metal.[12,13]

Theory on cyclic voltammograms of such surface-confined redox species has been developed with use of the Langmuir adsorption isotherm and the Nernst equation.[14] The self-assembling molecules possess a redox-active site, the adsorption process can be monitored in real time by electrochemical techniques, such as cyclic voltammetry (CV). A number of researchers have also employed various electrochemical techniques to study the stability, permeability, and uniformity of self-assembled films. The electrochemical techniques are very useful in monitoring the adsorption process; however, they can only monitor the amount of the adsorbed species having an electroactive site.[15] The redox reaction and EQCM response using viologen derivative showed influence of the anion.[16,17] Because the viologen molecules were finished in self-assembly on the gold electrode of the QCM, they became the reduction states. In this case, the anions were only adsorbed and desorbed, but the paper showed the property of the anion.

Since the first biosensor, based on enzyme, invented by Clark and Lyons,[18] there has been growing interest in the biosensor fields. Biosensor is usually an analytical device being used in biotechnology, medicine, industries and environmental monitoring. The detection of hydrogen peroxide (H_2O_2) is a great interesting object to the biochemists because of its validity in pharmaceutical, clinical, industrial, food area. Recently, an increasing number of redox-active proteins have been used to design biosensor for detecting H_2O_2 and some other target materials.[19,20] In this point of view, horse radish peroxidase (HRP) is the most commonly used enzyme.[21,22] But the HRP is not so stable in aqueous media and very expensive. Alternately, hemoglobin (Hb) is one of the promising biomolecules which can be used as a bioreceptor element. Hemoglobin, the main component in red blood cells, is one kind of protein that picks up oxygen in the lungs and delivers it to the body tissues. It has a molar mass of 67,000 g/mol and consists of four kinds of electroactive iron hemes. Usually, Hb can give the long-term stability to the H_2O_2 sensor as compared to other proteins. In addition, Hb is cheaper than commonly used HRP.

In order to make a biosensor with the protein, it is very important the electrochemical contact between redox-active protein, such as Hb and the transducer such as electrode. Normally, the fast electron transfer between protein, such as Hb, and electrode surface is not possible due to some reason. Among them, the deep entombing of the electroactive heme group in the protein structure is the main reason. There are numerous papers have been published on the basis of direct electrochemistry of Hb. In these circumstances, it is an intriguing object for biochemists to achieve efficient electrical contact between protein and electrode. There are numerous papers which have been published on the basis of direct electrochemistry of Hb. Among them, hemoglobin immobilized on egg-phosphatidycholine,[19] kieselgubr,[23] zirconium,[24] carbon nanotube[25] and DNA[26] have been described successfully.

Several methods such as chemical modification[27] and polymers modification[28] have been applied to electrically communicate redox protein with the electrode. Langmuir-Blodgett (LB) technique has also been used intensively for depositing the protein onto the electrode surface.[29] In addition, there are a lot of materials such as polyphenol, polypyrrole, polyaniline[30-32] have been used electrochemically with wide range to immobilize biomaterials. But these kinds of deposited films tend to be thick. Nonetheless, in the methods mentioned above, aloofness of the protein or conformational change may affect the functional activity of protein. As a result, the designed biosensor shows a lower sensitivity and short longevity. Therefore, researchers are paying their attention to fabricate the thin-films of a wide variety of materials which is just single molecule thick. Dong and Li[33] have published a paper based on viologen monolayer as the electron mediator of HRP. Viologen, redox-active material, plays an important role as electron relays in electrochemical process. It can enhance the electron transfer between the redox protein and electrode surface. There are only a few papers of SAMs of viologen derivatives for designing biosensor with redox protein, have been published till date. Therefore, it needs more investigation to understand the intrinsic redox behavior of the protein with SAMs.

The purpose of this study is to observe the electrochemical behavior of self-assembled viologen monolayers as electron transfer mediator and its application in a biosensor. An in situ quartz crystal microbalance (QCM) measurement indicated a similar adsorption trend and frequency change for the formation of self-assembled monolayers (SAMs) of the viologen.[1] A recently developed electrochemical quartz crystal microbalance (EQCM) method is used in ex situ experiments to measure mass changes at electrode surfaces after electrochemical deposition of a metal and offers one of the best ways to obtain this quantitative information.[34] We used a thiol-functionalized viologen derivative [VC$_8$SH] for increasing rate of electrons transfer of Hb. The method for modification of the electrode is very simple and convenient. We designed the sensor by self-assemble of viologen derivatives. To the best of our knowledge, for the first time, we have designed a hydrogen peroxide biosensor based on Hb-modified gold electrode where the thiol-functionalized viologen has been performed a very important role for enhancing electrons transfer rate of Hb.

2. Principle of QCM Technique

2.1. *Overview*

Since all solid materials have natural resonance frequencies of vibration, almost any solid made of an elastic material could be used like a crystal. For example, steel is very elastic and transmits sound with high-speed. It was often used in mechanical filters before quartz. The resonance frequency depends on size, shape, elasticity, and the speed of sound in the material. High-frequency crystals are typically cut in the shape of a simple, rectangular plate.

When a quartz crystal is properly cut and mounted, it gets deformed in an electric field by applying a voltage to both electrodes on the crystal. That is, mechanical strain generates electric polarization and electric field generates mechanical stress. This property is known as piezoelectricity. When the field is removed, the quartz will generate an electric field as it returns to its previous shape, and this can generate a voltage. Consequently, a quartz crystal behaves like a circuit composed of an inductor, capacitor and resistor, with a precise resonance frequency.

The quartz crystal has 32 symmetry groups of trigonal system. According to different cut angles, there are different kinds of quartz plates, for example, AT-, BT-, CT-, DT-, NT-, and GT-cut quartz plates. Each type of quartz cuts, indicated by a set of Euler angles, has different available elastic, piezoelectric and dielectric properties, which are the basic parameters for designing a crystal device. In this chapter, we used an AT-cut quartz crystal.[39]

Quartz has the further advantage that its elastic constants and its size do not change by the change of the temperature. The frequency dependence on temperature is also very low. The specific characteristics depend on the mode of vibration and the angle at which the quartz is cut. Therefore, the resonance frequency of the plate, which depends on its size and thickness, does not change much. An early use of this property of quartz crystals was in phonograph pickup. One of the most common piezoelectric uses of quartz today is as a crystal oscillator. The quartz clock is a familiar device using quartz. The resonance frequency of a quartz crystal oscillator is changed by mechanically loading mass, and this principle is used for very accurate measurements of very small mass changes in the QCM and in thin-film thickness monitors.

In this chapter, we introduce overall QCM, including the equivalent circuit, theory of a QCM and measurement principle of resonance parameters.

2.2. Theory and Equivalent Circuit of QCM

The frequency shift induced by mass loading to the quartz crystal or the electrode surface, is described by the Sauerbrey equation (1).[40-42] From Eq. (1), the ideal sensitivity can be calculated that a 9 MHz QCM has the mass sensitivity about 1.1 ng per 1 Hz change in the resonance frequency, where $\mu_q = 2.95 \times 10^{11}$ g/cm·s^2, $\rho_q = 2650$ kg/m^3

and $A = 0.2$ cm^2. This sensitivity is the reason the QCM has been widely used as a mass sensor.

$$\Delta f = -\frac{2f_0^2}{A \cdot \sqrt{\mu_q \rho_q}} \cdot \Delta m \tag{1}$$

where, Δf is the frequency change, f_0 is the fundamental resonance frequency of the quartz, Δm is the mass change, A is the electrode area, μ_q is the shear modulus of the quartz, and ρ_q is the density of the quartz.

Since the Sauerbrey equation is useful for rigid mass loading in air, Kanazawa and Gordon developed Eq. (2), which indicates the QCM can be performed in liquid.[35]

$$\Delta f = -f_0^{3/2} \left(\frac{\eta_l \rho_l}{\pi \mu_q \rho_q} \right)^{1/2} \tag{2}$$

where, η_l is the viscosity of the liquid and ρ_l is the density of the liquid.

The viscosity change of loading material also affects the resonance property of the QCM. To measure the resonance resistance, R_1, is useful to investigate the viscosity change on the QCM as shown in Eq. (3).[36] As the resonance resistance is unaffected by mass, this parameter is useful to analyze the rheological property on the QCM. After development of Eqs. (2) and (3), the QCM has progressed to biosensors.[37,38]

$$R_1 = -k_0^2 (2\pi \eta_l \rho_l)^{1/2} \cdot A \tag{3}$$

where, k_0 is the electromechanical compliance.

The electrical resonance and the mechanical resonance of the shear oscillation of the QCM are coupled each other. An AT-cut QCM is regarded to be electrically same as the equivalent circuit depicted in Fig. 1. Quantitatively, the left arm of the equivalent circuit represents the parasitic capacitance of the QCM while the right arm represents the mechanical oscillation of the QCM. Based on electromechanical analogy, L_1, C_1, and R_1 correspond to energy storage into kinetic energy (i.e., mass), energy storage into elastic energy (i.e., compliance), and energy dissipation (i.e., coefficient of viscous friction), respectively. When certain material, such as a film and/or viscous liquid, contacts with the

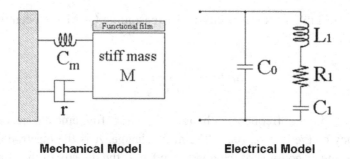

Fig. 1. The mechanical model and the equivalent circuit of the QCM. Both models are coupled electromechanically. In electrical model, the left arm stands for mechanical oscillation of the QCM and the right arm stands for the parasitic capacitance of the QCM.

QCM surface, the L_1, C_1, and R_1 change due to acoustic loading. For this reason, a QCM can be used as physical and chemical sensors.

2.3. *Measurement Principle of Resonance Parameters*

The resonance frequency and its changes are measured using and oscillator circuit. The electromechanical coupling provides a simple method to detect the resonance frequency by electrical means, such as a frequency counter and an impedance analyzer.

The resonance resistance is able to be measured by using the equivalent circuit. When the QCM is resonance,

$$\left(G - \frac{1}{2R_1}\right)^2 + (B - \omega C_0)^2 = \left(\frac{1}{2R_1}\right)^2 \quad (4)$$

where, G is conductance, B is susceptance and ω is angular frequency. According to Eq. (4), the curve of B versus G for various ω can be plotted. Based on this G-B curve, we can determine the serial resonance frequency $f = 2\pi/(L_1 C_1)^{1/2}$ and the resistance R_1 simultaneously and separately. The resonance resistance is obtained as the reciprocal expression of the maximum conductance value.

For the purpose of reliable measurement and the high-speed data acquisition, we used QCM 922 (Seiko EG&G, Japan)[47] whose measuring mechanism is based on Eq. (4).

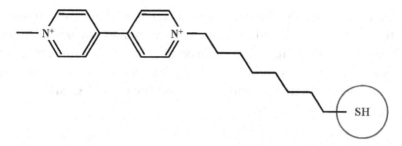

Fig. 2. Molecular structure of viologen derivative.

3. Experimental

3.1. Reagents

VC_8SH (viologen derivative) was synthesized by Qian *et al.* (Fudan University). Figure 2 shows the molecular structure of the viologen incorporated with a thiol group. Hemoglobin was obtained from Sigma. Their stock solutions were stored at a temperature of 4°C. Stock solutions of H_2O_2 were diluted from 30% solution. All other reagents were of analytical grade. The rest of the reagents used in this experiment were of analytical grade and were used without any purification. All solutions were prepared using Milli-Q water.

3.2. Electrode Modification

Self-assembly monolayers were similar on the gold electrode of the QCM which is AT-cut gold-coated quartz crystals with a resonant frequency of 9 MHz (5 mm diameter, Seiko EG&G, Japan). Prior to use, the gold electrode of the QCM was cleaned by piranha solution ($H_2SO_4:H_2O_2 = 3:1$) subsequently cleaned by cycling potential windows between 0 and 1.5 V in 0.05 M H_2SO_4 solution at a scan rate of 100 mV/s for nearly 25 min until stable scans were recorded. Then the electrode was thoroughly rinsed with the distilled water. After pretreatment, the electrode was immersed in ethanol-acetonitrile (1:1) solution containing 2 mM thiol-functionalized viologen with pure Ar gas.

After self-assembly, the electrode was removed from the deposition solution and rinsed with ethanol and water to remove weakly adsorbed viologen. And then, the viologen-modified electrode was immersed into phosphate buffer solution containing 3 mg/ml hemoglobin for 5 h. The modified electrodes will be abbreviated as VC_8SH/Au and $Hb/VC_8SH/Au$, respectively.

3.3. *Apparatus and Measurement System*

Figure 3 shows the measurement system of electrochemical 3-electrode set-up using the QCM. The frequency change and electrochemical behavior were operated by WinEchem software and measured using QCM 922 (Seiko EG&G, Japan) and potentiostat 263A (PerkinElmer, USA). For conventional CVs, a three-compartment electrochemical cell was employed using argon-purged electrolyte solutions. The self-assembled gold electrode onto the QCM was used as the working electrode. The Pt wire and Ag/AgCl (saturated KCl) were used as the

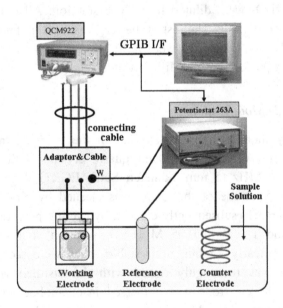

Fig. 3. Measurement system of electrochemical 3-electrode set-up using QCM.

counter and reference electrode, respectively. We observed the charge transfer properties of self-assembled viologen monolayers in 0.1 M $NaClO_4$ electrolyte solution. The scan range was from 0 to -1.0 V and the estimation of the peak current was based on the second cycle of the cyclic voltammetry (CV) curve. In biosensor experiment, the stock solutions were made by phosphate buffer solution (PBS, pH 7). HCl was used to maintain pH.

4. Results and Discussions

4.1. *Verification of Self-Assembly by Resonant Frequency Shift*

The self-assembling of viologen onto the gold electrode of the QCM was investigated by resonance properties. Figure 4 shows the time dependence to resonance frequency and resonance resistance shift of the QCM in ethanol-acetonitrile solution. In this case, the frequency decreased rapidly at first, and then the rate of decrease slowed. The assembling process was completely finished about 60 min. The measured frequency shift was about 386 Hz. From this value, we calculated that the adsorption mass was about 412 ng/cm^2, according to the Sauerbrey equation (Eq. (1)).[40-42]

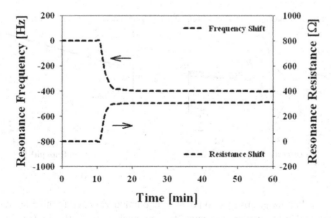

Fig. 4. Resonance properties during self-assembling process using QCM.

4.2. Redox Reaction and Electrochemical QCM Response

After the adsorption of viologen onto the gold electrode of the QCM was completed, the electrode was rinsed with ethanol:acetonitrile (1:1) solution and determined the electrochemical behavior. Figure 5(a) shows cyclic voltammograms of the self-assembled viologen mololayer onto gold-electrode of the QCM at different scan rates (50, 100, 200, and 400 mV) in 0.1 M NaClO$_4$. This is consequence of the first charge reaction, $V^{2+} \leftrightarrow V^{\cdot +}$. A pair of well-defined redox peaks determined. All curves indicated reversible broad redox reactions,[43] and the cathodic potential (E_{pc}) and anodic potential (E_{pa}) peaks were centered at -0.51 V and -0.47 V, respectively. In this case, the difference of E_{pc} and E_{pa} was below 59 mV ($|E_{pc} - E_{pa}| < 59/n$ mV). From this value, the redox reaction was completely reversible reaction according to the Nernst equation (5).[14]

$$E = E^0 + \frac{RT}{nF}\ln\frac{[ox]}{[red]} = E^0 + 0.059\ln\frac{[ox]}{[red]} \quad (5)$$

where, E_0 is formal potential, R is the gas constant (8.314 J/mol), T is absolute temperature, n is electron number in redox reaction and F is the Faraday's constant (96,520 C/mol).

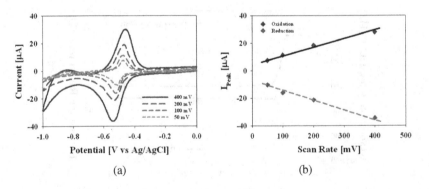

Fig. 5. Cyclic voltammograms reaction of self-assembled viologen monolayer at different scan rate (a) and relationship of scan rate and peak current (b) in 0.1 M NaClO$_4$ electrolyte solution.

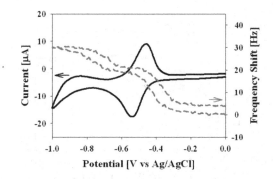

Fig. 6. Resonant frequency shift during the redox reaction in 0.1 M NaClO$_4$ electrolyte solution. (Scan rate is 100 mV/s).

Figure 5(b) shows the relationship between scan rates and peak currents. When the scan rate was increased, the redox peak current also increased linearly. This fact signified that the redox reaction was reversible.[17] Peak currents were increased linearly with potential scan rates with the square root of the scan rates, and the peaks were also symmetrical.[44]

Figure 6 shows the resonant frequency change which obtained during the CV for this same monolayer in NaClO$_4$ electrolyte solution. In CV, this was consequence of the first electron reaction, $V^{2+} + e^- \Leftrightarrow V^{\bullet +}$. The frequency change was about 18.1 Hz. From the data, the transferred mass was about 19.36 ng by Sauerbrey equation (1). Multiplying by Avogadro number, the number of shifted ions were about 2.33×10^{13}. We believe that this mass change was caused by the anions present for charge compensation of the viologen and some solvent. The EQCM frequency data reveal that mass is lost during the reduction process and that mass loss is reversed during the oxidation. These frequency changes have been previously interpreted as indicative of anion loss (gain) from the monolayer, which is consequent to the injection (removal) of electrons.

4.3. Electrochemistry of the Hb/Viologen-Modified Gold Electrode

Figure 7 shows the CVs recorded for the VC$_8$SH/Au (solid line) and Hb/VC$_8$SH/Au (dashed line) in 0.1 M PBS (pH 6.0). It can be seen that

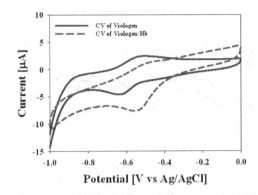

Fig. 7. CVs of VC$_8$SH/Au and Hb/VC$_8$SH/Au electrode in 0.1 M PBS (pH 6.0). (Scan rate: 200 mV/s).

only viologen-modified electrode gives a well-defined redox peaks. The peaks are located at -530 mV and -620 mV. But it is noteworthy, after the immersion of the VC$_8$SH/Au into the solution containing Hb, there is a new pair of peak appears at lower negative potential. For this new redox peaks, the formal potential E_0, was determined to be -135 mV form the average value of the cathodic and anodic peaks which are centered at -30 mV and -240 mV, respectively. This formal potential (-130 mV) is slightly negative compared with standard potential of Hb at gold colloid-thiol-modified gold electrode.[45] Therefore, it can be stated that new redox peaks arise from heme groups of Hb that immobilized onto the viologen-modified electrode. Normally only hemoglobin does not show any electrochemical characteristics as mentioned above. So, in case of this, redox-active viologen derivative is playing an important role to exchange electrons between Hb and electrode.

4.4. Catalytic Properties of the Sensor Towards H_2O_2

Figure 8 shows the CVs obtained from the Hb/VC$_8$SH-modified gold electrode in presence (dashed line) and absence (solid line) of H_2O_2 in 0.1 M PBS at pH 6.0. As shown in Fig. 5, when 1×10^{-4} M H_2O_2 was added into the electrolyte solution, an obvious increase of the cathodic peak current and the disappearance of the anodic peak current were observed. Therefore, it can be said that upon addition of H_2O_2 increased

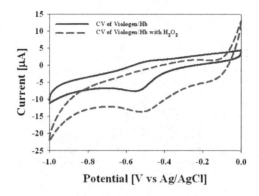

Fig. 8. Cyclic voltammograms Hb/VC$_8$SH/Au electrode in the presence of 1×10^{-4} M H_2O_2 and absence of H_2O_2.

reduction current at the electrode, indicated the electrocatalytic reduction property of H_2O_2. However, this obvious increase in the cathodic current indicated that the Hb keeps its natural structure after immobilizing onto the viologen SAM-modified electrode. The generic reaction for reduction of H_2O_2 by Hb could be as follows.

$$\text{Heme - Fe}^{2+} + 2H_2O_2 \rightarrow \text{Heme - Fe}^{3+} + 2H_2O + O_2$$
$$\text{Heme - Fe}^{2+} \rightarrow \text{Heme - Fe}^{3+} + e^-$$

4.5. Optimum Conditions of the Biosensor for Detecting H_2O_2

The pH value is one of the most important factors for working with biosensor that affect the response of the electrode to H_2O_2. So we investigated the pH effect on the sensor response in presence of H_2O_2. Figure 9 shows the cathodic current response of the sensor at different pH in presence of 1×10^{-4} M H_2O_2. According to this figure, it can be seen that the peak current which attains the maximum level at pH 6.0. When the pH was more than 6.0, the peak current started to decrease slowly. In the pH range 3.0 ~ 6.0, the peak current increases steeply. Especially for pH value < 4.0, the current response is very low. At these pH values, the low response may be due to the denaturation of Hb. Therefore, pH 6.0 was chosen to be used for working media solution to determine H_2O_2 of this sensor.

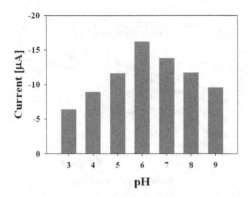

Fig. 9. Effect of pH in the presence of 1×10^{-4} M H_2O_2.

Influence of applied potential on amperometric response of the biosensor also is an important parameter. We studied the influence of applied potential on current response of the biosensor to H_2O_2. Figure 10 displays the attributed current at different applied potential in presence of 2.5×10^{-5} M H_2O_2. The resulting data shows that the steady-state current increases gradually as voltage increases from -100 mV to -300 mV. According to this experiment, the maximum current was achieved at -300 mV. Therefore, we have chosen applied potential -300 mV as the working potential for this sensor.

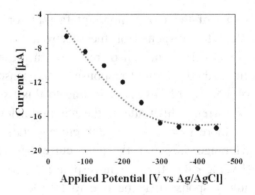

Fig. 10. Influence of applied potential on the amperometric response of the sensor in the presence of 2.5×10^{-5} M H_2O_2.

4.6. Amperometric Response of the Biosensor

Figure 11 illustrates the amperometric response of the Hb/VC$_8$SH-modified gold electrode using the chronoamperometric technique that consists of successive additions 1×10^{-5} M H$_2$O$_2$ in the PBS (pH 6.0) as recorded at operating potential of -300 mV. It can be observed that as soon as adding H$_2$O$_2$, the background current was changed and the reduction current rises steeply to reach value. Figure 12 displays calibration plot obtained for the Hb/VC$_8$SH-modified gold electrode using optimum experimental conditions. Under the optimized experimental

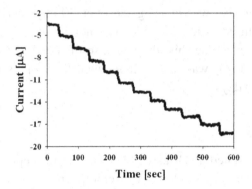

Fig. 11. Typical steady-state current-time response of the sensor with successive additions of 1×10^{-5} M H$_2$O$_2$ to the 0.1 M PBS (pH 6.0).

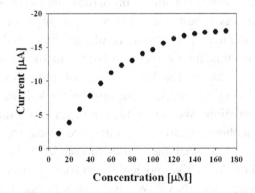

Fig. 12. Calibration curve of the biosensor.

parameters, this sensor shows the linear response within the range between 1×10^{-5} to 1.6×10^{-4} M concentrations of H_2O_2 with a correlation coefficient of 0.996 ($n = 12$). The detection limit was determined to be 3.0×10^{-7} M (based on the S/N = 3).[45] This detection limit is lower than the other biosensor.

4.7. *Stability of the Biosensor*

The repeatability of the sensor for current response was examined. It was found relative standard deviation (RSD) 3.75% for ten successive determinations at a H_2O_2 concentration of 20 μM. The long-term stability of this biosensor also was investigated over 30 days period. When the Hb/VC$_8$SH-modified gold electrode was not in use, it was stored in PBS (pH 7). To examine this, the response of the Hb/VC$_8$SH-modified electrode to the H_2O_2 was tested. It retained 90% of its initial current response after 30 days.

5. Conclusions

We have demonstrated a self-assembled method to prepare well-defined of a viologen derivative. The self-assembled process was determined by resonant frequency shift. The electrochemical property of viologen derivative was characterized in 0.1 M $NaClO_4$, Na_2SO_4 and Na_3PO_4 and electrolyte solutions using cyclic voltammetry. The modified electrode showed reversible electrochemical properties and high stability. The EQCM response was simultaneously determined by resonant frequency during cyclic voltammetry. From the result, the immersed viologen can act as an electron transfer mediator to access to the electrode surface because the mass change for SAMs was not only governed by the mobility of the ion in the viologen but also by the valence of the ion in the electrolyte solution. We know the viologen molecules have their own properties without regard to the influence of electrolyte solutions. Furthermore, we expect that the ability to contact metal nanoparticles electronically via redox-active molecules can form a basis for a range of nanoscale electronic switches.[46] We have also demonstrated a simple method for designing a H_2O_2 biosensor using Hb. It has proved that thiol-viologen provided a suitable biocompatible microenvironment for Hb. The catalytic properties of the designed sensor proved that Hb has been kept its natural structure and can retain its biological activity. The designed biosensor shows fast amperometric response, excellent linearity and low detection limit. In addition, it shows very high sensitivity, good reproducibility and stability.

Acknowledgment

This study was supported by research funds from Dong-A University.

References

1. N. Nakamura, H. X. Huang, D. J. Qian and J. Miyake, *Langmuir* 18, 5804 (2002).
2. A. Ulman, Academic Press, San Diego, Chap. 3 (1991).
3. R. Z. Nuzzo and D. L. Allara, *J. Am. Chem. Soc.* 105, 4481 (1983).
4. R. Maoz and J. Sagiv, *Langmuir* 3, 1034 (1987).
5. M. R. Deakin and D. A. Buttry, *Anal. Chem.* 61, 1147A (1989).
6. D. Y. Lee, S. H. Park, H. K. Shin, D. J. Qian and Y. S. Kwon, *Mol. Cryst. Liq. Cryst.* 445, 439 (2006).
7. J. Y. Ock, H. K. Shin, S. H. Song, S. M. Chang, D. J. Qian, J. Miyake and Y. S. Kwon, *Mol. Cryst. Liq. Cryst.* 407, 525 (2003).
8. J. Y. Ock, H. K. Shin, D. J. Qian, J. Miyake and Y. S. Kwon, *Jpn. J. Appl. Phys.* 43, 2376 (2004).
9. H.-X. Huang, D.-J. Qian, N. Nakamura, C. Nakamura, T. Wakayama and J. Miyake, *Electrochim. Acta* 49, 1491 (2004).
10. H. W. Kang, J. M. Kim, H. K. Shin and Y. S. Kwon, *J. Kor. Phys. Soc.* 32, 1750 (1998).
11. D. A. Buttry and M. D. Ward, *Chem. Rev.* 92, 1355 (1992).
12. S. Chen and F. Deng, *Langmuir* 18, 8942 (2002).
13. J.-Y. Ock, H.-K. Shin, Y.-S. Kwon and J. Miyake, *Coll. Surf. A* 257-258, 351 (2005).
14. A. J. Bard and L. R. Faulkner, J. Wiley, New York, Chap. 12 (1980).
15. K. Takada and H. D. Abruna, *J. Phys. Chem.* 100, 17909 (1996).
16. D.-Y. Lee, A. K. M. Kafi, S.-H. Park, D.-J. Qian and Y.-S. Kwon, *Jpn. J. Appl. Phys.* 45, 3772 (2006).
17. D.-Y. Lee, A. K. M. Kafi, S.-H. Park and Y.-S. Kwon, *J. Nanosci. Nanotechnol.* 6, 3657 (2006).
18. L. C. Clark and C. Lyons, *Ann. N. Y. Acad. Sci.* 102, 29 (1962).
19. X. Han, W. Huang, J. Jia, S. Dong and E. Wang, *Biosens. Bioelect.* 17, 741 (2002).
20. Y. Zhang, P. He and N. Hu, *Electrochim. Acta* 49, 1981 (2004).

21. J.-P. Guilla, *Electroanalysis* 3, 741 (1991).
22. Q. Deng and S. Dong, *J. Electroanal. Chem.* 377, 191 (1994).
23. H. Wang, R. Guan, C. Fan, D. Zhu and G. Li, *Sens. Actuators B* 84, 214 (2002).
24. S. Liu, Z. Dai, H. Chen and H. Ju, *Biosens. Bioelect.* 19, 963 (2004).
25. Y.-D. Zhao, Y.-H. Bi, W.-D. Zhang and Q.-M. Luo, *Talanta* 65, 489 (2005).
26. A. K. M. Kafi, F. Yin, H.-K. Shin and Y.-S. Kwon, *Thin Solid Films* 499, 420 (2006).
27. W. Schuhmann, T. J. Ohara, H.-L. Schmidt and A. Heller, *J. Am. Chem. Soc.* 113, 1394 (1991).
28. I. Willner, A. Riklin and N. Lapidot, *J. Am. Chem. Soc.* 112, 6438 (1990).
29. F. Yin, H.-K. Shin and Y.-S. Kwon, *Biosens. Bioelect.* 21, 21 (2005).
30. P. N. Bartlett and J. M. Cooper, *J. Electroanal. Chem.* 362, 1 (1993).
31. W. H. Scouten, J. H. T. Luong and R. S. Brown, *Trends Biotechnol.* 13, 178 (1995).
32. T. Wink, S. J. VanZuilen, A. Bult and W. P. VanBennekom, *Analyst* 122, 43 (1997).
33. S. Dong and J. Li, *Bioelectrochem. Bioenerg.* 42, 7 (1997).
34. W. K. Paik, S. Eu, K. Lee, S. Chon and M. Kim, *Langmuir* 16, 10198 (2000).
35. K. K. Kanazawa and J. G. Gordon, *Anal. Chim. Acta* 175, 99 (1985).
36. H. Muramatsu, E. Tamiya and I. Karube, *Anal. Chem.* 60, 2142 (1988).
37. H. W. Kang and H. Muramatsu, *Biosens. Bioelect.* 24, 1318 (2009).
38. H. W. Kang, K. Ida, Y. Yamamoto and H. Muramatsu, *Anal. Chim. Acta* 624, 154 (2008).
39. V. E. Bottom, *Introduction to Quartz Crystal Unit Design* (Van Nostrand, New York, 1982).
40. G. Sauerbey, *Z. Phys. Chem.* 155, 206 (1959).
41. A. Janshoff, C. Steinem, M. Sieber, A. el Bayâ, M. A. Schmidt and H.-J. Galla, *Eur. Biophys. J.* 26, 261 (1997).
42. D.-Y. Lee, S.-H. Park, H.-K. Shin, D.-J. Qian and Y.-S. Kwon, *Mol. Cryst. Liq. Cryst.* 445, 439 (2006).
43. H. Tatsumi, K. Takagi, M. Fujita, K. Kano and T. Ikeda, *Anal. Chem.* 71, 1753 (1999).
44. A. N. Khramov and M. M. Collinson, *Anal. Chem.* 72, 2943 (2000).
45. A. K. M. Kafi, D.-Y. Lee, S.-H. Park and Y.-S. Kwon, *Microchem. J.* 85, 308 (2007).
46. D. I. Gittins, D. Bethell, D. J. Schiffrin and R. J. Nichols, *Nature* 408, 67 (2000).
47. H. Muramatsu, A. Egawa and T. Ataka, *J. Electroanal. Chem.* 388, 89 (1995).

CHAPTER 5

ZINC (II), IRIDIUM (III) AND TIN (IV) COMPLEXES FOR NANOSCALE OLED DEVICES

Trinh Dac Hoanh and Burm-Jong Lee*

Department of Chemistry and Institute of Functional Materials,
Inje University, Gimhae 621-749, Korea
*E-mail: chemlbj@inje.ac.kr

In this chapter we present a review on the organometallic complexes which are employed for the organic light-emitting diodes (OLED) having nanoscale multilayer structures. The device fabrication and operation mechanism of OLED are followed by the ligands and metal complexes of Ir(III), Zn(II) and Sn(IV). The discussions have emphasized on their chemical structures and the electronic properties at the multilayer interfaces. The organometallic complexes are classified based on their colors and luminescent efficiencies.

1. Introduction

Organic and organometallic compounds applied in electroluminescent devices are of considerable interests because of their attractive characteristics and potential applications to flat panel displays.[1,2] The devices take advantages over their inorganic counterparts such as high luminous efficiency and fine-pixel formation.[3] Due to low driving voltage, high contrast, ease of fabrication, low cost, and wide viewing angle, OLED is believed to be the next-generation flat panel display. The phenomenon of organic electroluminescence (EL) is the electrically driven emission of light from non-crystalline organic materials.[2,4] In 1987, a research group of Kodak introduced a double layer device, which was used aluminum (III) tris(8-hydroxyquinoline) (Alq3) as the emitting layer by Tang and Van Slyke. It combined modern thin film deposition techniques with suitable materials and structure to give moderately low bias voltages and attractive luminance efficiency.[1] Since then, there have been increasing interests and research activities in this new field, and enormous progress has been made in the improvements of color gamut, luminance efficiency and device reliability. The growing interest is largely motivated by the promise of the use of this technology in flat panel displays. As a consequence, various OLED displays have been demonstrated.

The efficiency of an OLED is determined by charge balance, radiative decay of excitons, and light extraction. Significant progress has been made recently in developing phosphorescent emitters. In spite of the remarkable advance of OLED, it still remains some drawbacks. Charge injection and transport are the limiting factors in determining operating voltage and luminance efficiency. In OLEDs, the hole current is limited by injection, and the electron current is strongly influenced by the presence of traps owing to metal-organic interactions. In order to enhance carrier injection the selection of efficiently electron-injecting cathode materials and the use of appropriate surface treatments of anodes are of great importance. The most critical performance characteristic for OLEDs is the operational lifetime. Continuous operation of OLEDs generally leads to a steady loss of efficiency and a gradual rise in bias voltages. Despite OLEDs have achieved long operational stability, the material issues underlying the EL degradation are not fully understood.

1.1. *Structure and Operation of OLED Device*

An OLED has an organic EL medium consisting of extremely thin layers (< 0.2 mm in combined thickness) sandwiched by two electrodes. In a basic two-layer OLED structure, one organic layer is specifically chosen to transport holes and the other organic layer to transport electrons. The interface between the two layers provides an efficient site for the recombination of the injected hole-electron pair and resultant electroluminescence.

When an electrical potential difference is applied between the anode and the cathode such that the anode is at a more positive electrical potential with respect to the cathode, injection of holes occurs from the

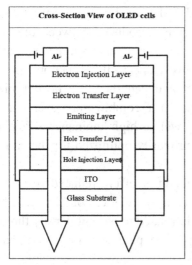
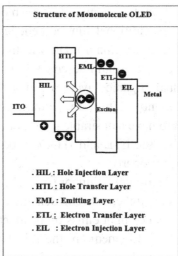

Fig. 1. Device structure and operation. Multilayer structure including glass substrate, anode and cathode, hole injection layer (HIL), hole transfer layer (HTL), emitting layer (EML), electron transfer layer (ETL), and electron injection layer (EIL).

anode into the hole transfer layer (HTL), while electrons are injected from the cathode into the electron transfer layer (ETL). The injected holes and electrons each migrate toward the oppositely charged electrode, and the recombination of electrons and holes occurs near the junction in the luminescent ETL. Upon recombination, energy is released as light, which is emitted from the light-transmissive anode and substrate.

The heterojunction should be designed to facilitate hole injection from the HTL into the ETL and to block electron injection in the opposite direction in order to enhance the probability of exciton formation and recombination near the interface region. As shown in Fig. 1, the highest occupied molecular orbital (HOMO) of the HTL is slightly above that of the ETL, so that holes can readily enter into the ETL, while the lowest unoccupied molecular orbital (LUMO) of the ETL is significantly below that of the HTL, so that electrons are confined in the ETL. The low hole mobility in the ETL causes a build up of hole density, and thus enhance the collision capture process. Furthermore, by spacing this interface at a sufficient distance from the contact, the probability of quenching near the metallic surface is greatly reduced.

The simple structure can be modified to a three-layer structure, in which an additional luminescent layer is introduced between the HTL and ETL to function primarily as the site for hole-electron recombination and thus electroluminescence. In this respect, the functions of the individual organic layers are distinct and can therefore be optimized independently. Thus, the luminescent or recombination layer can be chosen to have a desirable EL color as well as high luminance efficiency. Likewise, the ETL and HTL can be primarily optimized for the carrier-transport property.

The extremely thin organic EL medium offers reduced resistance, permitting higher current densities for a given level of electrical bias voltage. Since light emission is directly related to current density through the organic EL medium, the thin layers coupled with increased charge injection and transport efficiencies have allowed acceptable light emission to be achieved at low voltages.

1.2. *Emission Mechanism of OLED Device*

As presenting in the above section, the light is produced after the recombination of hole-electron in the emitting layer. Light consists of particles called photons and light emission occurs in a material when it is excited by energy sources. If the materials are excited by the heat source the emitted radiation is called thermal light, and analogue, by electron beam is cathodoluminescence, by mechanical mean is triboluminescence, by chemical reaction is chemiluminescence, by bio-organism is bioluminescence, and by light source is photoluminescence. In OLED the external source is electricity to emitting material, so it is electroluminescence (EL). Luminescence can be achieved by several physical or chemical mechanism which move the atoms and molecules of material to an excited state with energy higher than at their ground state, the lowest energy level of the atoms or molecules, in the luminance process, the decay from an excited state to the ground state will rise the emission of a photon, based on the principle of conservation of energy. In a molecular orbital, there are two electrons of opposite spins per orbital. But when an electron is excited to a higher energy level, its spins

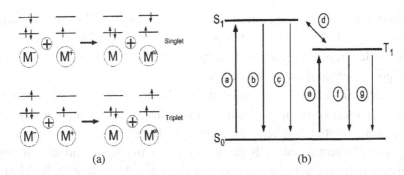

Fig. 2. Recombination of an electron and hole resulting in the formation of excited states (singlet or triplet) in organic material (Fig. 2(a)). Figure 2(b) illustrates different transitions states based on Jablonski diagram. (a) Singlet absorption; (b) fluorescence emission; (c) single quenching transition; (d) inter-recombination; (e) triple absorption; (f) phosphorescence emission; and (g) triplet quenching transition. Reproduced from Ref. 2.

can point either in opposite or in the same direction. In the case of opposite direction, we call singlet state or S, other case is triplet states or T. The transitions between these states with the ground state will be governed by the selection rule, which allows only those occurring between states of identical spin multiplicity. Jablonski diagram (Fig. 2(b)) summarized the different kinds of transitions. We can distinguish two main photoemission mechanisms, fluorescence and phosphorescence.

When the transition occurs from the singlet state S_1 to the ground state S_0 (arrow b in Fig. 2(b)), it is called fluorescence, this emission is spontaneous and the decay rate depends on the temperature. In the case of the triplet state T_1 to the ground state S_0 (arrow f in Fig. 2(b)), it is called phosphorescence. The decay rate is much slower than that of the fluorescence process.

To operate the OLED, the voltage must be applied efficiently high bias to the device. Because this bias enables a current moving within the diode and allows the charge carriers of electrons and holes to drain from the electrode to the emitting layer to recombine there, then the photons produced can go out of the diode. The process can be simply understood as a conversion of electrical energy to the light energy. The conversion rate can be defined for the electroluminescence. In general, the overall mechanisms of light emission in the device can be described in five-step sequence such as charge injection, transport of carrier through the emitting layer, recombination of charge carrier, emission of photons, and transport of photons out of the device.[5]

Charge injection: When we consider the injection of charge carriers from the electrodes, it is required that charges overcome or tunnel through a potential barrier at the interface of emitter and electrode. The height of the barrier depends on the nature of the metal and that of the semiconductor (organic or polymer compounds). According to the Mott-Schottky model,[6] the vacuum and Fermi levels of the electrode and the semiconductor are continuous to across the interface, leading the three types of contact: neutral, ohmic, and blocking. If we call Ψ_m and Ψ_s is the work function of the metal and the semiconductor, when $\Psi_m = \Psi_s$, it is neutral. There is no charge transfer between the materials and the conduction band remains flat up to the interface. When $\Psi_m > \Psi_s$, this type is ohmic. Under the applied voltage the electrode can supply charges to compensate for those flowing to the semiconductor. At thermal equilibrium, electrons are injected from the electrode into the conduction band of conducting material, creating a negative space charge region of width λ_0. This type of contact can supply as many electrons to the semiconductor as are required by bias conditions.

When $\Psi_m < \Psi_s$, this type of contact is blocking, electrons flow from the conduction band of the semiconductor to the metal, forming the negative space charge region. The depletion region has a low electron density, therefore an applied voltage to the contact will drop almost entirely across it. Figure 3 shows the electrons injected into the semiconductor via different mechanism.

The field emission mechanism is described by direct tunneling through the base of the barrier from Fermi level E_{Fm} at low temperature, provide that the barrier is thin enough to allow the electrons to cross through it.

This mechanism was first treated by Fowler and Nordheim[6] and the model was modified by many researchers for adaptation to semiconductor. Assuming a triangular shape for the barrier, the current density can be written as follow:

$$J = \frac{A^*T^2}{\phi_B}\left(\frac{eF}{\alpha k_B T}\right)\exp\left(\frac{2\alpha\phi_B^{3/2}}{3eF}\right) \quad (1)$$

where $A^* = (4\pi e m^* k_b^2)/h^3$, ϕ_B is the potential barrier and

$$\alpha = \frac{4\pi(2m^*)^{1/2}}{h}.$$

The thermionic field emission is described by tunneling from an energy level $E_m > E_{Fm}$ from the metal to the semiconductor at a relatively high temperature range. In this case, the barrier is thinner for the carriers because of their higher energy. The mechanism is understood that when a semiconductor-metal contact is brought to a sufficiently high temperature, electrons are thermally excited over the potential barrier into the conduction band of the semiconductor. The current density is given by the Richardson thermionic emission equation.

$$J = A^*T^2 \exp\left(-\frac{\phi_B}{k_B T}\right) \quad (2)$$

where A^* takes the value 120 when J is expressed in Acm^{-2}.

Thermionic or Schottky emission: By emission over the potential barrier at a high temperature range. The energy of the electrons is higher than the barrier height φ_n, allowing them to flow to the conduction band of the semiconductor.

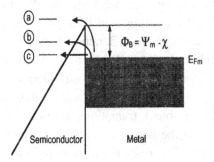

Fig. 3. Different injection mechanism of electrons from the cathode into the semiconductor under the applied voltage: (a) thermionic emission; (b) thermionic field emission; and (c) direct tunneling.

Charge transport: After being injected into the emitter, electrons and holes are transported inside the material under the applied electric field. They can contact each other and recombine, but they can also go through the emitting layer and be extracted at the opposite electrode. The moving of the carrier inside the organic layer depends on some parameters inherent to the nature of the material. It can occur in the conduction band or the band gap via the localized states. The former mechanism is possible when the charge carrier have efficiently high energy to enter the conduction band while the latter is involved when the carriers are confined in the band gap on localized states but can jump along these with less energy because of the proximity of these states. When the number of injected charges to a semiconductor is higher than that the material has in thermal equilibrium without carrier injection, the excess of injected charges will form a space charge inducing an electric field inside the semiconductor, which, in turn, will reduce the charge injection from the electrode. Therefore the current flow is no longer limited by the charge injection from the electrode but by the bulk of the semiconductor. The condition for the mechanism to occur is that one of the injecting contacts should be able to provide as many carriers as needed.

Charge recombination: After the charge carriers are injected to the semiconductor, they may decay by recombination to produce a photon, but they may also move along several different paths to reach the opposite electrode. In conventional semiconductors, which contain both the donor and acceptor, the recombination of the electrons and holes can occur through several mechanisms like direct band-to-band recombination, indirect recombination through a recombination center-example, donor to valence band, conduction band to acceptor, and donor to acceptor. Recombination can lead either to an electron-hole pair that decays to the ground state by the emission of photon (radiative recombination) or to exclusively photon emission (non-radiative recombination). Band-to-band transitions occur at energies less than equal to or greater than the band gap and free electrons and holes may become bound via the Coulomb interaction to form free excitons. In contrast, indirect transitions occur at energies less than the band gap.

OLED fabrication: There are many methods to fabricate OLED such as vacuum deposition, Langmuir-Blodgett, spin-coating, inkjet-printing, dye-diffusion and silk-screen. Depending on the material we can choose a proper method. In this section we introduce about vacuum deposition method, which widely used for fabrication of OLED of the small molecular materials.

Vacuum deposition: Organic materials are evaporated on to an indium-tin-oxide (ITO)-coated glass substrate. Organic materials are previously purified by vacuum sublimation. The EL cells are generally constructed by a conventional thermal evaporation method at a chamber pressure of about 1×10^{-5} Torr. This technique consists of heating the materials under reduce pressure. The material is put into a metal boat or crucible, which is heated by Joule effect, or with an electron gun. Emitting layers are sometimes deposited each one of two metal sources. The cathode is deposited onto the organic layer. The cathode is generally selected from low work function metals (Table 1). For example, Mg:Al (10:1) alloy as the cathode is also deposited from each one of two metal sources.[2]

Table 1. The work function of metal.

Metal	eV
ITO	4.7
Ga	2.87-3
MgAg	3.7
Al	4.06-4.41
Ag	4.26-4.74
Au	5.1-5.46
In	4.12-4.2
Yb	2.6

2. Ligands Coordinated to Ir(III), Zn(II) and Sn(IV)

Small molecules as the metal complexes with ligands are very important to fabrication for OLED devices due to their luminance efficiency and other advance properties. The main characteristic of those is that they can be evaporated under vacuum to form thin films. It enables their use in the field of small area devices, especially to making colored pixels for display application. This deposition technique allows precise control of the film thickness, so for this reason, multilayer diodes for multicolor emission are usually made using these materials. Therefore the ligands are received much attention to make complexes with metal for the application of this area. We introduce three main types of ligands which have popular complexes with Ir(III), Zn(II), and Sn(IV) presented in this review.

8-Hydroxyquinone and its derivatives: 8-Hydroxyquinone (**1**) has become the well-known ligand since it was successfully synthesized with aluminum to form Alq3 at the first time in 1955 by Freeman and White.[7] Then Tang *et al.* utilized Alq3 for the fabrication of the first OLED operating at low voltage (< 20 V), green photoluminescence (PL) with a quantum efficiency around 32%.[8] And also the material is subjected to an electrical field.[9] On the other hand, it is easily sublimated under vacuum to form a smooth thin layers.

The 8-hydroxyquinone was first prepared from o-aminophenol, glycerol and H_2SO_4 in 1882 by Skraup.[10] The structure of 8-hydorxyquinone (**1**) shows in Fig. 4. It is the basic ligand due to the capacity to produce the chelate compounds with metals. Researchers have been tried to modify the chemical structure of 8-hydroxyquinon to investigate the EL and PL of the complexes by adding the donor or withdraw functional group to the structure of the 8-hydroxyquinone.

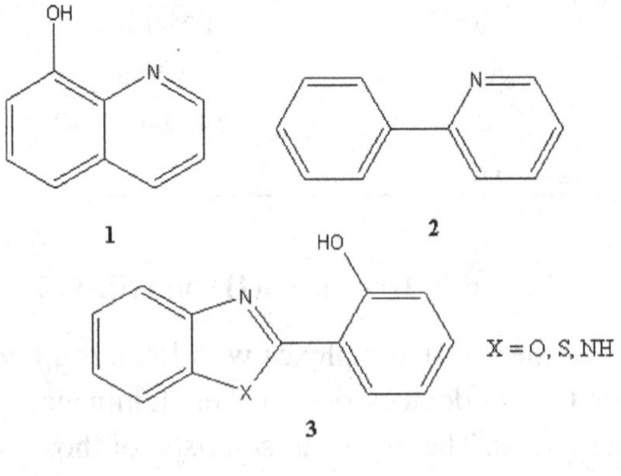

Fig. 4. The basic ligands for the coordination with Ir(III), Zn(II), Sn(IV). **1**: 8-hydroxyquilone; **2**: 2-phenyl pyridine; **3**: benzo-compounds.

The complexes of 8-hydroxyquinone with aluminum (Alq3) was first applied for OLED in 1987 by Tang and Van Slyke but until 1993 the complexes with zinc metal (Znq2) was first reported to apply for OLED by Hamada,[11] the studies of zinc complexes as active materials for OLEDs have focused on improving electrons mobility or producing a blue shift. Besides the Al and Zn, other metal complexes, the 8-hydroxyquinone were also investigated with tin or iridium complexes.

Benzo-compounds: Including 2-2(hydroxylphenyl)benzoxazole, 2-2(hydroxylphenyl)benzothiazole, 2-2(hydroxylphenyl)benzoimidazole and their derivatives.

The general structures of the benzo-compounds (**3**) are shown in Fig. 4, where X can be oxygen, sulfur, or hydronitrogen. Like the 8-hydroxyquinone, these ligands also can coordinate to metals, especially Zn. The zinc complexes with these ligands usually have the EL and PL of blue or green emission with the wavelength around 400-520 nm. For examples, zinc 2-2(hydroxyphenyl) benzoxazole shows efficient blue material, we have indicated that it is also good for the performance of device if the material is applied as the hole blocking layer. While bluish-green emitting layer of bis 2-2[(hydroxyphenyl)benzothiazolate] zinc $(Zn(BTZ)_2)$, doped with other materials to make white light emitting diode device.

2-Phenyl pyridine: This ligand was synthesized with iridium (III), $Ir(ppy)_3$, at the first time in 1985 by King.[12] It opened the new method to synthesized and new materials which later successfully applied to OLED as the emitting material. After the success of these researches, a serial of reports based on the structure of pyridine and the (C^N) ligands were investigated to support the good materials for the OLED. Most of the iridium complexes in this field are based on 2-phenylpyridine and its derivative. In addition, in the last decade 2-phenylpyrine was known as to have several potent azapeptide Human Immunodeficiency Virus (HIV) protease inhibitors, a moiety showing anti-HIV activity such as BMS-232632a.[13] Strategies for the synthesis of the 2-phenylpyridine nucleus has varied from Chichibabin condensation reactions to Heck substitution reaction, and a new route for the synthesis of 2-phenylpyridines over molecular sieve catalysts was report by Radha Rani.[14]

3. Light Emitting Ir(III), Zn(II), and Sn(IV) Materials

3.1. Blue Emitting Materials

Blue emitting materials are the one of the main color for fabricating the white OLED device, they are also one of the most challenges of OLED's drawback because of the lifetime limitation. Therefore, blue emitting materials have been received much attention from researchers (Fig. 5). In the previous section, Ir(ppy)$_3$ was introduced as the emitting materials for OLED, then in 2001, Lamansky et al.,[15] modified the structure to Ir(ppy)$_2$(acac) (acac = acetylacetonate) and Ir(dfppy)$_2$(acac), compounds (4) and (5) where R$_1$ = R$_2$ = CH$_3$, and they showed the PL at green and blue emission at 519 and 543 nm. Later Liu et al.,[16] prepared the compounds (4') and (5') (where R$_1$ = -t-Bu, R$_2$ = 9-ethyl-9H- carbazole). The PL of 4 = 4' and 5 = 5' but in the device the maximum luminence efficiency of compound 4' (4.54 lm/W) and 5' (0.51 lm/W) are higher than that of 4 (0.53 lm/W) and 5 (0.06 lm/W). Compounds (6) and (7) were synthesized by Ragni et al.,[17] with the presence of electron-withdrawing fluorine substituents in the chelating ligands, which should decrease the HOMO energy and, as the consequence, increase the HOMO-LUMO energy gap. These materials showed the blue emitting with emission quantum yields in solution as high as 53%. The other serials of Iridium complexes (8), (9), (10), (11), based on diazine were prepared by Ge and Guo,[18] showed EL around blue-green region and the best performance in the device of compound (10) with the configuration of the device ITO/NPB (40 mn)/CBP: (10) (30 nm)/TPBi (15 nm)/Alq$_3$ (50 nm)/Mg:Ag (150 nm, 10:1)/Ag (10 nm). With turn-on voltage of 3.2 V has its highest brightness ($L_{max,\ cd/m^2}$) of 120,450 cd/m^2 at 18 V, a maximum external quantum efficiency ($\eta_{ext.max}$) of 13.9%, a highest power efficiency ($\eta_{max,\ lm/W}$) of 48.6 lm/W and a highest current efficiency ($\eta_{max,\ cd/A}$) of 58.2 cd/A.

Since zinc complexes are promised as blue emitting materials, we also investigated the zinc complex with other ligands (12), Zn(II) 2-(2-hydroxyphenyl)benzoxazole,[19] which showed EL of blue emission at 447 nm when (12) as the emitting layer, turn-on voltage at 4 V and the L_{max} of 2800 cd/m^2.

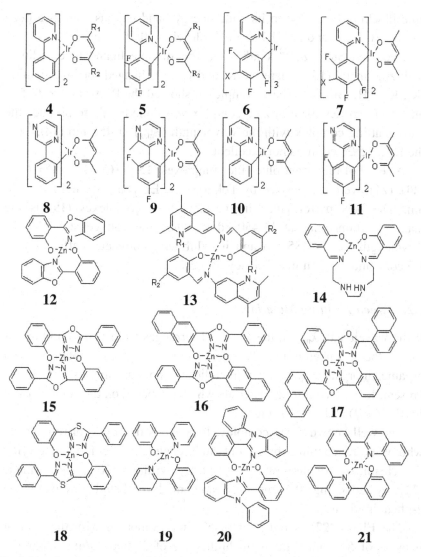

Fig. 5. The chemical structures of blue emitting materials based on the Zn- and Ir-containing complexes.

In this research also indicated that when (**12**) was applied as the hole blocking layer in the device structure of ITO/NPB (40 nm)/Alq$_3$ (60 nm)/ 12 (5 nm)/ LiF/Al, the L$_{max}$ was improved immediately of 15,000 cd/m^2 when the voltage at 10 V. The (**12**) complex was suggested to use as the hole blocking layer due to its wide energy band gap, as well as a deep HOMO (6.5 eV) and LUMO (2.8 eV) closely with that on ETL.

Recently, Su et al.,[20] synthesized the suitable ligand to coordinate with zinc metal (**13**), where the R_1 and R_2 are changed for investigation (R_1, R_2 = H, ^{t}Bu, Br). These complexes showed the PL from 476 to 578 nm. An other new type of zinc complex was done by Yu to produce the blue emitting complex with the wavelength (λ_{max}) of 455 nm (**14**),[21] but the L_{max} = 5.26 cd/m^2 is not excellent.

A serial of important blue zinc complexes (**15**), (**16**), (**17**), (**18**), (**19**), (**20**), (**21**), were synthesized by Tokito,[22] the EL peaks from 447 to 508 nm. The best performance in the device of complexes (**19**) is the maximum luminance at 11,450 cd/m^2 when the voltage at 15 V, the wavelength λ_{max} = 485 nm and it exhibited photometric efficiency of 1.3 cd/A and a quantum efficiency of 1.2%.

3.2. Green Emitting Materials

If the wavelength λ_{max} in the EL or PL arranges from 500 to 550 nm, it shows the green emission. The green emission can be obtained from organic, inorganic, and organometallic compounds. In this section we present the green emitting materials which are based on the complexes of Ir(III), Zn(II) and Sn(IV) (Fig. 6).

The well-known green complex, Ir(ppy)$_3$, is the topic for researchers who want to continue break out and renew this field by modifying the chemical structure of ppy (2-phenylpyridine). Xu et al.[23] made (**22**) by attaching the methoxy and methoxyphenyl group to the 2-phenylpyridine.

The EL of (**22**) is red shift to 531 nm compared with the data of Ir(ppy)$_3$ of 512 nm. Maximum luminance of 35,000 cd/m^2, maximum quantum efficiency of 9.0% and luminance efficiency of 36 cd/A were achieved, which are higher than that of Ir(ppy)$_3$. By adding trimethylsilyl group to replace the hydrogen in the structure of 2-phenylpyridine, Jung et al.[24] got the compound (**23**).

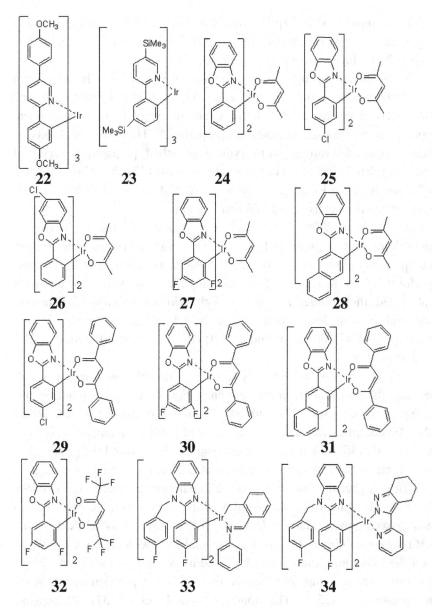

Fig. 6. The chemical structures of blue emitting materials based on the Zn- and Ir-containing complexes.

To compare with Ir(ppy)$_3$ the EL have changed not much but significant improvement of the power efficiency (17.3%), which is nearly 200% higher than that of Ir(ppy)$_3$.

In the blue emitting material section, we have already mentioned the benzoxazole complex with zinc (**12**). Because benzoxazole and the derivatives have the high quantum yield,[25] high non-linear optical effectiveness, photostable properties.[26] Here several kinds of benzoxazole derivatives were synthesized then prepared a series of cyclometalated iridium (III) complexes (**24**)-(**32**) by Chen.[27] Most of them show the PL of green emission at 530 nm but the (**28**) and (**31**) show the yellow emission at 568 and 579 nm.

In the year 2009, Li et al.[28] has synthesized the high-efficient phosphorescent iridium (III) complexes with benzimidazole ligand, compounds (**33**) and (**34**). Both of them showed the promise performance in the devices. Devices using (**33**) and (**34**) as dopants were fabricated, they emitted strong cyan light with an emission maximum at 498 and 492 nm, low turn-on voltage at 3.4 and 3.1 V, high maximum brightness of 47,100 cd/m^2 and 41,150 cd/m^2, current efficiency at 21.7 and 35.3 cd/A.

Zinc complexes are not only known as the blue emitting materials but also the excellent green emitting ones (Fig. 7). The green emitting zinc complexes (**35**), (**36**), (**37**), and (**38**)[29-32] are the examples of that. Where the (**35**) complex was done by Sharma et al., (**36**) by Wang et al., (**37**) by Chen et al., (**38**) by Park et al. The complex (**35**) showed the green EL in the OLED device at the wavelength of 545 nm, the turn-on voltage at 7 V (it is higher than normal) and the maximum brightness of 363 cd/m^2 at 15 V. The same (**35**), the complex (**36**) also based on the derivative of 8-hydroxyquinone. The single layer OLED device was fabricated with (**36**), which exhibited the green-yellow emission with the peak at 557 nm and the maximum luminance of 450 cd/m^2 at 20 V. In addition, (**36**) is very heat stable compound that the thermal decomposition occurs at the temperature of 435°C. The abundant complexes of (**37**) are possible when using the different groups of R in the structure. Most of them showed the blue-green PL, and the best performance in the device where R = carbazole, the maximum brightness at 2249 cd/m^2, and EL at 552 nm.

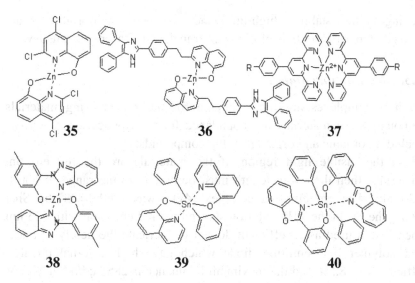

Fig. 7. The chemical structures of green emitting materials based on Zn- and Sn-containing complexes.

The complex (**38**) was successfully synthesized in our laboratory and applied to OLED. In the both PL and EL properties, it exhibited the green emission at 530-540 nm, high thermal stability up to 400°C. When it was used as the emitting layer the maximum brightness was 9500 cd/m^2 at 9.5 V, but when it played as the role of hole blocking layer the maximum of brightness is twice higher than that of emitting layer at 18,700 cd/m^2, when applied the voltage of 10.5 V. And the same as in (**12**), it also suggested to be used as the hole blocking layer in the OLED devices.

Tin (IV) organometallic compounds have a lot of applications as biocides like fungicides or antifouling agents,[33] but in the field of OLED it seems to be new and there are not so many literatures about this topic.

Tin (IV) complexes of (**39**) and (**40**)[34] were the first time synthesized and applied for OLED in our laboratory. The EL of (**40**) is blue emission, and green emission in case of (**39**). The (**40**) also showed the good performance when it was applied in the hole blocking layer with the maximum luminance of 17,600 cd/m^2 and maximum EL efficiency reached 1.4 lm/W at 34 mA/cm^2, while the (**39**) showed L_{max} = 720 cd/m^2 at 7.5 V. From these results we could observe the good properties of benzoxazole ligand when it can coordinate with various metals to form metal complexes having the heat stability, high luminance efficiency, high brightness, and also good material for hole blocking layer in the multilayer structure devices.

3.3. Red Emitting Materials

Iridium complexes are most suitable for the red emitting materials, although these materials are abundance for the application of OLED, including organic and organometallic compounds.

In the visible light region, if the materials are the red emitting materials, then their wavelength must be belong to the range of 600 to 720 nm. The red emitting complex (**41**)[35] was synthesized by Shen *et al.*, the PL in the THF solution showed the red emission with 623 nm, then it was used as an efficient dopant to fabricate the nearly saturated red polymer light-emitting diode which gave high external quantum efficiency of 8.5% and the maximum luminance is 2858 cd/m^2 at 16.5 V. The series of compounds (**42**)[36] were obtained when changing the different X_1 and X_2 (where X_1 and X_2 = H, F, CF$_3$) by Park. The PL data of them showed in the peaks around 600 nm. Gao *et al.*, synthesized (**43**)[37] which shows the PL of pure red light at 675 nm, when fabricated the diode with blending 3% in CBP, the EL spectrum shows the wavelength at 677 nm, L$_{max}$ = 2800 cd/m^2 at 22 V. The (**44**), (**45**), (**46**), and (**47**) were synthesized by different authors, Chuang,[38] Kim,[39] Huang,[40] but they can be classified as the same group, because changing the chemical structure a little, all of them show the emission in the red region.

It is rarely to see the zinc complex shows the red emission but in case of (**48**)[41] when Zn(II) in the center of porphyrins it exhibited the PL in red emission of 720 nm, because of the PL of the ligand also in the red region. This work was investigated by Paul-Roth.[43]

To modify the structure of 2-phenylpyridine (ppy), Yao *et al.*[44] designed and synthesized the crown ligand instead of ppy, (**49**), (**50**), (**51**), and (**52**) when acac was used as the second ligand after the synthesis of the chloro bridge dimmer. In summary of EL and device characteristics the (**52**) has the EL of red emission at 609 nm when the rest have the yellow emission around 550 nm, and the (**49**) showed the best LE$_{max}$ of 36.4 cd/A with the L$_{max}$ = 30,956 cd/m^2. There are a lot of other red emitting materials but in the restricted area of this review we just introduced some of them which are shown in Fig. 8.

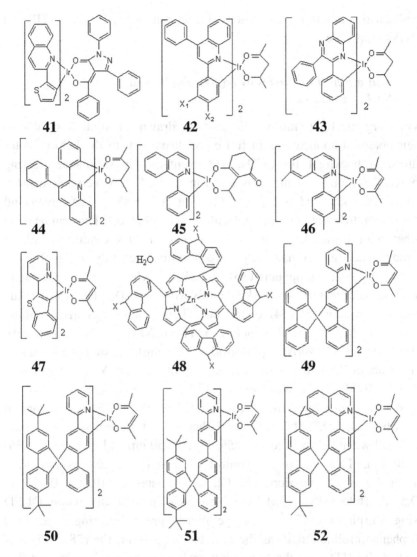

Fig. 8. The chemical structures of red emitting materials based on the Ir- and Zn-containing complexes.

Red emitting materials are very important when fabricating white OLED, the efficiency and high luminance of the diode depend on the blending, combination, and the nature of the monochromatic of red, blue and green color. In the next section we will discuss the white emitting diodes which based on main color emitting materials of Ir(III), Zn(II) and Sn(IV) complexes.

3.4. White Emitting Diodes by Ir(III), Zn(II) and Sn(IV) Complexes

White organic light-emitting diodes have drawn much attention due to their potential applications in full color displays with the help of color filters, in backlight for LCDs, and eventually in solid-state lighting sources.[43] To obtain white emission from an OLED, two complementary colors (ex., blue and orange) or three primary colors (red, green and blue) from different emitting molecules are combined.[44] This can be done either using a multilayer structure with two or more emitting layers or doping an active host material with several fluorescent dyes (Fig. 9).

Starting with compound (**53**) and (**54**),[45] which are the two main emitting color materials to make the this white OLED, where (**53**) is blue emission, meanwhile (**54**) is red one. This work was carried out by Lim et al. A device with structure of ITO/α-NPD (50 nm)/53 (4 nm)/ 53:54 (10 nm, 0.8%)/Alq$_3$ (30 nm)/LiF (0.5 nm)/ Al shows an external QE of about 0.5%, a luminous efficiency of 0.27 lm/W. The luminance of about 4000 cd/m^2 was achieved at 17 V and 580 mA/cm^2.

We have combined the compound (**55**-blue) and (**56**-red) to obtain the white OLED.[46] But we found that the white emission was only achieved when the thickness of (**55**) layer is 30 nm and that of (**55**):(**56**) is 10 nm. That means the control of device structure is important to get the white emission. The CIE coordinates are (0.304, 0.332) at 10.5 V. Before this work, we also characterized the white OLED using Zn(phen)q (**57**)[47] as a yellowish green emitting layer and 10-phenanthroline (BCP) as the hole blocking layer, the (**58**) NPB was used as the HTL and the blue emitting layer. The result indicated that when adjust the thickness of BCP layer from 0 to 10 nm, the white emission was achieved at 5 nm of thickness. And the blue emission increases in proportion to applied voltage. The CIE coordinates are (0.26, 0.33) at an applied voltage of 10 V.

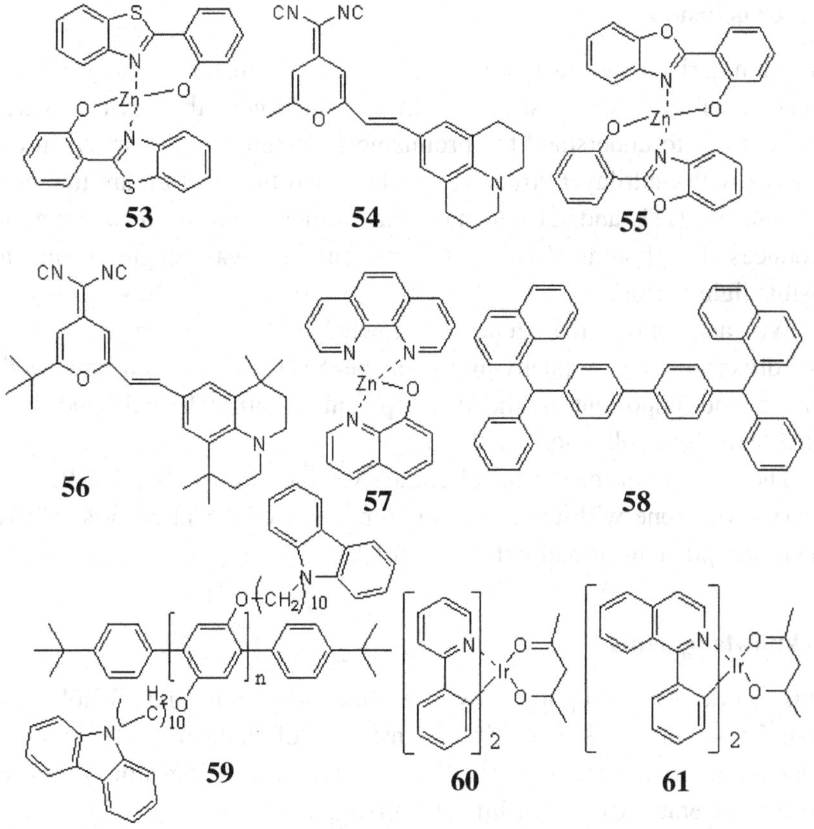

Fig. 9. The chemical structures of color materials for fabricating white OLED.

Huang et al.[48] prepared a high-efficiency white polymer light-emitting diode (WPLED) based on carbazole (**59**) doped with green (**60**) and red emitting (**61**) iridium complexes as the emitting layer. The device exhibits pure white light yet with stable sharp blue (430 nm), green (512 nm), and red (613 nm) emissions and has the maximum current and external quantum efficiencies of 16.8 cd/A and 8.65%, respectively. The CIE coordinates are (0.32, 0.33) at 20 V, quiet close to that of pure white light (0.33, 0.33).

4. Conclusion

We summarized the background and the basic phenomenon of OLED together with the device structures in Introduction of this review, which is necessary to understand the promising next-generation display. In the devices with multilayer structure, the hole and the electron are tunneled through the HTL and ETL to meet each others, and the recombination produces the photons (light emission) with the wavelength belong to visible light region.

We also gave the general knowledge for the basic ligands, 8-hydroxyquinone, 2-phenylpyridine, benzo-ligands which are well-known and important to the development of other ligands and their complexes for application to OLED.

The description of the metal complexes based on Ir(III), Zn(II), and Sn(IV) was done with the selective complexes of recent reports to give the new, update information for the OLED.

Acknowledgments

This work was supported by the Inje Research and Scholarship Foundation in 2008 and Korea Institute of Industrial Technology Evaluation and Planning (ITEP) through the Biohealth Products Research Center (BPRC) of Inje University.

References

1. C. W. Tang, S. A. Van Slyke, *Appl. Phys. Lett.* 51 (1987) 913.
2. H. S. Nalwa, L. S. Rohwer, *Handbook of Luminescence, Display Materials, and Devices*, Vol 1. American Scientific Publisher. 2003.
3. M. A. Baldo, D. F. O'Brien, M. E. Thomson, S. R. Forest, *Phys. Rev. B* 60 (1999) 14422.
4. L. S. Hung, C. H. Chen, *Mater. Sci. Eng.* R39 (2002) 143.
5. M. Schott, C. R. Acad, *Sci.* 1, 381 (2000).
6. R. H. Fowler, L. Nordheim, *Proc. R. Soc.* 119A (1928) 173.
7. D. C. Freeman, Jr., C. E. White, *J. Appl. Am. Chem. Soc.* 78 (1956) 2678.
8. C. W. Tang, S. A. Van Slyke, C. H. Chen, *Appl. Phys. Lett.* 65 (1989) 3610.
9. Z. Shen, P. E. Burrow, V. Bulovic, D. M. McCarthy, M. E. Thomson, S. R. Forest, *Jpn. J. Appl. Phys.* 35 (1996) L401.

10. Z. P. Skraup, *Monatsh.* 1 (1880) 316.
11. Y. Hamada, T. Sano, M. Fujita, T. Fujii, Y. Nishio, K. Shibata, *Jpn. J. Appl. Phys.* 32 (1993) L514.
12. K. A. King, P. J. Spellane, R. J. Watts, *J. Am. Chem. Soc.* 107 (1985) 1431.
13. M. F. Siddiqui, A. I. Levey, *Drugs Fut.* 24 (1999) 375.
14. V. Radha Rani et al., *Catal. Commun.* 6 (2005) 7174.
15. S. Lamansky, P. Djurovich, D. Murphy, F. Abdel-Razzaq, H. E. Lee, C. Adachi, P. E. Burrows, S. R. Forrest, M. E. Thompson, *J. Am. Chem. Soc.* 123 (2001) 4304.
16. Z. Liu, Z. Bian, *Organic Electron.* 9 (2008) 171.
17. R. Ragni et al., *J. Mater. Chem.* 16 (2006) 1161.
18. G. Ge, H. Guo, *J. Organomet. Chem.* 694 (2009) 3050.
19. W. S. Kim, B. J. Lee, *Thin Solid Films* 515 (2007) 5070.
20. Q. Su et al., *Polyhedron* 26 (2007) 5053.
21. G. Yu, Y. Liu, *Synth. Met.* 117 (2001) 211.
22. S. Tokito, *Synth. Met.* 111-112 (2000) 393.
23. M. L. Xu et al., *Thin Solid Films* 497 (2006) 239.
24. S. O. Jung et al., *Organic Electron.* 10 (2009) 1066.
25. A. Rzeska, J. Malicka, K. Guzow, M. Szabelski, W. Wicz, *J. Photochem. Photobiol. A* 146 (2001) 9.
26. A. F. Tarek et al., *J. Photochem. Photobiol. A* 121 (1999) 17.
27. T.-R. Chen, *J. Organomet. Chem.* 693 (2008) 3117.
28. C. Li et al., *J. Organomet. Chem.* 694 (2009) 2415.
29. A. Sharma, *Mater. Lett.* 61 (2007) 4614.
30. T.-T. Wang, *Tetrahedron* 65 (2009) 6325.
31. X. Chen, *J. Luminescence* 126 (2007) 81.
32. J. K. Park, D. E. Kim, T. D. Hoanh, Y. S. Kwon, B.-J. Lee, *J. Nanosci. Nanotechnol.* 8 (2008) 5071.
33. A. Matsuno-Yagi, Y. J. Hate, *Biol. Chem.* 268 (1993) 1539.
34. J. K. Park et al., *Coll. Surf. A: Physicochem. Eng. Asp.* 321 (2008) 266.
35. L. Shen et al., *Inorganic Chem. Commun.* 9 (2006) 620.
36. G. Y. Park, Y. S. Kim, Y. K. Ha, *Thin Solid Films* 515 (2007) 5090.
37. J. Gao et al., *Synth. Met.* 155 (2005) 168.
38. T.-H. Chuang et al., *Inorganica Chim. Acta* 362 (2009) 5017.
39. D.-E. Kim et al., *Curr. Appl. Phys.* 6 (2006) 805.
40. H.-H. Huang et al., *Thin Solid Films* 517 (2009) 3788.
41. C. O. Paul-Roth, *C. R. Chimie* 9 (2006) 1277.
42. J. H. Yao et al., *Tetrahedron* 64 (2008) 10814.
43. C. Hosokawa, M. Eida, M. Matsuura, K. Fukuoka, H. Nakamura, T. Kusumoto, *Synth. Met.* 91 (1997) 3.
44. E. H. Rhoderick, *Metal-Semiconductor Contacts*, Clarendon Press, Oxford, 1980.

45. J. T. Lim, *Curr. Appl. Phys.* 2 (2002) 295.
46. D.-E. Kim et al., *Thin Solid Films* 516 (2008) 3637.
47. D.-E. Kim et al., *Coll. Surf. A: Physicochem. Eng. Asp.* 313-314 (2008) 320.
48. S. P. Huang, H. H. Lu, S. A. Chen, *Synth. Met.* 159 (2009) 1940.

CHAPTER 6

STRUCTURE OPTIMIZATION FOR HIGH EFFICIENCY WHITE ORGANIC LIGHT-EMITTING DIODES

Ji Hoon Seo, Ji Hyun Seo and Young Kwan Kim*

Department of Information Display, Hongik University, Sangsu-dong, Mapo-ku, Seoul 121-791, Korea
*E-mail: kimyk@hongik.ac.kr

White organic light-emitting diodes (WOLEDs) have proven to have great utility in solid-state light sources (SSLs) as well as backlights in liquid-crystal displays (LCDs) and full-color OLEDs, due to their light weight, low operating voltage, thinness, diffusiveness, and harmless-light source for the restriction of certain hazardous substances directive/waste electrical and electronic equipment. In this article, advances in the understanding of WOLEDs are reviewed with dry- and wet-process and structure optimization for high efficiency WOLEDs. This article also reviews the highlights of the last few years and summarizes recent trends on WOLEDs. The technical discussions explore the question of whether WOLEDs are a viable alternative component in SSLs and backlights in LCD and full-color OLEDs.

1. Introduction

The first observations of electroluminescence from organic materials were made in the 1950s.[1] Interest in this phenomenon was fueled by the work of Pope *et al.*,[2] who observed electroluminescence from single crystals of anthracene. A voltage was applied between silver paste electrodes that were placed on the opposite sides of an anthracene crystal. A bright blue emission was observed. However, these devices were impractical for commercial applications because of the high voltages required for their operation and the need for exceptionally pure crystals. Thanks to innovations in the fields of vacuum and thin film coating technologies, Vincett et al.[3] were able to fabricate light-emitting devices based on evaporated thin films of anthracene in 1982. These were an order of magnitude thinner than the single crystals used by Pope et al. By using very thin vapor-deposited films, high fields were generated across the devices at much lower voltages, thereby substantially improving the device efficiency.

In recent years, the optimization of organic light-emitting diodes (OLEDs) as well as the synthesis of electroluminescent materials have been attracting great attention. In particular, white OLEDs (WOLEDs) have many advantages that can apply to illuminating light sources, backlights for liquid-crystal displays, and full-color, flat-panel displays with a color filter.[4,5] Many attempts to achieve WOLEDs have been reported, including the use of small organic molecules,[6] specially synthesized polymers,[7] polymer blends,[8-10] and polymer bilayers,[11] mixing the above with phosphorescent iridium complexes,[12] as well as luminescent semiconducting polymers blended with organometallic emitters.[13]

WOLEDs can be fabricated by vapor deposition and a solution process using fluorescent materials and phosphorescent materials. Therefore, this chapter is organized in the following manner. Following is very brief introduction into the structure optimization of high efficiency WOLEDs. Firstly, in Sec. 2, we introduce the different types of WOLED fabricated by thermal evaporation, which treat only small molecules because polymers cannot be thermally evaporated in a vacuum chamber. Polymers generally crosslink or decompose upon heating. In Sec. 3, we present various types of WOLEDs fabricated by wet-process for both small molecules and polymers. Finally, we offer a short conclusion on this topic. (Sec. 4)

2. WOLEDs with Small Molecules Based on Dry-Process

2.1. *Two Complementary Colors versus Three Primary Colors*

White light emission can be obtained from two complementary colors (sky blue and yellow-orange) or three primary colors (blue, green, and red).[14,15] The electroluminescence spectra of organic materials generally are broader than that of inorganic materials. Therefore, two complementary colors can produce white light emission. In general, WOLEDs with two complementary colors can be used as solid-state lighting (SSL). However, its weak point is a low color rendering index (CRI) because it lacks green emitting material. To serve as a backlight in thin film transistor liquid-crystal displays or full-color OLEDs, WOLEDs must be produced with organic emitting materials of three primary colors. WOLEDs with three primary colors had lower efficiency, more complex process, and more difficulty in controlling the location of the exciton recombination zone than WOLEDs with two complementary colors. So, the development of one or two emitting materials with three primary colors is necessary due to simple process. However, despite their potential capabilities, no one emitting material with three primary colors have yet been considered as proper WOLEDs, due to their extremely low efficiency and difficulty in finding a proper host material.

Seo *et al.* recently reported highly efficient WOLEDs using two emitting materials for three primary colors (red, green, and blue).[16] The 1,4-bis[2-(7-N-diphenyamino-2-(9,9-diethyl-9H-fluoren-2-yl)))vinyl] benzene used in the study had blue and green emissions at a peak of 478 and 510 nm, respectively. Iridium(III) bis(5-acetyl-2-phenylpyridinato-N,C2') acetylacetonate also had red emissions at the peak of 568 nm. Figure 1 shows the electroluminescence (EL) spectrum of WOLEDs at the driving voltage of 8, 10, and 12 V.

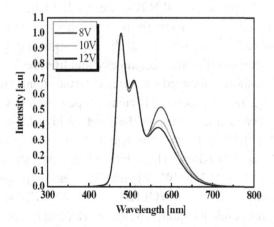

Fig. 1. EL spectra of WOLEDs at driving voltages from 8 to 12 V.

2.2. All Fluorescent- versus All Phosphorescent- versus Hybrid-WOLEDs

After the first fluorescent WOLEDs were reported with multilayer structure in 1994 by Kido *et al.*, WOLEDs have drawn increasing attention as a potential source of SSLs and backlights in liquid-crystal displays and full-color OLEDs.[6,17-22] There are three ways of realizing WOLEDs as SSLs, depending on the materials used. One type uses all fluorescent WOLEDs. The other two types are phosphorescent WOLEDs (PHWOLEDs), which consist solely of phosphorescent emitters, and hybrid WOLEDs (HWOLEDs), which use both fluorescent blue and phosphorescent red-green emitters. In 2002, Forrest *et al.*[23] developed entirely phosphorescent WOLEDs in a multilayered structure. Researchers have also conducted intense investigations of PHWOLEDs, because PHWOLEDs allow them to harvest both singlet and triplet excitons.[24-28] Unfortunately, despite the possibility of achieving 100% internal quantum efficiency (IQE) with PHWOLEDs, researchers have not yet found a phosphorescent blue emitter with a sufficiently high energy gap, long lifetime, and host materials. However, a few researchers have recently reported data concerning highly efficient phosphorescent deep-blue OLEDs and also investigated the synthesis of novel blue Ir- and Pt-complexes. Sun *et al.* have demonstrated highly efficient HWOLEDs using fluorescent blue and phosphorescent red-green emitters, where a spacer of 4,4'-N,N'-dicarbazole-biphenyl was used to prevent singlet energy transfer from a fluorescent blue emitter to phosphorescent red-green emitters and the exchange energy losses were minimized to increase efficiency, brightness, and lifetime.[29]

Figure 2(a) shows forward-viewing external quantum efficiency (EQE) (filled squares) and power efficiency (open circles) versus current density of the HWOLED shown in the inset. White device showed an EQE of 11.0% ± 0.3% at 1.0 mA/cm^2 and a slight decrease to an EQE of 10.8% ± 0.3% at 500 cd/m^2. The white device also showed a power efficiency of 22.1 ± 0.3 lm/W. Figure 2(b) shows proposed energy transfer mechanisms in the HWOLED. The HWOLED was harvested completely in independent channels as the singlet and tripelt excitons and the energy transfer from host to dopant could be separately optimized.

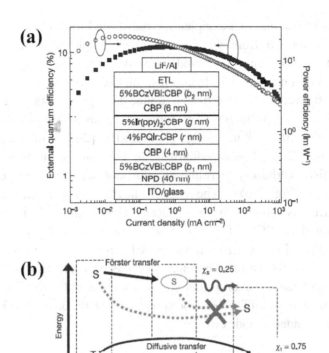

Fig. 2. (a) Forward-viewing EQE (filled squares) and power efficiency (open circles) versus current density of the HWOLED shown in the inset. (b) Proposed energy transfer mechanisms in the fluorescent/phosphorescent WOLED.

Therefore, the efficiency of WOLEDs was demonstrated with fluorescent dopants with IQE of 25% and phosphorescent dopant with IQE of 100%, resulting in a total WOLED IQE of 100%.

2.3. *Excimer- versus Exciplex-WOLEDs*

WOLEDs are obtained from a blue emitter and its excimer- and exciplex-based emissions with multiple-wavelength emissions.[30,31] The emissions of excimer and exciplex are red-shifted from blue emitter. An excimer is formed from the formation of complexes formed by the combination between an excited-state molecule and a ground-state molecule into one molecule. In contrast, an exciplex is a donor-acceptor complex between an excited-state donor and ground-state acceptrors in two molecules. Williams *et al.* reported excimer-based PHWOLEDs with nearly 100% IQE using platinum(II) [2-(4′,6′-difluorophenyl) pyridinato-N,C2′)](2,4-pentanedionato) as phosphorescent blue dopant and 2,6-bis(N-carbazolyl)pyridine as phosphorescent host, respectively, as shown in Fig. 3(a).[32] The energy diagram of the excimer-based PHWOLED is shown in Fig. 3(a). The optimized PHWOLED demonstrated an EQE of 15.9% and a power efficiency of 29 lm/W at 500 cd/m^2. Figure 3(b) shows the EL spectrum of excimer-based PHWOLED at 500 cd/m^2. The monomer was observed from 470 nm to 500 nm and the excimer peaks showed around 600 nm. The device showed an emission of CIE$_{x,y}$ coordinates of (0.46, 0.47) at 500 cd/m^2 and a color rendering index of 69.

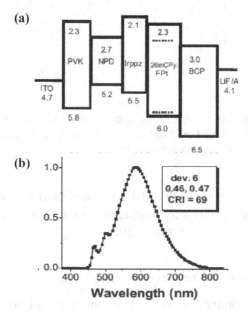

Fig. 3. (a) Energy diagram of excimer-based WOLEDs. (b) EL spectrum of excimer-based WOLED at 500 cd/m^2.

An exciplex-based WOLED was reported by Kim et al. which demonstrated exciplex between a charge transport layer and an EML.[33] The authors developed a WOLED using exciplex formation between active layers and a charge transport layer consisting of near ultraviolet and yellow emission. The white device showed a current efficiency of 12 cd/A at 1000 cd/m^2 and Commission Internationale de l'Eclairage (CIE$_{x,y}$) coordinates of (0.34, 0.34) at 8.6 mA/cm^2, respectively.

2.4. PIN- and Tandem-WOLEDs

The highly conductive p-doped material in the hole transporting layer and n-doped material in the electron transporting layer could improve the charge injection from the metals (anode and cathode) reduce the drive voltage in OLEDs.[34-36] The operating voltages of OLEDs are close to the thermodynamic limit. The device concept from LEDs has generalized to OLEDs. Ho et al. demonstrated PIN WOLEDs with high power efficiency and stable color which were comprised of 50% v/v tungsten oxide (WO$_3$)-doped N,N'-bis(naphthalen-1-yl)-N,N'-bis(phenyl)-benzidine (NPB) as the p-doped transport layer and 2% cesium carbonate (Cs2CO$_3$)-doped 4,7-diphenyl-1,10-phenanthroline (BPhen) as the n-doped transport layer.[37]

Tandem WOLEDs were first introduced in 1996 with stacked multiple emitting units in a single OLED.[38] The charge generation layers (CGL) are formed by contact of a n-doped ETL and a p-doped HTL or transparent inorganic conductors. The current efficiency of the tandem device with N units is N times as high as that of the single-unit device, normally. Kanno et al. developed tandem WOLEDs based on a combination of fluorescent and phosphorescent emitters.[39] CGL consisted of MoO$_3$ as p-type material and Li-doped BPhen as n-type material, respectively. Figure 4 shows EQE and power efficiencies of one- (circles), two- (squares), and three- (triangles) element tandem WOLEDs as functions of current density. 1-, 2-, and 3-element white devices showed a peak EQE of 12 ± 1, 23 ± 2, and 33 ± 3% at 0.78, 1.0, and 1.1 mA/cm^2, respectively.

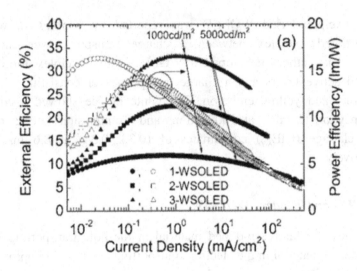

Fig. 4. EQE and power efficiencies of one- (circles), two- (squares), and three- (triangles) element tandem WOLEDs as functions of current density.

2.5. Sensitizer- and Microcavity-WOLEDs

WOLEDs can be obtained from phosphorescent emitters as sensitizer and fluorescent emitter.[40,41] OLEDs that use a phosphorescent sensitizer were first introduced in 2000 reported by Baldo et al.[42] The device with a phosphorescent sensitizer showed an EQE three times higher than the control device without a phosphorescent sensitizer. Lei et al. reported blue phosphorescent dye as an effective sensitizer and emitter for WOLEDs, consisting of iridium(III) bis[(4,6-di-fluoropheny)-pyridinato-N,C2] picolinate as the blue emitter and sensitizer and 4-(dicyanomethylene)-2-t-butyl-6-(1,1,7,7-tetramethyljulolidyl-9-enyl) as the fluorescent emitter.[40] The mechanism showed that a blue phosphorescent sensitizer can harvest both singlet and triplet excitons, then transfer energy to the fluorescent emitter and emit blue phosphorescence for white emissions; Fig. 5 illustrates energy transfer mechanism in the white light device. The white device showed a maximum luminance of 18200 cd/m^2 and a peak current efficiency of 9.2 cd/A.

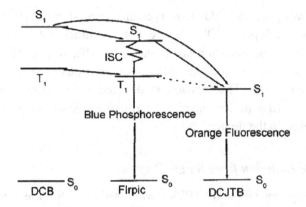

Fig. 5. Energy transfer mechanism in the sensitizer WOLEDs.

A microcavity is one of the most effective methods to improve the luminance and enable brightness. A microcavity resonator for OLEDs consists of a pair of mirrors facing each other across an EML. Typically, one of the mirrors is a metal electrode and the other is a half mirror. WOLEDs can also be obtained from a microcavity mechanism. Multiwavelength resonant cavities for WOLEDs were reported by Shiga *et al.* in 2003.[43] The authors designed WOLEDs with a multiwavelength resonant cavity (MWRC) by using optical simulation. The white device with MWRC was 1.3 times as bright as that of a control device.

3. WOLEDs Based on Wet-Process

Thermal evaporation process has been used as a deposition method for the organic materials, because it can process relatively uniform layers as compared with wet-process.[30,44-47] However, its high fabrication cost (due to wasting material) and its limited substrate size are weak points. On the other hand, the fabrication of OLEDs by wet-process is well-known to be low cost, as it does not require vacuum processing.[48-53] It has been reported that solution processes such as spin-coating,[54,55] doctor blade,[56] ink-jet printing,[57] gravure printing,[58,59] and screen printing[60] remain cost efficient because of large-area OLED fabrication.

In the wet-process, WOLEDs typically consist of a white emissive single-emission layer. WOLEDs are not manufactured with separated red, green, and blue emission layer since it is difficult to form multilayer structures because of solvent erosion of previously deposited layers during spin-coating. This session is devoted to exploring methods to improve WOLEDs performance by optimizing the device structure and materials used in the device.

3.1. *White Emission from Single Polymer*

Poly(*p*-phenylene vinylene) (PPV) was used first for the operation of polymer organic light-emitting diodes (PLEDs); increasing attention has been paid to using poly(fluorine) (PF) because it is possible to develop the blue emission and easily control the emission spectrum by bringing the comonomer in a PF molecule. Various white emissive polymers can be also synthesized based on the copolymerization of the PF backbone. Some examples are shown in Fig. 1, where the PF with the blue emission was copolymerized with red and green emissive structure for the white emission.[61-65] Phenylenevinylene oligomer (Fig. 6(a)), naphthalimide (Figs. 6(b)-6(d)), and fluorine-benzothiadiazole oligomer (Figs. 6(e) and 6(f)) with the triphenylamine are introduced the for the green emission, while the 4-(dicyanomethylene)-2-methyl-6-(*p*-dimethylaminostyryl)-4H-pyran (DCM) derivative (Fig. 6(a)) and thiophene-benzothiadiazole oligomer (Figs. 6(b)-6(f)) are induced for the red emission in polymer. In contrast, Fig. 7 shows the polymers with two-color white emission using the complementary color relation.[66-71] The PF with blue emission is copolymerized with reddish-yellow emitting comonomers, naphthalimide (Fig. 7(a)), triphenylamine-benzothiadiazole oligomer (Figs. 7(b) and 7(c)), DCM derivative (Fig. 7(d)), quinacridone derivative (Fig. 7(e)), and cyanovinylene complex (Fig. 7(f)) based on phenothiazine. The triphenylamine and oxadiazole are introduced in almost all chemical structures for the efficient transport of hole and electrons, respectively, as well as color tuning of PF. Even if naphthalimide moiety is the same, the emission color can be changed as its connected location, as shown in Figs. 6(b)-6(d) and Fig. 7(a). Each naphthalimide moiety connected to the conjugated

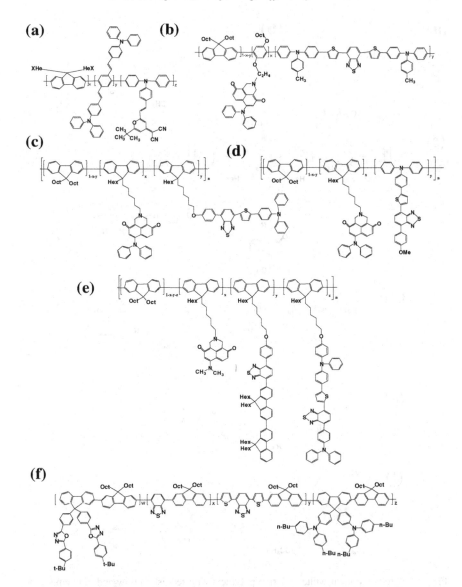

Fig. 6. Single-component white emissive polymers with red, green, and blue emission. (a) $CIE_{x,y}(0.33, 0.35)$, 0.10 cd/A, x = 0.95, y = 0.03, z = 0.02, (b) $CIE_{x,y}(0.31, 0.34)$, 1.59 cd/A, 0.83 lm/W, x = 0.0002, y = 0.0003, (c) $CIE_{x,y}(0.31, 0.32)$, 7.30 cd/A, 4.17 lm/W, x = 0.0002, y = 0.0002, (d) $CIE_{x,y}(0.30, 0.31)$, 3.80 cd/A, 1.99 lm/W, x = 0.0002, y = 0.0002, (e) $CIE_{x,y}(0.33, 0.36)$, 8.6 cd/A, 5.4 lm/W, x = 0.0005, y = 0.0005, z = 0.0002, (f) $CIE_{x,y}(0.37, 0.36)$, 4.87 cd/A, w = z = 0.4938 n, x = 0.0080 n, y = 0.0044 n.

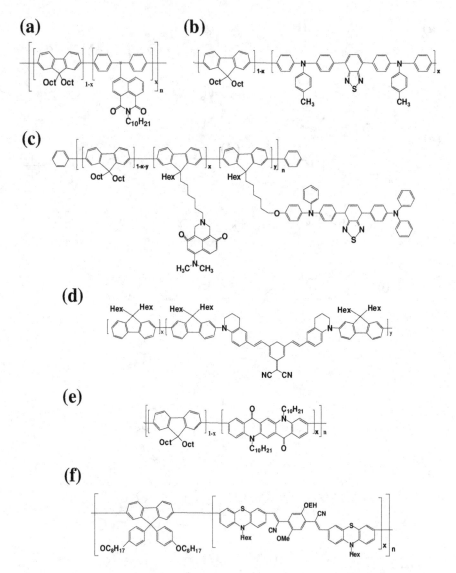

Fig. 7. Single-component white emissive polymers with reddish-yellow and blue emission. (a) $CIE_{x,y}$(0.25, 0.35), 5.3 cd/A, 2.8 lm/W, x = 0.0005, (b) $CIE_{x,y}$(0.35, 0.34), 8.99 cd/A, 5.75 lm/W, x = 0.0003, (c) $CIE_{x,y}$(0.31, 0.36), 12.8 cd/A, 8.51 lm/W, x = 0.0005, y = 0.0003, (d) $CIE_{x,y}$(0.33, 0.31), 0.60 cd/A, x = 0.995, y = 0.005, (e) $CIE_{x,y}$(0.27, 0.35), 3.47 cd/A, 2.18 lm/W, x = 0.0003, (f) $CIE_{x,y}$(0.33, 0.39), 1.95 cd/A, 0.97 lm/W, x = 0.005.

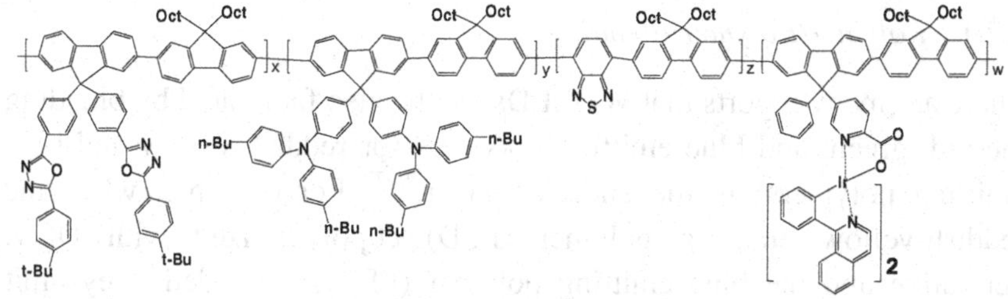

Fig. 8. Single-component white emissive polymers with phosphorescent emitter. (a) $CIE_{x,y}$(0.31, 0.36), 4.6 cd/A, x = 0.0003, y = 0.003, (b) $CIE_{x,y}$(0.35, 0.38), 8.2 cd/A, 7.2 lm/W, x = y = 0.2496 n, z = 0.004 n, w = 0.0004 n.

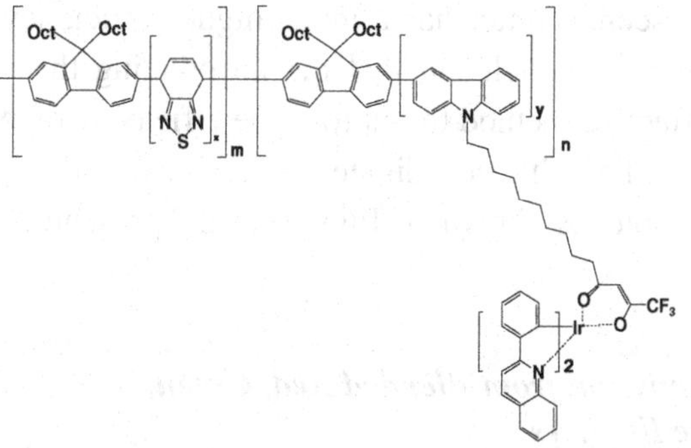

main-chain and non-conjugated side-chain of PF emits the reddish-yellow and green color, respectively. Moreover, as shown in Figs. 6(e) and 7(c), the dimethylamine-substituted naphthalimide, when used instead of diphenylamine, is functionalized like a blue dopant and improving the WOLED efficiency because its blue emission is more efficient than fluorescent emission efficiency from PF.

Fluorescent material limits the realization of efficient devices due to its low quantum yield; thus, the efficiency of WOLEDs can be enhanced by using the phosphorescent emitters as comonomers. All iridium complexes shown in Fig. 8 are used as red phosphorescence emitters.[72-75] A remarkable thing is that the complexes of Figs. 6(f) and 8(b) were reported by same researchers and their chemical structures are the same except for the red emitting site. The device efficiency with fluorescent emitter shows a maximum efficiency of 4.87 cd/A, whereas the device with phosphorescent emitter has a much higher efficiency of 8.20 cd/A. Therefore, it can be concluded that the introducing the phosphorescent emitter is an effective method to improve the efficiency of WOLEDs based on the polymer. The CIE coordinates, efficiency, and copolymerization ratio of all complexes shown in Figs. 6-8 are presented in each figure caption.

3.2. White Emission from Blended Red, Green, and Blue Emitters

3.2.1. Polymer/Polymer Blending

There are many reports that WOLEDs can be also fabricated by blending the red, green, and blue emitting polymers (or reddish-yellow and blue emitting polymers) in the emission layer.[76-79] For example, when the reddish-yellow emitting polymer (PFD) copolymerized with DCM derivative and the blue emitting polymer (PF) are blended, they emit the white color of CIE (0.33, 0.35) and demonstrate an efficiency of 0.26 cd/A.[79] However, these blended polymers lead to relatively low efficiency and high driving voltage, due to their difficult purification and wide energy band gap. Therefore, most contemporary researchers prefer to use phosphorescent small molecular red, green, and blue dopants rather than a polymer emitter.

3.2.2. Polymer Host/Phosphorescent Small Molecular Dopants Blending

In order to improve the device efficiency, it is advisable to apply the host-dopant system with phosphorescent small molecular dopants. Recently, poly(N-vinyl-carbazole) (PVK) and poly(9,9-dioctylfluorene) (PFO) shown in Fig. 9 have been used as polymer hosts for the fabrication of OLEDs by wet-process.[80-84] Wu *et al.* fabricated a highly efficient WOLED based on a PVK host.[80] The device consisted of bis(2-(4,6-difluorophenyl)-pyridinato-N,C2') picolinate (Firpic), iridium tris(2-(4-tolyl)pyridinato-N,C2') (Ir(mppy)$_3$), and iridium bis(1-phenylisoquinoline) (acetylacetonate) (Ir(piq)$_2$(acac)) as blue, green, and red phosphorescent dopants, respectively, in a PVK host. This device had

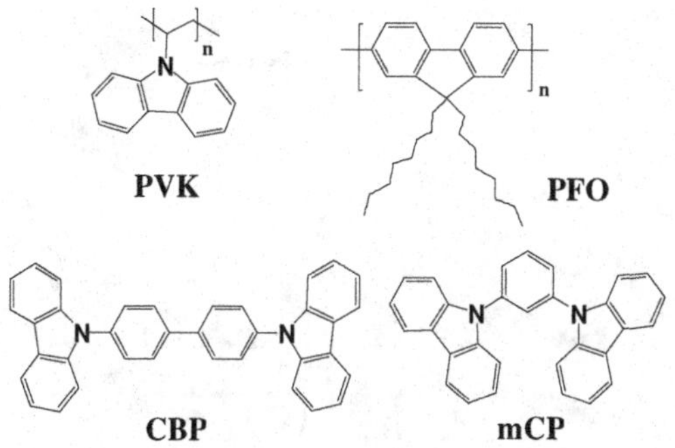

Fig. 9. Polymers and small molecules generally used as host in wet-process OLEDs.

current efficiency, EQE, and power efficiency of 24.3 cd/A, 14.4%, and 9.5 lm/W, respectively, with CIE$_{x,y}$ coordinates of (0.34, 0.46). Although this device emitted a reddish-white color, these efficiency values are the best among wet-processed WOLEDs reported so far. PVK is a very suitable polymer host, due to its excellent film-forming properties, high glass transition temperature (T_g) (~160°C) and hole transport characteristics.[85] However, PVK has a relatively low-lying triplet state (T_1) (ca. 2.5 eV). When using dyes to tune the emission color of a device, the host material should not act as a quencher of the dye emission via energy transfer from the dye to the host. When the dye is a phosphorescent emitter, its emission can be quenched by energy transfer to the triplet excited-state of the host.[86,87] To prevent this, the triplet energy of the host must be higher than that of the dopant. In general, red, green, and blue phosphorescent emitters have T_1 of approximately 2.0 eV, 2.5 eV, 2.7 eV, respectively. Thus, even if PVK is an excellent host material for the red and green emitters, it is not suitable for the blue emitter in terms of energy transfer, because it leads to endorthermic energy transfer and exchange energy loss.[85] To obtain highly efficient WOLEDs, it is necessary to develop a host material with high T_1.

3.2.3. *Small Molecular Host/Phosphorescent Small Molecular Dopants/Polymer Binder Blending*

In recent years, it has been shown that the small molecular hosts of carbazole-type shown in Fig. 9 are suitable for the phosphorescent

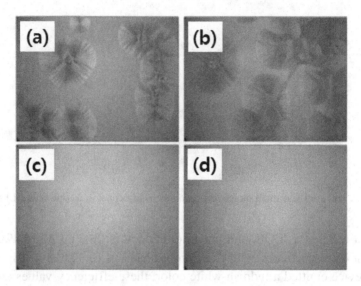

Fig. 10. Optical microscopy images of thin films stored at room temperature for 24 h after spin-coating with toluene solution of (a) CBP, (b) CBP and Firpic, (c) PS, CBP, and Firpic, and (d) PVK, CBP, and Firpic.

dopant in wet-process OLEDs as well as thermal-evaporated OLEDs.[88-90] This is because they have relatively high T_1, narrow energy band gap, and excellent charge transport ability in comparison with polymer hosts. The CBP (T_1; 2.6 eV) and mCP (T_1; 2.9 eV) are the most generally used host materials in phosphorescent OLEDs. However, although high efficiency can be obtained by using mCP or CBP, crystallization often occurs during the annealing process (Figs. 10(a) and 10(b)) required by wet-process for the exclusion of solvents remaining in the film.[88] As such, small molecular hosts with high T_1 and T_g need to be developed for higher efficiency and device stability. However, Lee *et al.* reported that crystallization of small molecules can be suppressed by adding a polymer binder into emission layer (Figs. 10(c) and 10(d)).[88] They added a PVK or polystyrene (PS) binder into the emission layer, consisting of CBP, Firpic, and iridium bis(2-phenylbenzothiozolato-N,C2')(acetylacetonate) ($Bt_2Ir(acac)$) as host, blue emitter, and red emitter, respectively. The introduction of polymer binders resulted in good film formation properties as well as better device performance than that of the PVK conventionally used as a host material. When the 23% PS binder was added, the device emits the white light of $CIE_{x,y}$(0.33, 0.39) and has an EQE and a current efficiency of 6.1% and 13 cd/A, respectively, at 100 cd/m^2.

3.3. WOLEDs Based on Color Conversion Method

Generally, the emission color of WOLEDs is easily changed from white to reddish-white as driving voltage is increased. The problem is very serious in WOLEDs with blended red, green, and blue emitters, as well as those with a white emissive single polymer. In order to overcome this problem, the color conversion method was developed by GE (Fig. 11).[91] At that time, PLED was fabricated with a blue emitting PF derivative from CDT, and then the color conversion materials — organic (perylene orange, perylene red) and inorganic (Y(Gd)AG:Ce) fluorescent materials — were sequentially deposited on the reverse side of a glass substrate. Subsequently, OSRAM also fabricated a WOLED with a similar method, in which a non-conjugated polymer host and Firpic are blended in the emission layer and the blue emission color due to Firpic is converted by nitridosilicate fluorescent material.[92] These color conversion methods are also a core issue for the realization of highly efficient WOLEDs by wet-process.

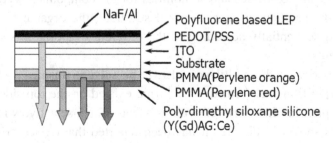

Fig. 11. Down conversion WOLED structure by GE.

3.4. Wet-Coating Techniques for the Multilayer Formation

Polymers cannot be thermally evaporated in a vacuum chamber because they generally crosslink or decompose upon heating. This limitation imposes restrictions on the nature of the polymers and the sidegroups attached to the polymer backbone since the polymer must be soluble. For example, unsubstituted poly(*p*-phenylenevinylene) (PPV) is insoluble, thus, it is generally fabricated by coating a soluble precursor polymer onto the desired anode substrate. The precursor polymer film is then converted to PPV by annealing at a temperature $150 \leq T \leq 250°C$ for up to ~ 24 h.[55,93,94] As this conversion process yields an insoluble layer of PPV, additional layers may be deposited on it by wet-process.

On the other hand, when soluble PPV derivatives such as 2,5-dialkoxy PPVs are coated onto the substrate, only solvents which would not redissolve the deposited film can be used to deposit additional layers. For example, Gustafsson *et al.*[95] fabricated flexible PLEDs by sequentially spin-coating an aqueous solution of water-soluble, conducting transparent polyaniline onto a transparency, and a xylene solution of poly(2-methoxy-5-(2'-ethyl)-hexoxy-1,4-phenylene vinylene) (MEH-PPV).

In order to form the multilayer, S-OLED Co. from Korea developed a self-assembled graded junction made by simple wet-process.[96] In this case, it is necessary to prepare the predissolving organic charge-transporting and light-emitting compounds in mixed organic solvents that have different volatilities and solubilities for the compounds used. Upon coating the solution on to a prepared substrate, the organic layers are deposited sequentially according to the different evaporation speeds of the solvents.

GE fabricated the multilayer OLEDs by solvent rinsing.[97] After the PVK:Firpic film was formed, only Firpic existed on the top side of the PVK:Firpic film. It was rinsed with *p*-xylene, which selectively removes Firpic, as shown in Fig. 12. It has been reported that device efficiency can be improved after the rinsing process.[97]

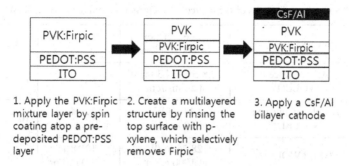

Fig. 12. Multilayer formation process through the solvent rinsing process.

In wet-process, these multilayer formation techniques can be helpful in improving the efficiency, lifetime, and stability of the device because each functionalized layer can enhance the balance of hole and electron in emission layer and suppress the exciton leakage out of the emission layer.

4. Conclusions and Outlook

Highlights from WOLED development over the past few years were reviewed, and recent trends in dry- and wet-process and structure optimization for high efficiency WOLEDs were reviewed.

Table 1 summarizes the device characteristics of some representative results regarding the dry- and wet-process WOLEDs. It has been established that WOLEDs fabricated by dry-process exhibit higher device performance than those developed by wet-process. Because of the simple process and low cost, WOLEDs will have to change from the wet-process to the dry-process-type in future. WOLEDs were assessed as a SSL source and backlight in liquid-crystal displays and full-color OLEDs. WOLEDs will not be practical in most SSLs because they are less efficient (~ 50 lm/W) than a fluorescent lamp (~ 70 lm/W). In addition, WOLEDs currently have short lifetimes in comparison with other displays or SSLs. Although WOLEDs have many weak points, this technology will soon be practical in manufactured goods such as SSLs and backlights in liquid-crystal displays and full-color OLEDs.

Table 1. Summarized results of the device characteristics from representative WOLEDs.

	Process type	Power efficiency or Current efficiency	EQE	CIE$_{(x,y)}$
D R Y	All fluorescent WOLED[19]	12.8 cd/A, 4.3 lm/W at 20 mA/cm^2	-	(0.31, 0.38)
	All phosphorescent WOLED[21]	44 lm/W at 1000 cd/m^2	25% at 1000 cd/m^2	(0.34, 0.40)
	Hybrid WOLED[18]	28.75 lm/W at 1000 cd/m^2	13.18% at 1000 cd/m^2	(0.48, 0.43)
	Excimer WOLED[32]	37.8 cd/A, 12.6 lm/W at 500 cd/m^2	15.9% at 1000 cd/m^2	(0.46, 0.47)
	Exciplex WOLED[33]	12 cd/A, 2 lm/W at 1000 cd/m^2	-	(0.34, 0.34)
	PIN WOLED[37]	10 cd/A, 9.3 lm/W at 1000 cd/m^2	-	(0.32, 0.43)
	Tandem (2 unit) WOLED[39]	19 lm/W at 1000 cd/m^2	38% at 1000 cd/m^2	(0.38, 0.44)
	Sensitizer WOLED[41]	20.2 lm/W at 800 cd/m^2	13.1% at 800 cd/m^2	(0.38, 0.42)
	Microcavity[43]	12.8 cd/A at 300 cd/m^2	-	(0.32, 0.40)
W E T	PF + PFD blending[77]	Max. 0.26 cd/A	-	(0.33, 0.35)
	PVK + Firpic + Ir(mppy)$_3$ + Ir(piq)$_2$(acac) blending[78]	Max. 9.5 lm/W, Max. 24.3 cd/A	Max. 14.4%	(0.34, 0.46)
	CBP + PS + Firpic + Bt$_2$Ir(acac) blending[86]	13 cd/A at 100 cd/m^2	6.1% at 100 cd/m^2	(0.33, 0.39)
	Single-component RGB white polymer[62]	Max. 5.4 lm/W, Max. 8.6 cd/A	-	(0.33, 0.36)
	Single-component RB white polymer[66]	Max. 8.51 lm/W, Max. 12.8 cd/A	-	(0.31, 0.36)
	Single-component white phosphorescent polymer[71]	Max. 7.21 lm/W, Max. 8.2 cd/A	-	(0.35, 0.38)

Acknowledgments

This work was supported by Energy Resources Technology Development program (2007-E-CM11-P-07), Strategy Technology Development program (10030834) from Ministry of Knowledge Economy (MKE) and the ERC program of the Korea Science and Engineering Foundation (KOSEF) grant funded by the Korea Ministry of Education, Science and Technology (MEST) (No. R11-2007-045-03001-0).

References

1. A. Bernanose, *Br. J. Appl. Phys.* **4**, S54 (1955).
2. M. Pope, H. P. Kallmann and P. J. Magnante, *J. Chem. Phys.* **38**, 2042 (1963).
3. P. S. Vincett, W. A. Barlow, R. H. Hann and G. G. Roberts, *Thin Solid Films* **94**, 171 (1982).
4. J. Kido, M. Kimura and K. Nagai, *Science* **267**, 1332 (1995).
5. H. Spreitzer, H. Vestweber, P. Stobel and H. Becker, *Proc. SPIE-Int. Soc. Opt. Ent.* **125**, 4105 (2005).
6. C. W. Ko and T. Tao, *Appl. Phys. Lett.* **79**, 4234 (2001).
7. M.-L. Tsai, C.-Y. Liu, Y.-Y. Wang, J.-Y. Chen, T.-C. Chou, H.-M. Lin, S.-H. Tsai and T. J. Chow, *Chem. Mater.* **16**, 3373 (2004).
8. M. Granstrom and O. Inganas, *Appl. Phys. Lett.* **68**, 147 (1996).
9. B. Hu and F. Karasz, *J. Appl. Phys.* **93**, 1995 (2003).
10. J. I. Lee, H. Y. Chu, S. H. Kim, L. M. Do, T. Zyung and D. H. Hwang, *Opt. Mater.* **21**, 205 (2002).
11. J. H. Park, T.-W. Lee, Y. C. Kim, O. O. Park and J. K. Kim, *Chem. Phys. Lett.* **403**, 293 (2005).
12. I. Tanaka, M. Suzuki and S. Tokoto, *Jpn. J. Appl. Phys.* **42**, 2737 (2003).
13. X. Gong, S. H. Lim, J. C. Ostrowski, D. Moses, C. J. Bardeen and G. C. Bazan, *J. Appl. Phys.* **95**, 948 (2004).
14. M. F. Lin, L. Wang, W. K. Wong, K. W. Cheah, H. L. Tam, M. T. Lee, M. H. Ho and C. H. Chen, *Appl. Phys. Lett.* **91**, 073517 (2007).
15. Y. Sun and S. R. Forrest, *Appl. Phys. Lett.* **91**, 263503 (2007).
16. J. H. Seo, J. H. Seo, J. H. Park, J. H. Kim, G. W. Hyung, K. H. Lee, S. S. Yoon and Y. K. Kim, *Appl. Phys. Lett.* **90**, 203507 (2007).
17. J. Kido, J. Hongawa, K. Okuyama and K. Nagai, *Appl. Phys. Lett.* **64**, 815 (1994).
18. J. H. Seo, J. S. Park, J. H. Kim, J. R. Koo, B. M. Seo, K. H. Lee, J. K. Park, J. T. Je, S. S. Yoon and Y. K. Kim, *IMID 2009*, 34-3 (2009).
19. T. H. Liu, Y. S. Wu, M. T. Lee, H. H. Chen, C. H. Liao and C. H. Chen, *Appl. Phys. Lett.* **85**, 4304 (2004).

20. H. Kanno, R. J. Holmes, Y. Sun, S. K. Cohen and S. R. Forrest, *Adv. Matert.* **18**, 339 (2006).
21. S. J. Su, E. Gonmori, H. Sasabe and J. Kido, *Adv. Mater.* **20**, 4189 (2008).
22. H. I. Baek and C. H. Lee, *J. Appl. Phys.* **103**, 124504 (2008).
23. B. W. D'Andrade, M. E. Thompson and S. R. Forrest, *Adv. Mater.* **14**, 147 (2002).
24. B. P. Yan, C. C. C. Cheung, S. C. F. Kui, H. F. Xiang, V. A. Roy, S. J. Xu and C. M. Che, *Adv. Mater.* **19**, 3599 (2007).
25. K. S. Yook and J. Y. Lee, *Appl. Phys. Lett.* **92**, 193308 (2008).
26. S. Seidel, R. Krause, A. Hunze, G. Schmid, F. Kozlowski, T. Dobbertin and A. Winnacker, *J. Appl. Phys.* **104**, 064505 (2008).
27. R. Meerheim, R. Nitsche and K. Leo, *Appl. Phys. Lett.* **93**, 043310 (2008).
28. Q. Wang, J. Ding, D. Ma, Y. Cheng, L. Wang and F. Wang, *Adv. Func. Mater.* **19**, 84 (2008).
29. Y. Sun, N. C. Giebink, H. Kanno, B. Ma, M. E. Thompson and S. R. Forrest, *Nature* **440**, 908 (2006).
30. B. W. D'Andrade, J. Brooks, V. Adamovich, M. E. Thompson and S. R. Forrest, *Adv. Mater.* **14**, 1032 (2002).
31. J. Y. Li, D. Liu, C. Ma, O. Lengyel, C. S. Lee, C. H. Tung and S. Lee, *Adv. Mater.* **16**, 1538 (2004).
32. E. L. Williams, K. Haavisto, J. Li and G. E. Jabbour, *Adv. Mater.* **19**, 197 (2007).
33. Y. M. Kim, Y. W. Park, J. W. Huh, J. H. Choi, B. W. Ju, J. H. Jung and J. K. Kim, *Society for Information Display 07 Digest*, 796 (2007).
34. Y. Tomita, C. May, M. Toerker, J. Amelung, M. Eritt, F. Loeffler, C. Luber and K. Leo, *Appl. Phys. Lett.* **91**, 063510 (2007).
35. J. Huang, M. Pfeiffer, A. Werner, J. Blochwitz, K. Leo and S. Liu, *Appl. Phys. Lett.* **80**, 139 (2002).
36. M. H. Ho, T. M. Chen, P. C. Yeh, S. W. Hwang and C. H. Chen, *Appl. Phys. Lett.* **91**, 233507 (2007).
37. M. H. Ho, S. F. Hsu, J. W. Ma, S. W. Hwang, P. C. Yeh and C. H. Chen, *Appl. Phys. Lett.* **91**, 113518 (2007).
38. V. Bulovic, G. Gu, P. E. Burrows, M. E. Thompson and S. R. Forrest, *Nature* **380**, 29 (1996).
39. H. Kanno, N. C. Giebink, Y. Sun and S. R. Forres, *Appl. Phys. Lett.* **89**, 023503 (2006).
40. G. Lei, L. Wang and Y. Qiu, *Appl. Phys. Lett.* **85**, 5403 (2004).
41. H. Kanno, Y. Sun and S. R. Forrest, *Appl. Phys. Lett.* **89**, 143516 (2006).
42. M. A. Baldo, M. E. Thompson and S. R. Forrest, *Nature* **403**, 750 (2000).
43. T. Shiga, H. Fujikawa and Y. Taga, *J. Appl. Phys.* **93**, 19 (2003).
44. B. W. D'Andrade, R. J. Holmes and S. R. Forrest, *Adv. Mater.* **16**, 624 (2004).
45. C. H. Chuen and Y. T. Tao, *Appl. Phys. Lett.* **81**, 4499 (2002).
46. Y. S. Huang, J. H. Jou, W. K. Weng and J. M. Liu, *Appl. Phys. Lett.* **80**, 2782 (2002).

47. H. G. Lee, J. H. Seo, Y. K. Kim, J. H. Kim, J. R. Koo, K. H. Lee and S. S. Yoon, *J. Kor. Phys. Soc.* **49**, 1052 (2006).
48. C. A. Landis, S. R. Parkin and J. E. Anthony, *Jpn. J. Appl. Phys.* **6A**, 3921 (2005).
49. S. C. Lo, N. A. H. Male, J. P. J. Markham, S. W. Magennis, P. L. Burn, O. V. Salata and I. D. W. Samuel, *Adv. Mater.* **14**, 975 (2002).
50. H. Kim, Y. Byun, R. R. Das, B. K. Choi and P. S. Ahn, *Appl. Phys. Lett.* **91**, 093512 (2007).
51. N. Rehmann, D. Hertel, K. Meerholz, H. Becker and S. Heun, *Appl. Phys. Lett.* **91**, 103507 (2007).
52. Y. Hino, H. Kajii and Y. Ohmori, *Organic Electron.* **5**, 265 (2004).
53. M. Ooe, S. Naka, H. Okada and H. Onnagawa, *Jpn. J. Appl. Phys.* **45**, 250 (2006).
54. J. P. J. Markham, S.-C. Lo, S. W. Magennis, P. L. Burn and I. D. W. Samuel, *Appl. Phys. Lett.* **80**, 2645 (2002).
55. N. C. Greenham, S. C. Moratti, D. D. C. Bradley, A. B. Holmes and R. H. Friend, *Nature* **365**, 628 (1993).
56. W. Reiss, in *Organic Electroluminescent Materials and Devices*, eds. S. Miyata and H. S. Nalwa, Chap. 2 (Gordon and Breach, NY, 1997).
57. T. R. Hebner, C. C. Wu, D. Marcy, M. H. Lu and J. C. Sturm, *Appl. Phys. Lett.* **72**, 519 (1998).
58. J. J. Michels, S. H. P. M. de Winter and L. H. G. Symonds, *Organic Electron.* **10**, 1495 (2009).
59. P. Kopola, M. Tuomikoski, R. Suhonen and A. Maaninen, *Thin Solid Films* **517**, 5757 (2009).
60. D. A. Pardo, G. E. Jabbour and N. Peyghambarian, *Adv. Mater.* **12**, 1249 (2000).
61. S. K. Lee, D. H. Hwang, B. J. Jung, N. S. Cho, J. Lee, J. D. Lee and H. K. Shim, *Adv. Funct. Mater.* **15**, 1647 (2005).
62. J. Liu, Q. Zhou, Y. Cheng, Y. Geng, L. Wang, D. Ma, X. Jing and F. Wang, *Adv. Mater.* **17**, 2974 (2005).
63. J. Liu, Z. Xie, Y. Cheng, Y. Geng, L. Wang, X. Jing and F. Wang, *Adv. Mater.* **19**, 531 (2007).
64. J. Liu, L. Chen, S. Shao, Z. Xie, Y. Cheng, Y. Geng, L. Wang, X. Jing and F. Wang, *Adv. Mater.* **19**, 4224 (2007).
65. C. Y. Chung, P. I. Shih, C. H. Chien, F. I. Wu and C. F. Shu, *Macromolecules* **40**, 247 (2007).
66. G. Tu, Q. Zhou, Y. Cheng, L. Wang, D. Ma, X. Jing and F. Wang, *Appl. Phys. Lett.* **85**, 2172 (2004).
67. J. Liu, Q. Zhou, Y. Cheng, Y. Geng, L. Wang, D. Ma, X. Jing and F. Wang, *Adv. Funct. Mater.* **16**, 957 (2006).
68. J. Liu, S. Shao, L. Chen, Z. Xie, Y. Cheng, Y. Geng, L. Wang, X. Jing and F. Wang, *Adv. Mater.* **19**, 1859 (2007).
69. S. K. Lee, B. J. Jung, T. Ahn, Y. K. Jung, J. I. Lee, I. Kang, J. Lee, J. H. Park and H. K. Shim, *J. Polym. Sci. Polym. Chem.* **45**, 3380 (2007).

70. J. Liu, B. Gao, Y. Cheng, Z. Xie, Y. Geng, L. Wang, X. Jing and F. Wang, *Macromolecules* **41**, 1162 (2008).
71. M. J. Park, J. Lee, J. H. Park, S. K. Lee, J. I. Lee, H. Y. Chu, D. H. Hwang and H. K. Shim, *Macromolecules* **41**, 3063 (2008).
72. J. Jiang, Y. Xu, W. Yang, R. Guan, Z. Liu, H. Zhen and Y. Cao, *Adv. Mater.* **18**, 1769 (2006).
73. F. I. Wu, X. H. Yang, D. Neher, R. Dodda, Y. H. Tseng and C. F. Shu, *Adv. Funct. Mater.* **17**, 1085 (2007).
74. K. Zhang, Z. Chen, C. Yang, Y. Tao, Y. Zou, J. Qui and Y. Cao, *J. Mater. Chem.* **18**, 291 (2008).
75. P. I. Lee, S. L. C. Hsu and J. F. Lee, *J. Polym. Sci. Polym. Chem.* **46**, 464 (2008).
76. P. I. Shih, Y. H. Tseng, F. I. Wu, A. K. Dixit and C. F. Shu, *Adv. Funct. Mater.* **16**, 1582 (2006).
77. F. I. Wu, P. I. Shih, Y. H. Tseng, C. F. Shu, Y. L. Tung and Y. Chi, *J. Mater. Chem.* **17**, 167 (2007).
78. S. K. Lee, T. Ahn, N. S. Cho, J. I. Lee, Y. K. Jung, J. Lee and H. K. Shim, *J. Polym. Sci. Polym. Chem.* **45**, 1199 (2007).
79. B. Y. Hsieh and Y. Chen, *Macromolecules* **40**, 8913 (2007).
80. H. Wu, J. Zou, F. Liu, L. Wang, A. Mikhailovsky, G. C. Bazan, Q. Yang and Y. Cao, *Adv. Mater.* **20**, 696 (2008).
81. Y. Xu, J. Peng, J. Jiang, W. Xu, W. Yang and Y. Cao, *Appl. Phys. Lett.* **87**, 193502 (2005).
82. X. Gong, W. Ma, J. C. Ostrowski, G. C. Bazan, D. Moses and A. J. Heeger, *Adv. Mater.* **16**, 615 (2004).
83. X. Gong, D. Moses and A. J. Heeger, *J. Phys. Chem. B* **108**, 8601 (2004).
84. T. H. Kim, H. K. Lee, O. O. Park, B. D. Chin, S. H. Lee and J. K. Kim, *Adv. Funct. Mater.* **16**, 611 (2006).
85. Y. Kawamura, S. Yanagida and S. R. Forrest, *J. Appl. Phys.* **92**, 87 (2002).
86. M. Sudhakar, P. I. Djurovich, T. E. Hogen-Eash and M. E. Thompson, *J. Am. Chem. Soc.* **125**, 7796 (2003).
87. F.-C. Chen, S.-C. Chang, G. He, S. Pyo, Y. Yang, M. Kurotaki and J. Kido, *J. Polym. Sci. Pt. B: Polym. Phys.* **41**, 2681 (2003).
88. J. I. Lee, H. Y. Chu, Y. S. Yang, L. M. Do, S. M. Chung, S. H. K. Park and C. S. Hwang, *Jpn. J. Appl. Phys.* **45**, 9231 (2006).
89. J. J. Park, T. J. Park, W. S. Jeon, R. Pode, J. Jang, J. H. Kwon, E. S. Yu and M. Y. Chae, *Organic Electron.* **10**, 189 (2009).
90. S. Reineke, G. Schwartz, K. Walzer and K. Leo, *Appl. Phys. Lett.* **91**, 123508 (2007).
91. A. R. Duggal, J. J. Shiang, C. M. Heller and D. F. Foust, *Appl. Phys. Lett.* **80**, 3470 (2002).
92. B. C. Krummacher, M. Mathai, F. So, S. Choulis and V. E. Choong, *J. Display Tech.* **3**, 200 (2007).

93. J. H. Burroughs, D. D. C. Bradley, A. R. Brown, R. N. Marks, K. Mackay, R. H. Friend, P. L. Burns and A. B. Holmes, *Nature* **347**, 539 (1990).
94. R. H. Friend, R. W. Gymer, A. B. Holmes, J. H. Burroughes, R. N. Marks, C. Taliani, D. D. C. Bradley, D. A. Dos Santos, J. L. Bredas, M. Logdlund and W. R. Salaneck, *Nature* **397**, 121 (1999).
95. G. Gustafsson, Y. Cao, G. M. Treacy, F. Klavetter, N. Colaneri and A. J. Heeger, *Nature* **357**, 477 (1992).
96. B. Park, The International Symposium on Super-Functionality Organic Devices, Program Schedule and Abstract Book, 38p (2004).
97. J. Liu, Q. Ye, L. N. Lewis and A. R. Duggal, *Proc. SPIE* **6695**, 66550Y (2007).

Part 2
Molecular Electronics

CHAPTER 7

STATISTICAL ANALYSIS OF ELECTRONIC TRANSPORT PROPERTIES OF ALKANETHIOL MOLECULAR JUNCTIONS

Tae-Wook Kim, Gunuk Wang, Hyunwook Song and Takhee Lee*

*Department of Materials Science and Engineering,
Gwangju Institute of Science and Technology, Gwangju 500-712, Korea
E-mail: tlee@gist.ac.kr

We present a review on the electronic transport of alkanethiols molecular junctions in microscale or nanoscale via-hole structure. The statistical analysis provides criteria for defining "working" molecular electronic devices and selecting "representative" devices. Based on statistical analysis, we proposed a multi-barrier tunneling (MBT) model. The charge conduction mechanisms in molecular junctions are studied. These statistical analyses on the electronic transport properties provide an insight of metal-molecule contact effect in the conduction mechanism.

1. Introduction

Due to the merits such as low cost, high density, and less heat problems for using functional molecules as nanoscale building-blocks in miniaturized electronic devices, molecular electronics is currently undergoing rapid development, although poor reproducibility and low device yield still remains a challenge.[1-6] Extensive efforts have been made to understand charge transport in molecular layers.[7,8] Alkanethiol ($CH_3(CH_2)_{n-1}SH$) self-assembled monolayers (SAMs) on Au surfaces are one of the most extensively studied molecular systems because of the robust formation of monolayers of alkanethiols on a Au surface.[3,9] The yield of molecular electronic devices of even these robust alkanethiol molecular systems, however, is very low, mainly because of electrical shorts caused by the penetration of the top electrode through the molecular layer and making contact with the bottom electrode.[10,11] A recent study, with the objective of preventing electrical shorts by using a layer of a highly conducting polymer resulted in a significant improvement in the yield of molecular electronic devices.[7] However, studies on the device yield of simple metal-molecule-metal (M-M-M) junctions have not been extensive. In particular, systematic studies with the goal of defining "working" molecular devices, device yield, and even selecting "representative" devices have not been investigated thoroughly. Furthermore, determining the average transport parameters from a statistically meaningful number of molecular working devices is important because the statistically averaged transport parameters can provide more accurate and meaningful characteristics of molecular systems. Statistical measurement has been performed, for example, to extract the electrical conductance of single molecules using mechanically controllable break junctions.[12]

In this chapter, we summarize the fabrication of a large number of alkanethiol molecular electronic devices as vertical M-M-M structures without using any intermediate external polymer layer which might cause an additional interface to be produced in the molecular junctions and the results of their electronic transport properties. Gaussian distribution functions were used to statistically analyze the mass-fabricated molecular devices and a simple criterion for the statistical determination of working devices and representative devices is proposed. Average transport parameters such as current density, transport barrier height, effective electron mass, and tunneling decay coefficient were obtained from the statistically defined working molecular electronic devices. In addition, the statistical criterion was employed to demonstrate that determining working molecular electronic devices should be done prior to further analysis such as temperature-variable characterization on the devices. Also, the statistical analysis would be useful for comparing the transport parameters of different molecular systems.

In addition, we explain a statistical method to investigate the electronic transport of nanoscale molecular junction. For this, comprehensive temperature-variable current-voltage ($I(V,T)$) characterization was performed with subsequent statistical analysis, using mass-fabricated molecular devices with nanometer-scale junction diameter. The $I(V,T)$ characterization can play a critical role in determining the transport mechanism which makes it possible, for example, to distinguish electronic tunneling transport from thermally activated conduction such as impurity-mediated transport.[13] Study based on the statistical approach would give impartiality in determining the intrinsic molecular transport properties.

Molecular electronics utilizing functional molecules as the ultimate nanoscale electronic components have demonstrated their potential in device applications in variety functional electronic device components for ultrahigh density future electronics.[1,2,5,14-17] However, despite the numerous potential advantages of molecular electronics as compared to traditional silicon-based electronics, there are many issues and challenges that need to be overcome to apply molecules to actual electronic circuits. Among those, metal-molecule contact is important not only for understanding the transport properties of molecular devices,[18-24] but also for realizing reproducible molecular electronic devices, due to its role in controlling metal-molecule interfaces.[11,22-25]

Here, we explain the influence of metal-molecule contacts in molecular junctions and the essential charge transport mechanisms using a proposed multi-barrier tunneling (MBT) model where the metal-molecule-metal junction can be divided into three parts: the molecular-chain body with metal-molecule contacts on either side of molecule. The MBT model will help introduce a new insight for studying charge transport mechanisms, one focused on the metal-molecule contacts in molecular electronic devices or other nanoscale devices.

2. Fabrication of Molecular Devices

The alkanethiol M-M-M junction devices were fabricated on a p-type (100) Si substrate covered with a thermally grown 3000 Å thick layer of SiO_2. As schematically illustrated in Fig. 1, the conventional optical lithography method was used to pattern bottom electrodes that were prepared with Au (1000 Å)/Ti (50 Å) using an electron beam evaporator. A SiO_2 layer (700 Å thick) was deposited on the patterned bottom by plasma enhanced chemical vapor deposition (PECVD). Reactive ion

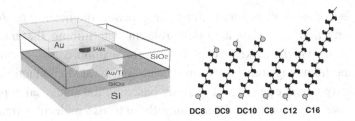

Fig. 1. A schematic diagram of a metal-alkanethiol-metal junction device and molecular structures of octanedithiol (DC8), nonanedithiol (DC9), decanedithol (DC10), octanethiol (C8), dodecanethiol (C12), and hexadecanethiol (C16).

etching (RIE) was then performed to produce microscale via-holes of 2 μm diameter through SiO_2 layer to expose the Au surfaces of the bottom electrodes. Several different ~5 mM alkanethiol solutions were prepared by adding ~10 μL alkanethiols to ~10 mL anhydrous ethanol. The chips were left in the solution for 24-48 h for the alkanethiol SAMs to assemble on the Au surfaces exposed by RIE in a nitrogen-filled glove box with an oxygen level of less than ~10 ppm.

Alkanemonothiols (Aldrich Chem. Co) and alkanedithiols (Aldrich Chem. Co and Tokyo Chem. Industry) of different molecular lengths: octanemonothiol ($CH_3(CH_2)_7SH$, C8), dodecanemonothiol ($CH_3(CH_2)_{11}SH$, C12), hexadecanemonothiol ($CH_3(CH_2)_{15}SH$, C16), octanedithiol ($HS(CH_2)_8SH$, DC8), nonanedithiol ($HS(CH_2)_9SH$, DC9), and decanedithol ($HS(CH_2)_{10}SH$, DC10) were used to form the active molecular components in M-M-M junctions. After the alkanethiol SAMs were formed on the exposed Au surfaces, a top Au electrode was produced by thermal evaporation to form M-M-M junctions. This evaporation was done with a shadow mask on the chips with a liquid nitrogen cooled cold stage in order to minimize thermal damage to the active molecular component under a pressure of ~10^{-6} torr. For the same reason, the deposition rate for the top Au electrode was kept very low, typically ~0.1 Å/s until the total thickness of the top Au electrode reached ~500 Å. Figure 1 shows a schematic diagram of a microscale M-M-M junction device and molecular structures of different alkanethiols. The room temperature current-voltage (I-V) characteristics of the fabricated molecular devices were evaluated using a HP4155A

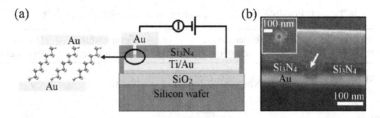

Fig. 2. (a) Schematic of the molecular device structure. (b) An SEM image of a nanowell (marked by the arrow) in cross-sectional view. The inset shows a nanowell in top view. The scale bars are 100 nm.

semiconductor parameter analyzer. The fabricated chips were packaged and loaded into a cryostat (from Janis. Co.). The temperature was varied from 300 to 77 K by flowing liquid nitrogen into the sample holder in the vacuum chamber.

In order to fabricate nanoscale molecular junction, alkanethiol SAMs are sandwiched between two metallic contacts through a nanowell. The junction diameter is estimated to be ~50 nm from a cross-sectional scanning electron microscope (SEM) image of the nanowell as shown in Fig. 2.

3. Theoretical Basis

3.1. *Possible Conduction Mechanisms*

In Table 1, possible conduction mechanisms are listed with their characteristic current, temperature- and voltage-dependencies.[26] Based on whether thermal activation is involved, the conduction mechanisms fall into two distinct categories: (i) thermionic or hopping conduction which has temperature-dependent $I(V)$ behavior and (ii) direct tunneling or Fowler-Nordheim tunneling which does not have temperature-dependent $I(V)$ behavior. For example, thermionic and hopping conductions have been observed for 4-thioacetylbiphenyl SAMs[28] and 1,4-phenelyene diisocyanide SAMs.[29] On the other hand, the conduction mechanism is expected to be tunneling when the Fermi levels of contacts lie within the large energy gap for short length molecule, as for the case of alkanethiol molecular system.[30-32]

Table 1. Possible conduction mechanisms. Adapted from Ref. 26.

Conduction Mechanism	Characteristic Behavior	Temperature Dependence	Voltage Dependence
Direct Tunneling*	$J \sim V \exp\left(-\dfrac{2d}{\hbar}\sqrt{2m\Phi}\right)$	none	$J \sim V$
Fowler-Nordheim Tunneling	$J \sim V^2 \exp\left(-\dfrac{4d\sqrt{2m}\,\Phi^{3/2}}{3q\hbar V}\right)$	none	$\ln\left(\dfrac{J}{V^2}\right) \sim \dfrac{1}{V}$
Thermionic Emission	$J \sim T^2 \exp\left(-\dfrac{\Phi - q\sqrt{qV/4\pi\varepsilon d}}{kT}\right)$	$\ln\left(\dfrac{J}{T^2}\right) \sim \dfrac{1}{T}$	$\ln(J) \sim V^{1/2}$
Hopping Conduction	$J \sim V \exp\left(-\dfrac{\Phi}{kT}\right)$	$\ln\left(\dfrac{J}{V}\right) \sim \dfrac{1}{T}$	$J \sim V$

3.2. Tunneling Models

When the Fermi level of the metal is aligned close enough to one energy level (either highest occupied molecular orbital (HOMO) or lowest unoccupied molecular orbital (LUMO)), the effect of the other distant energy level on the tunneling transport is negligible, and the widely used Simmons model[33] is an excellent approximation. Simmons model expressed the tunneling current density through a barrier in the tunneling regime of $V < \Phi_B/e$ as[33,34]

$$J = \left(\frac{e}{4\pi^2 \hbar d^2}\right)\left\{\left(\Phi_B - \frac{eV}{2}\right)\exp\left[-\frac{2(2m)^{1/2}}{\hbar}\alpha\left(\Phi_B - \frac{eV}{2}\right)^{1/2} d\right] - \left(\Phi_B + \frac{eV}{2}\right)\exp\left[-\frac{2(2m)^{1/2}}{\hbar}\alpha\left(\Phi_B + \frac{eV}{2}\right)^{1/2} d\right]\right\} \quad (1)$$

where m is the electron mass, d is the barrier width, Φ_B is the barrier height, and V is the applied bias. For molecular systems, the Simmons model has been modified with a parameter α.[34,36] α is a unitless adjustable parameter that is introduced to provide either a way of applying the tunneling model of a rectangular barrier to tunneling through a non-rectangular barrier,[34] or an adjustment to account for the effective mass (m^*) of the tunneling electrons through a rectangular barrier,[34-37] or both. $\alpha = 1$ corresponds to the case for a rectangular barrier and bare electron mass. By fitting individual $I(V)$ data using Eq. (1), Φ_B and α values can be obtained.

Equation (1) can be approximated in two limits: low bias and high bias as compared with the barrier height Φ_B. For the low bias range, Eq. (1) can be approximated as[33]

$$J \approx \left(\frac{(2m\Phi_B)^{1/2} e^2 \alpha}{h^2 d}\right) V \exp\left[-\frac{2(2m)^{1/2}}{\hbar}\alpha(\Phi_B)^{1/2} d\right]. \quad (2a)$$

To determine the high bias limit, we compare the relative magnitudes of the first and second exponential terms in Eq. (1). At high bias, the first term is dominant and thus the current density can be approximated as

$$J \approx \left(\frac{e}{4\pi^2 \hbar d^2}\right)\left(\Phi_B - \frac{eV}{2}\right)\exp\left[-\frac{2(2m)^{1/2}}{\hbar}\alpha\left(\Phi_B - \frac{eV}{2}\right)^{1/2} d\right]. \quad (2b)$$

The tunneling currents in both bias regimes are exponentially dependent on the barrier width d. In the low bias regime the tunneling current density is $J \propto (1/d)\exp(-\beta_0 d)$, where β_0 is bias-independent decay coefficient:

$$\beta_0 = \frac{2(2m)^{1/2}}{\hbar}\alpha(\Phi_B)^{1/2} \quad (3a)$$

while in the high bias regime, $J \propto (1/d^2)\exp(-\beta_V d)$, where β_V is bias-dependent decay coefficient:

$$\beta_V = \frac{2(2m)^{1/2}}{\hbar}\alpha\left(\Phi_B - \frac{eV}{2}\right)^{1/2} = \beta_0\left(1 - \frac{eV}{2\Phi_B}\right)^{1/2}. \quad (3b)$$

At high bias β_V decreases as bias increases, which results from barrier lowering effect due to the applied bias.

4. Statistical Approach on Charge Transport Through Alkanethiols

4.1. *Statistical Analysis of Electronic Properties of Alkanethiols in Metal-Molecule-Metal Junctions*

As mentioned before, the yields of the molecular electronic devices are very low, mainly due to electrical short problems.[11,38,39] However, what "working" devices are and the yields of the molecular electronic devices have not been studied thoroughly. Typically, working devices might be defined as a device showing non-linear *I-V* behavior and not being electrical open and short. Electrical open and short devices can be readily recognized. Open devices are noisy with a current level typically in the pA range and short devices show Ohmic *I-V* characteristics with a current level larger than a few mA.[40] However, scientific criteria are needed for determining working devices more precisely. Although the choice of such criterion is not universal, current density can be a good criterion for determining working devices, because *I-V* data are major characteristics that are measured initially and the current directly reflects the conductivity of different lengths of alkanethiols or different molecular systems.

For this purpose, we fabricated a statistically sufficient number of molecular devices and characterized their electronic properties to obtain reasonable criteria for defining working devices and device yield. Specifically, we first fabricated 13,440 molecular electronic devices of alkanethiol SAMs (C8, C12, and C16 SAMs) with a microscale via-hole structure, as shown in Fig. 1. We then statistically analyzed all of the fabricated devices. From the *I-V* characterizations of all 13,440 devices, 11,744 showed electrical shorts. The devices with an electrical short showed short-circuit Ohmic *I-V* characteristics and a current typically larger than 10 mA at 1.0 V. Fabrication failure (392 devices) and electrical open devices (1103 devices) occurred mainly because of failures during the fabrication process. The electrical open devices show noisy and open circuit *I-V* behaviors. We then performed a statistical analysis on the remaining 201 "candidate" working devices as follows. First, we plotted histograms of the logarithmic current densities (log *J*) of the C8, 12, and C16 candidate working devices and then performed

Gaussian fittings on the histograms using the normal distribution equation,

$$f(x) = \frac{1}{\sigma\sqrt{2\pi}} \exp\left[-\frac{(x-\mu)^2}{2\sigma^2}\right] \quad (4)$$

where μ is the average and σ is the standard deviation. We selected the 99.7% of the devices from the overall population, which are included in the interval of the 3σ range between $\mu + 3\sigma$ and $\mu - 3\sigma$. This 3σ range was chosen arbitrarily to include as many devices as possible. When current densities are within the 3σ range (indicated as dotted lines in Fig. 3(a)), they are defined as working molecular electronic devices whereas the others are defined as "non-working" devices when the current densities are out of this range. Figure 3(a) shows an example of a histogram plot for logarithmic current densities of all C8 candidate devices. One can see that some of the non-working devices are located out of the 3σ range. Similarly, we are able to define the working device ranges of C12 and C16 devices. Figure 3(b) shows a summarized histogram plot of the current densities at 1 V for all of the working C8, C12, and C16 devices, based on the statistical analysis. This graph shows the difference in current density levels for different lengths of alkanethiols. Here, we

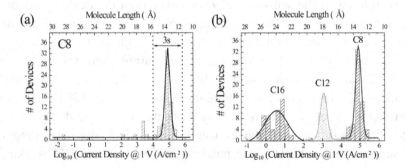

Fig. 3. (a) Histogram of logarithmic current densities at 1 V for "candidate" C8 molecular electronic devices. (b) Histograms of current densities at 1 V for "working" C8, C12, and C16 devices. (See text for the definition of candidate and working devices.) Solid lines are Gaussian fitting curves. Reprinted with permission from [27], T. W. Kim et al., *Nanotechnology* 18, 315204 (2007), @ Institute of Physics.

consider that the charge carriers tunnel via "through-bond" pathway. Tunneling rate is independent of the chain tilt with respect to the substrate.[41,42] Therefore, it is possible to think that the tunneling length is length of molecule than film thickness. In spite of a slight overlap around the tails of the intervals of three different alkanethiols, each alkanethiol shows a unique current density range. By using the relationship $\ln(J) \propto -\beta d$ which means the known exponential dependence of tunneling current through alkanethiols[11,21,36,39,43,44] and assigning the known molecular lengths for C8 (13.3 Å) and C16 (23.2 Å) at the mean current densities of C8 and C16 devices,[11] we deduced the relationship between the logarithmic current at the entire bottom axis and the molecular length at the top axis, shown in Fig. 3(b). Then, the molecular length at the mean current density of C12 devices was determined as 18.2 Å from Fig. 3(b), the known molecular length of C12.[11] Thus, Fig. 3 shows that the logarithmic current density is linearly dependent on molecular length, suggesting the exponential length-dependent charge transport through the alkanethiols.[11,21,36,39,43,44] Therefore, the current density is significantly affected by a slight change in molecular length. One can note that the histograms in Fig. 3 show the distribution of the logarithmic current densities, indicating the existence of a fluctuation factor causing the exponential distribution in the current densities. This fluctuation factor could be the tunneling distance, indicating that fluctuations in molecular configurations in the self-assembled monolayers in the device junctions are possible, such as molecular tilting angle, surface flatness of the Au bottom electrode on which the molecules are assembled.[45]

The values of $\{\mu, \sigma\}$ for the logarithmic current densities at 1 V for C8, C12, and C16 were found to be {4.87, 0.23}, {3.15, 0.29}, and {0.533, 0.527}, respectively. Among the above mentioned 201 candidate devices, 45 were found to be non-working devices and 156 were determined to be working devices, using the statistical criteria (3σ range). As summarized in Table 2, the numbers of C8, C12, and C16 working devices were 63, 33, and 60, respectively, among the total 13,440 fabricated devices. Thus, the device yield is ~1.2% (156/13,440). This device yield ~1.2% was determined using the 3σ range criterion. If more narrow ranges such as the 2σ range or the 1σ range are used, the device yield is reduced to ~1.1% (142/13,440) or ~1.0% (132/13,440),

Table 2. Summary of results for the fabricated micro via-hole devices.

	# of fabricated devices	Fab. Failure	Short	Open	Non-working	Working				Device Yield
						DC8	C8	C12	C16	
Monothiol	13,440 (100%)	392 (2.9%)	11,744 (87.4%)	1103 (8.2%)	45 (0.3%)		63 (1.41%)	33 (0.69%)	60 (1.44%)	1.2%
Dithiol	4800 (100%)	192 (4%)	4080 (85%)	428 (8.9%)	16 (0.3%)	84 (1.75%)				1.75%

Note: Working and non-working devices were defined by statistical analysis after Gaussian fitting on histograms of the logarithmic scale current densities (see text). Reprinted with permission from [27], T. W. Kim et al., Nanotechnology 18, 315204 (2007), @ Institute of Physics.

respectively. The similar statistical analysis was also performed for the octhanedithiol (DC8) molecule to demonstrate the device yield for different molecular devices, in this case dithiol has both chemisorbed contacts [Au-S] at both side to metal electrodes whereas monothiol has one chemisorbed contact and the other physisorbed contact [Au-CH$_3$]. As shown in Table 2, the device yield (~1.75%) of DC8 dithiol devices is not so much different from that of C8 monothiol device. This result may suggest that device yield is not much affected by the metal-molecular contact, but rather affected more by the device structures, fabrication condition, and quality of the self-assembled monolayer.

As mentioned above, the main reason for such a very low device yield is because the top Au contacts penetrate the thin molecular monolayer and make contact with the bottom electrode.[11,38,39] It should be noted that the device yield of ~1.2% applies to the vertical structures of microscale molecular electronic devices, and may not be the same for the vertical structures of nanoscale devices[36,46] or horizontal structures such as break junction[47] and electromigration nanogap devices.[14]

A number of groups have demonstrated that the charge transport through alkanethiol SAMs is tunneling and can be explained by the Simmons tunneling model (Eq. (1)).[33,36,48,49] This Simmons tunneling fitting was done on all the working C8, C12 and C16 devices (total 156 devices) to obtain the statistical Φ_B and α values. The molecular lengths used in this work are 13.3, 18.2, and 23.2 Å for C8, C12 and C16, respectively, determined by adding an Au-thiol bond length to the length

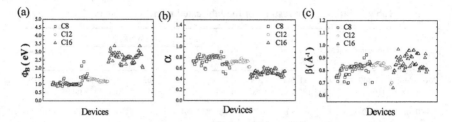

Fig. 4. Distribution in transport barrier height Φ_B (a), parameter α (b), and decay coefficient β (c) for all of the working C8, C12, and C16 devices. These values were determined by Simmons fitting using 156 working devices.

of the original molecule.[50] Figures 4(a) and 4(b) show the distribution for the Simmons fitting results, Φ_B and α values of all of the individual working C8, C12, and C16 devices. The current densities for the different length alkanethiols exhibit exponential length-dependent transport, characterized by a tunneling decay coefficient.[1,21,36,38,43,44] The β value (Eq. (3a)) in the low bias range can be defined from the Simmons equation (Eq. (1)).

The β values were calculated for all of the individual working devices and are summarized in Fig. 4(c). The Φ_B (Fig. 4(a)) and α (Fig. 4(b)) values obtained from Simmons fitting increase and decrease with increasing molecular length, respectively. The dependence of effective electron mass ($m^* = \alpha^2 m$, where m = electron rest mass) on the molecular length has been studied previously.[36,43] However, probably due to a lack of analytical data on statistically meaningful devices, the trend for length dependence for Φ_B and α values has not been explained well. In our study, Figs. 4(a) and 4(b) show the distributions of Φ_B and α values from the statistically acceptable 156 working devices and clearly show molecular length dependencies. The dependence of Φ_B and α values on the tunneling barrier has been extensively studied for SiO_2 materials. Ng et al. reported that the Φ_B and effective electron mass values decreased and increased with decreasing oxide thickness, respectively, which is the same dependence trend found for our alkanethiol molecular systems.[51] The decrease in barrier height was predicted for smaller oxide thickness due to the abrupt nature of the SiO_2/Si interface.[47] The enhancement in effective electron mass with decreasing oxide thickness is presumably due to the modification of the configuration of the Si-O-Si bond in the compressively strained oxide layer near the SiO_2/Si interface,[51,52] or several combined effects of the graded potential drop across the SiO_2/Si interface, the presence of defect-assisted tunneling, and an image force effect.[53] Although our molecular system is not the same as the inorganic SiO_2 layer, the decrease in Φ_B and increase in α (or effective electron mass) with decreasing molecular length may be attributed to a combination of effects such as the different molecular configuration, potential drops at the metal-molecule interface, and defect-assisted transport through the molecular layers.

Although a slight increase in β values with molecular length can be seen (Fig. 4(c)), the individual β values C8, C12, and C16 devices distributed in the range of 0.7 to 1.0 Å^{-1}, which are in agreement with previous reported β values.[34,49,50] It should be noted that the α values were less than unity in our experimental results. A lower α value indicates that the potential barrier shape is not rectangular and that tunneling through alkanethiols may be different from tunneling through a vacuum. Therefore, the molecular length dependence for α values indicates that the potential barrier shape is also molecular length-dependent. Engelkes et al. also reported that the Simmons model is inadequate for molecular systems, because of the simplistic model that approximates a single rectangular energy barrier with height Φ_B between two metal electrodes.[21] Hence, the interpretation of potential barrier height Φ_B may be thought of as an effective barrier to charge transport, which may not be the same as the difference in Fermi energy and molecular orbital energy (HOMO in this case).[21]

Table 3 summarizes the electrical transport parameters for Φ_B, α, β values, and the current densities at 1 V for the C8, C12, and C16 alkanethiols, statistically averaged over all the working devices. The average transport parameters do not change significantly with the different criterion ranges (3σ, 2σ, or 1σ ranges) for determining acceptable working devices.

When showing device data and doing further analysis such as length-dependent analysis, as in Fig. 6(a), the task would be tremendous if one were to use all the working devices, 156 devices in our case. Thus, it is necessary to choose a few devices that represent the different molecules.

Table 3. Summary of the statistical average transport parameters of alkanethiol SAMs from all of the working devices in micro via-hole structure.

Alkanethiol	J at 1 V (A/cm^2)	Φ_B (eV)	α	β (Å^{-1})
C8	~7.4 × 10^4	1.14 ± 0.28	0.76 ± 0.08	0.81 ± 0.05
C12	-4.4 × 10^3	1.26 ± 0.08	0.72 ± 0.04	0.83 ± 0.04
C16	~3.4	2.66 ± 0.28	0.52 ± 0.05	0.86 ± 0.06

Note: These parameters were obtained by taking statistical averages from individual parameters of all of the working devices (see text).

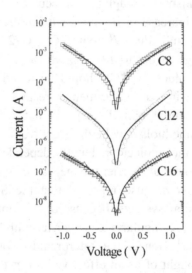

Fig. 5. *I-V* characteristics of three representative devices for three different length alkanethiols. (See text regarding the method used to select these representative devices). Symbols are experimental data and solid lines are curves fitted with the Simmons equation. Reprinted with permission from [27], T. W. Kim *et al.*, *Nanotechnology* 18, 315204 (2007), @ Institute of Physics.

Such representative devices can be chosen from the positions of the mean values in the histograms in Fig. 3 that were used as the criterion for determining working devices. Figure 5 summarizes the *I-V* characteristics for three representative C8, C12, and C16 devices chosen in this manner, which shows the length-dependent transport properties. Using these representative devices, one can plot a length-dependent

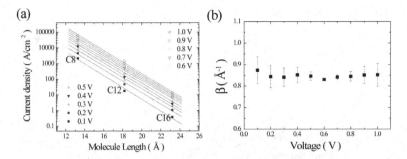

Fig. 6. (a) A semilog plot of tunneling current densities at different biases for three representative alkanethiol devices versus molecular length. The lines through the data points are exponential fittings. (b) Decay coefficient β values obtained from the slope of the exponential fittings as a function of the applied bias. Reprinted with permission from [27], T. W. Kim et al., *Nanotechnology* 18, 315204 (2007), @ Institute of Physics.

tunneling analysis, a semilog plot of tunneling current densities at various voltages as a function of the molecular length of the different alkanethiols, as shown in Fig. 6(a). The tunneling current densities show an exponential dependence $[J \propto \exp(-\beta d)]$ for molecular length. The decay coefficient β values can be determined from the slopes of the line fittings at different biases in Fig. 6(a) and are plotted in Fig. 6(b) as a function of bias. The β values obtained here are in the range of 0.83-0.87 Å^{-1} and are in good agreement with previously reported values for alkanethiols.[11] Both bias dependence[38] and bias independence[34,49] of β values have been reported. However, we did not observe a bias dependence of decay coefficients as shown in Fig. 6(b), which may indicate that the barrier lowing effect with applied bias is relatively weak for microscale junction devices, compared with nanometer-scale junction devices.[36]

The statistical approach to molecular electronic devices can provide a useful way to distinguish the transport property of different molecular systems. As an example, we performed the statistical analysis on octanedithiol (DC8) devices and compared the charge transport property of DC8 devices with that of octanethiol (C8) devices. As mentioned above, DC8 has thiols [-SH] at both ends and can have chemisorbed contacts [As-S] on the both sides of the metal electrodes whereas C8 has a thiol at only one end and thus has only one chemisorbed contact and

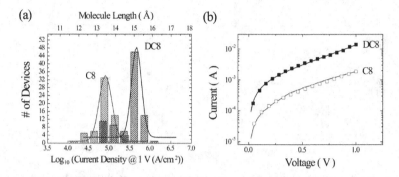

Fig. 7. (a) Histogram of logarithmic current densities at 1 V for C8 and DC8 working devices. (b) *I-V* characteristics of C8 and DC8 representative devices. Symbols are experimental data and solid lines are fitting curves with Simmons equation. Reprinted with permission from [27], T. W. Kim *et al.*, *Nanotechnology* 18, 315204 (2007), @ Institute of Physics.

the other a physisorbed contact [Au-CH$_3$]. Figure 7(a) shows a histogram of the logarithmic current densities at 1 V for C8 working (63 devices) and DC8 (84 devices) molecular devices after considering the criterion of the working devices to be the 3σ range. The statistical mean current densities of the C8 and DC8 devices were found to be ~74,000 and ~355,000 A/cm^2 at 1 V, respectively. The current density of DC8 is larger than that of C8 by a factor ~5. Figure 7(b) shows the *I-V* data for the C8 and DC8 representative devices, chosen from the positions of the mean values in the histogram of Fig. 7(a). As shown in Fig. 7(a), the histogram of C8 and DC8 has some overlap range in current densities. In this range, one may make a mistake in data selection of C8 and DC8. Therefore, statistical analysis is necessary to determine the intrinsic property of molecular electronic devices. In this point of view, this statistical analysis will be a good guide for studying the electronic property of alkanethiol molecules.

4.2. *Statistical Method for Determining Intrinsic Electronic Properties of Alkanethiols in Nanoscale Molecular Junctions*

We examined a total of 6745 molecular devices fabricated employing the alkanethiol SAMs of various chain lengths at room temperature using nanowell structure as shown in Fig. 2. Of these nanoscale devices examined, we found 6244 devices (92.6%) with linear $I(V)$ and current in the milliampere (mA) range. This indicates a metallic short caused by the penetration of the vapor-deposited top electrode through the molecular layer. Also, we found 19 devices (0.3%) with no detectable current in the subpicoampere (pA) range, which is likely due to a failure during the device fabrication such as an incomplete etching of the Si_3N_4 insulating layer. In addition, 482 devices (7.1%) showed non-linear $I(V)$ characteristics and current in the nanoampere (nA) range, which corresponds to the general characteristics of metal-molecule-metal junctions under investigation. To investigate electronic transport properties from this last group of candidate molecular devices, we performed temperature-variable current-voltage ($I(V,T)$) characterization, through which we identified the transports in four different categories: (i) direct tunneling, (ii) Fowler-Nordheim tunneling, (iii) thermally activated conduction, and (iv) Coulomb blockade phenomenon.

Figures 8(a)-8(g) show the characteristic behaviors of different transports over ±1 V range observed from four representative devices. In

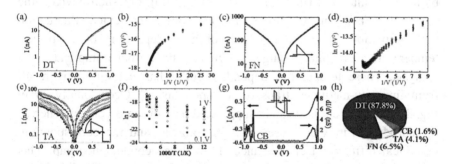

Fig. 8. (a)-(g) show the characteristic behaviors of various transports observed from four representative devices. (a) $I(V,T)$ data of a direct tunneling (DT) device in 300-80 K. (b) Plots of ln (I/V^2) versus $1/V$ of (a). (c) $I(V,T)$ data of a Fowler-Nordheim tunneling (FN) device in 300-80 K. (d) Plots of ln (I/V^2) versus $1/V$ of (c). (e) $I(V,T)$ data of a thermal-activated conduction (TA) device in 300-80 K. (f) Arrhenius plots (ln I versus $1/T$) of (e). (g) $I(V)$ data (upper line) of a Coulomb blockade (CB) device at 80 K and corresponding numerical differential conductance (dI/dV) (lower line). (h) A pie chart summarizing the statistics of various transports observed in this study. Reprinted with permission from [54], H. Song et al., Appl. Phys. Lett. 91, 253116 (2007), @ American Institute of Physics.

direct tunneling, no significant temperature dependence of the transport characteristics is observed [Figs. 8(a) and 8(b)]. Based on the applied bias as compared to barrier height (Φ_B), the tunneling transports can be divided into either direct ($V < \Phi_B/e$) or Fowler-Nordheim ($V > \Phi_B/e$) tunneling. These two tunneling mechanisms can be distinguished by their distinct current-voltage dependencies.[26] The molecular devices in direct tunneling do not exhibit an inflection point on a plot of ln (I/V^2) versus $1/V$ as shown in Fig. 8(b). This is consistent with tunneling through a trapezoidal barrier when the applied bias is less than the barrier height. Fowler-Nordheim tunneling also shows no temperature-dependent characteristics (Figs. 8(c) and 8(d)), which is analogous to direct tunneling. However, at a high voltage regime when the applied bias exceeds the barrier height, the barrier shape is changed from trapezoidal to triangular barrier, and the conduction mechanism causes a transition on a plot of ln (I/V^2) versus $1/V$. This gives rise to a linear decay region at the high bias tail as shown in Fig. 8(d).[55] Figures 8(e) and 8(f) show the charge transport in which thermal activation is involved. It has an obvious temperature-dependent $I(V)$ behavior, which can be identified by an Arrhenius plot (ln I versus $1/T$) [Fig. 8(f)]. For such devices, the conduction mechanism shows a large device-to-device fluctuation with activation energy in hundreds of meV depending on a specific temperature and voltage range, which thereby further complicates the analysis. These temperature-dependent $I(V)$ characteristics may result from impurity-mediated transport components.[13,56] Occasionally as in Fig. 8(g), the current is strongly suppressed near the zero voltage, whereas the current abruptly increases at high voltages, and the corresponding numerical differential conductance shows large conductance gap. This can be due to the Coulomb blockade effect. In these samples, it is likely that the Au atoms which were separated from the vapor-deposited top electrode migrated into the SAM, forming a localized state in molecular junctions.[57] This results in forming a double-barrier tunnel junction incorporating a Au nanoparticle in the SAM. The width of the zero conductance region can be approximately determined by e/C (C is the capacitance).[58] The capacitance formed between the Au particle and the electrode can be estimated from the normalized capacitance $C/4\pi\varepsilon r$ with respect to the dependence of the normalized position (z/r) in a mirror image point-charge model of a charged sphere (radius r) at the distance z from the electrode.[59] By considering the relative permittivity of alkanethiol as 2.6,[60] the molecular length, and the position of Au particle (z) from the electrode, it is roughly estimated that the Au particle radius (r) is less than ~0.3 nm. The estimated particle radius corresponds to the size of the Coulomb island formed in the junction and is enough small to be incorporated into the alkanehtiol SAMs.

Figure 8(h) summarizes the percentage of statistical distribution for which we observe each of the various transports based on the comprehensive $I(V,T)$ characterizations. We obtained a total of 123 $I(V,T)$ data in a complete temperature range of 300-80 K.[61] Among them, 108 devices (87.8%) showed direct tunneling characteristics in accordance with temperature-independent $I(V)$ characteristics and no transition on a plot of ln (I/V^2) versus $1/V$. Thus, as the most probable occurrence, the statistical assessment demonstrates that direct tunneling is indeed the dominant charge transport mechanism in the alkanethiol molecular devices. The dominance of direct tunneling in alkanethiol SAMs is in good agreement with previous reports[13,36] and can be reasonably anticipated due to their large HOMO–LUMO gap (~8 eV). However, as shown in Fig. 8(h), an uncontrolled device-to-device variation in transport mechanisms indicates the importance of a statistical study.

Repeated measurements give a statistical picture of molecular transport properties, typically presented as histograms. Figure 10 shows histograms for the conductance measured at the Ohmic region inside ±0.1 V. The data of devices governed by all different transport mechanisms (Fig. 8) exhibit the linear $I(V)$ characteristics inside ±0.1 V and this linear portion of individual $I(V)$ curve is used to estimate a junction conductance from the slope. For creating the histograms, we take all the $I(V)$ curves measured without any data selection or processing. Thereby, the histograms faithfully exhibit the intact device statistics. In Fig. 9, black-filled columns represent the conductance histograms for all the fabricated 6745 devices measured at room temperature and gray-patterned columns represent those for 108 direct tunneling devices confirmed from the acquired 123 $I(V,T)$ data as explained above. The histograms exhibit the appearance of an intermediate regime excluding typical non-working devices such as short or open junctions. The appearance of such an intermediate regime in the histograms is unambiguously different in the absence of molecules. As an example, the histogram for short devices intentionally made without molecules is plotted in inset (a) of Fig. 9. Therefore, we ascribe this intermediate regime to the formation of a molecular junction in which, depending on the molecular length of alkanethiols (marked as C16, C12, and C8), the conductance values vary over orders of magnitude and appear to be distributed log-normally with well-defined conductance peaks highlighted by the Gaussian curves for each alkanethiol. Note that the log-normal distribution of the conductance values stems from a parameter that affects the conductance exponentially and is therefore likely due to a variation in tunneling distance.[27,62] This effect is not observed in the absence of molecules where the devices show linearly normal distribution in the histogram [see inset (a) in Fig. 9].[63] The variation in tunneling distance could be attributed to detailed microscopic configurations of metal-molecule junctions in numerous degrees of freedom such as contact distance, molecular binding site, surface structure of electrodes, and molecular conformations and orientations.[64,65]

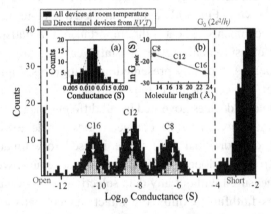

Fig. 9. Conductance histograms for all devices (black-filled columns) measured at room temperature and the direct tunneling devices (white columns) confirmed by $I(V,T)$ characterizations. The Gaussian curves (dashed lines) highlight the conductance peaks for each alkanethol. Inset (a) is a conductance histogram for the intentional short devices in the absence of molecules. Inset (b) shows a logarithmic plot of conductance peak values (marked by the arrows) versus molecular length. Reprinted with permission from [54], H. Song et al., Appl. Phys. Lett. 91, 253116 (2007), @ American Institute of Physics.

5. Analysis of Metal-Molecular Junctions by Multi-Barrier Tunneling Model

5.1. *Statistical Analysis of Electronic Properties of Alkanethiols and Alkanedithols*

For this study, we characterized a significantly large number of such molecular devices (27,840 devices in total) to statistically analyze the molecular electronic properties of a sufficient number of working molecular electronic devices (427 devices). The working molecular electronic devices were extracted from devices showing a majority of current densities in the statistical distribution (Figs. 10(a)-10(f)) following the aforementioned criterion for determining working devices. The numbers of C8, C12, C16, DC8, DC9 and DC10 working devices were found as 63, 33, 60, 84, 94 and 93, respectively, among the total 27,840 fabricated devices. Then, the device yields were found as ~1.2% (156/13,440) for monothiol and ~1.9% (271/14,440) for dithiol devices. Since the device yield (~1.75%) of DC8 dithiol devices is not much different from that of C8 monothiol devices (~1.41%), this result may suggest that device yield is not much affected by metal-molecule contact, but rather affected more by the device structures, fabrication condition, and quality of the self-assembled monolayers. In this study, the use of a

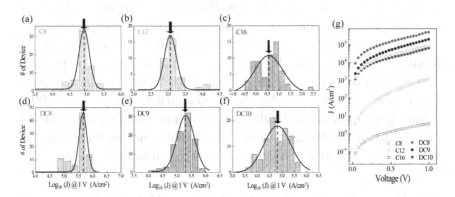

Fig. 10. The statistical histograms of log (*J*) measured at 1.0 V for C8 (a), C12 (b), C16 (c), DC8 (d), DC9 (e), and DC10 (f). The line curves are fitting results obtained from the histograms with Gaussian functions, and the mean positions are indicated with arrows. (g) Current density-voltage characteristics of representative devices chosen from the mean positions of the fitted Gaussian functions.[68]

statistical approach is very significant, as the analysis of a large number of devices increases the ability to develop more accurate and meaningful characteristics of molecular systems.[27,66,67] Note that the main conduction through alkyl molecular devices is tunneling mechanism, which can be checked by temperature-independent current-voltage characteristics.[33,69]

Figures 10(a) to 10(f) present the statistical histograms of current densities in logarithmic scale for different lengths of alkanemonothiols (C8, C12, and C16) and alkanedithiols (DC8, DC9, and DC10) at 1.0 V with the mean positions as representative devices indicated with arrows from the fitting results by Gaussian functions. The current densities for these representative devices were found to be $\sim 8.3 \times 10^4$, 1.2×10^3, 3.5, 4.9×10^5, 2.0×10^5, and 6.3×10^4 A/cm^2 at 1.0 V for C8, C12, C16, DC8, DC9, and DC10, respectively. The current densities-voltage (J-V) characteristics for these six representative devices are plotted in Fig. 10(g). The conductance and J-V characteristics are clearly dependent on the molecular length and metal-molecular contacts (i.e., monothiol versus dithiol). This observation is supported by previous reports of M-M-M junctions that have shown that the current density for alkanedithiol is higher than that for alkanemonothiol due to their different natures of metal-molecule contact properties (chemisorbed versus physisorbed contact) at Au-molecule contacts.[21,70] The histograms in Figs. 10(a) to 10(f) show the distribution of the logarithmic current densities, indicating the existence of fluctuation factors causing the exponential distribution in the current densities.[27] The variation of junction area may exist, but the area fluctuation does not produce exponential distribution in current, instead fluctuation in the tunneling path is probably responsible for the distribution data of Figs. 10(a)-10(f). Some fluctuations in molecular configurations in the self-assembled monolayers in the device junctions are possible, such as fluctuations in molecular configuration or microstructures in metal-molecule contacts.[64,65]

5.2. Length-Dependent Decay Coefficients by Multi-Barrier Tunneling Model

To investigate the effect of metal-molecule contacts on the electronic transport, we propose a multi-barrier tunneling (MBT) model, which generalizes the Simmons tunneling model.[33] As compared to the Simmons tunneling model where the tunneling barrier is represented by a single barrier, the M-M-M junction in MBT model can be divided into three parts: a molecular-chain body and metal-molecule contacts on either side of molecule, represented as three individual conduction barriers, as schematically illustrated in Fig. 11(a). In the alkanedithiol M-M-M junction, there is one molecular-chain body barrier [$(CH_2)_n$] (n is the number of carbon units), and two chemisorbed contact barriers [Au-S-C] on either side. Conversely, the alkanemonothiol M-M-M junction with the same a molecular-chain body barrier [$(CH_2)_n$] as the n-alkanedithiol junction has one chemisorbed contact barrier [Au-S-C] and one physisorbed contact barrier [CH_3/Au]. This approach of separation of metal-molecule contact and molecular body from alkanethiol M-M-M junction is reasonable, since hybridization of the metal-molecule wavefunction decays rapidly into the junction for alkanethiol devices.[22,23]

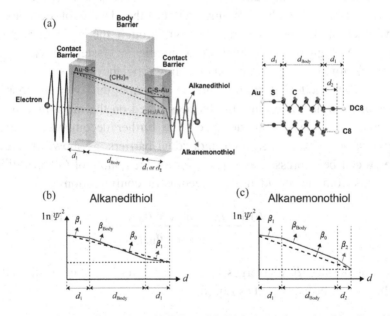

Fig. 11. (a) Left is an illustration of MBT model. Right is a schematic of barrier widths for C8 and DC8. Schematics of MBT model for an alkanedithiol M-M-M junction (b) and for an alkanemonothiol M-M-M junction (c).[68]

In Fig. 11(a), the widths of the barriers for d_1, d_{Body}, and d_2 represent the length of the chemisorbed contact on the molecule [Au-S-C], a molecular-chain body region [$(CH_2)_n$], and the physisorbed contact on the molecule [CH_3/Au], respectively. Here, d_1 ([Au-S-C]) is ~3.80 Å, and d_2 ([CH_3/Au]) is ~2 Å.[50] d_{Body} is the projected length along the molecular-chain, and the incremental length per carbon atom (Δd_{Body} [CH_2]) is ~1.25 Å; all of which were obtained using Chem3D software following the method in previous literature.[50] The length d_{Body} is identical for alkanemonothiol and alkanedithiol with the same n value; for example, octanemonothiol (C8) and octanedithiol (DC8) have an identical length, d_{Body} [$(CH_2)_8$, ~7.46 Å]. The total width of the barriers in alkanemonothiol (alkanedithiol) is $d = d_1 + d_{Body} + d_2$ (d_1).

For small length molecules with a large HOMO-LUMO energy gap, such as alkyl chain molecules, coherent tunneling is the main conduction mechanism of the electronic charge transport at relatively low bias regime.[11,13,71] In low bias regime, the tunneling current density can be approximated as Eq. (2a).[11,13,21,33] And, β_0 (Eq. (3a)), the decay coefficient in a low bias regime, reflects the degree of decrease in wavefunction of the tunneling electron through the molecular tunnel barrier. A higher decay coefficient implies a faster decay of the wavefunction, i.e., lower electron tunneling efficiency.

In MBT model, it is possible to describe the overall slope of wavefunction decay through the barriers based on the magnitude of the β_0 value, and this overall decay can be further decomposed to three individual decays through three individual barriers, as shown in Fig. 3. The β_0 can be expressed as Eq. (5) for alkanemonothiol (alkanedithiol) junctions from the consideration of geometric configurations

$$\beta_0 = \frac{\beta_1 d_1 + \beta_{Body} d_{Body} + \beta_{1(2)} d_{1(2)}}{d_1 + d_{Body} + d_{1(2)}}. \tag{5}$$

One can see that β_0 converges to β_{Body} for a very long molecule. Also, $\alpha(\Phi_B)^{1/2}$ can be expressed as Eq. (6).

$$\alpha(\Phi_B)^{1/2} = \frac{\hbar}{2(2m)^{1/2}} \frac{\beta_1 d_1 + \beta_{Body} d_{Body} + \beta_{1(2)} d_{1(2)}}{d_1 + d_{Body} + d_{1(2)}}. \tag{6}$$

As mentioned above, because the main conduction mechanism is coherent (elastic) tunneling at low bias regime (and at room temperature), it is assumed that the energy of electron tunneling through the molecular barriers does not decrease, as expressed by the horizontal bottom dashed line in Fig. 11(a). Furthermore, due to the different nature of the metal-molecule contact properties, electron transmission for the chemisorbed contact [Au-S-C] is found to be more efficient than that for the physisorbed contact [CH_3/Au]. As a result, the slope (β_0) for alkanemonothiol junctions is steeper than that for alkanedithiol junctions, as illustrated by the dashed lines in Figs. 11(b) and 11(c). In this MBT model, it was possible to define d_1 (d_2) as the components of the decay coefficients corresponding to the chemisorbed (physisorbed) contact barrier width d_1 (d_2) in Fig. 11. Similarly, β_{Body} is the decay coefficient component for the molecular-chain body barrier.

Figure 12(a) shows the statistical distribution of β_0 values obtained for different length alkanemonothiol and alkanedithiol M-M-M devices. In this plot, β_0 values were determined from fitting the I-V data of all the statistically-defined-working molecular electronic devices (total 427 devices) with the Simmons tunneling model. The values for the mean and standard deviation of β_0 are presented as 0.81 ± 0.05, 0.83 ± 0.03, and 0.87 ± 0.05 Å^{-1} for C8, C12, and C16 alkanemonothiols, and 0.55 ± 0.06, 0.57 ± 0.06, and 0.58 ± 0.08 Å^{-1} for DC8, DC9, and DC10

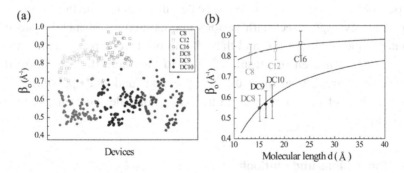

Fig. 12. The overall decay coefficient β_0 for different length alkanemonothiol (C8, C12 and C16) and alkanedithiol (DC8, DC9 and DC10) junctions. (b) The mean (symbols) and standard deviations (error bars) of β_0 versus molecular length d. The black solid lines were calculated from MBT model.[68]

alkanedithiols, respectively. As previously mentioned, the β_0 values for alkanemonothiol devices appear to be larger than those for alkanedithiol devices due to the poor tunneling rate of physisorbed contact [CH_3/Au] in alkanemonothiol junctions, as compared to alkanedithiol junctions. Also, a slight increase of β_0 values in Figs. 12(a) and 12(b) can be seen as the molecular length increases, which reflects the different tunneling rates for different lengths of alkanethiols, i.e., the wavefunction of the tunneling electron decays further when it tunnels through longer molecules. The solid lines in Fig. 12(b) are the results calculated using the estimated β_{Body}, β_1, and β_2 values determined from MBT model. Moreover, Fig. 12(b) shows that the difference in β_0 values between monothiol and dithiol becomes larger as the molecular length decreases. This phenomenon explains that the metal-molecular contact effect becomes relatively more important than the molecular-chain body effect in electronic transport for shorter molecules. On the contrary, if the molecular length increases, the molecular-chain body effect becomes more important and the β_0 values of monothiol and dithiol molecular systems become closer and eventually converge to body decay coefficient (β_{Body}), as seen in Fig. 12(b).

Note that our analysis with MBT model does not consider the details of the Fermi level alignment and molecular binding sites, which will generally influence the charge transport of molecular devices.[72] Furthermore, the transport properties values obtained from our experimental results with microscale molecular junctions are an ensemble average effect with various microstructures of metal-molecule contacts and binding sites, and thus should not be compared with single-molecular measurement results,[23] due to the contribution from the probability amplitude of multiple-reflection and the possibility of cooperative effects between individual molecules in ensemble of molecules.

6. Conclusions and Outlook

We summarized detailed statistical analysis on the electronic transport properties of molecular electronic devices in micro and nanoscale via-hole structured alkanethiol molecular devices. The comprehensive temperature-variable current-voltage characterization for candidate molecular devices monitors the full categories of transport properties. Our statistical analysis shows that direct tunneling is indeed the dominant conduction mechanism responsible for the intrinsic transport properties of alkanethiol molecular devices. The introduction of the statistical consideration of determining the working molecular devices and representative devices can be a meaningful concept to understand electronic transport properties for expanding over organic conducting molecular devices or other devices of nanoscale elements which have typically low device yields and poor reproducibility.

Based on statistical analysis of molecular electronic devices, we proposed multi-barrier tunneling (MBT) model and studied the effect of metal-molecule contacts and molecular structures on the electronic transport properties in metal-molecule-metal junctions. The decay coefficients were obtained and compared for these molecules with different molecular lengths, based on a proposed MBT model where the molecular junction was decomposed into three individual barriers through the molecular body and metal-molecule contacts on either side of the molecule.

There have been many proposals and promise of various device concepts. However reliable data, characterization techniques are still necessary. In this point of view, statistical analysis is necessary to determine the intrinsic property of molecular electronic devices and will be good guide for studying the electronic property of molecular devices.

Acknowledgment

This work was supported by the National Research Laboratory (NRL) Program from the Korean Ministry of Education, Science and Technology.

References

1. M. A. Reed and T. Lee (eds.), *Molecular Nanoelectronics* (American Scientific Publishers, Stevenson Ranch, 2003).
2. A. Nitzan and M. A. Ratner, *Science* 300, 1384 (2003).

3. A. Ulman, *An Introduction to Ultrathin Organic Films from Langmuir-Blodgett to Self-Assembly* (Academic Press, Boston, 1991).
4. G. Cuniberti, G. Fagas and K. Richter, *Introducing Molecular Electronics* (Springer, Berlin, 2005).
5. J. E. Green, J. W. Choi, A. Boukai, Y. Bunimovich, E. Johnston-Halperin, E. DeIonno, Y. Luo, B. A. Sheriff, K. Xu, Y. S. Shin, H. Tseng, J. F. Stoddart and J. R. Heath, *Nature* 445, 414 (2007).
6. E. Lörtscher, J. W. Ciszek, J. Tour and H. Riel, *Small* 2, 973 (2006).
7. H. B. Akkerman, P. W. M. Blom, D. M. de Leeuw and B. de Boer, *Nature* 441, 69 (2006).
8. J. M. Beebe, B. Kim, J. W. Gadzuk, C. D. Frisbie and J. G. Kushmerick, *Phys. Rev. Lett.* 97, 026801 (2006).
9. J. Jiang, W. Lu and Y. Luo, *Chem. Phys. Lett.* 400, 336 (2004).
10. J. O. Lee, G. Lientschnig, F. Wiertz, M. Struijk, R. A. J. Janssen, R. Egberink, D. N. Reinhoudt, P. Hadley and C. Dekker, *Nano Lett.* 3, 113 (2003).
11. T. Lee, W. Wang, J. F. Klemic, J. J. Zhang, J. Su and M. A. Reed, *J. Phys. Chem. B* 108, 8742 (2004).
12. M. T. González, S. Wu, R. Huber, S. J. van der Molen, C. Schönenberger and M. Calame, *Nano Lett.* 6, 2238 (2006).
13. W. Wang, T. Lee and M. A. Reed, *Rep. Prog. Phys.* 68, 523 (2005).
14. J. Park, A. N. Pasupathy, J. I. Goldsmith, C. Chang, Y. Yaish, J. R. Petta, M. Rinkoski, J. P. Sethna, H. D. Abruña, P. L. McEuen and D. C. Ralph, *Nature* 417, 722 (2002).
15. J. Chen, M. A. Reed, A. M. Rawlett and J. M. Tour, *Science* 286, 1550 (1999).
16. C. P. Collier, E. W. Wong, M. Belohradský, F. M. Raymo, J. F. Stoddart, P. J. Kuekes, R. S. Williams and J. R. Heath, *Science* 285, 391 (1999).
17. Y. Chen, G. Y. Jung, D. A. A. Ohlberg, X. Li, D. R. Stewart, J. O. Jeppesen, K. A. Nielsen, J. F. Stoddart and R. S. Williams, *Nanotechnology* 14, 462 (2003).
18. S. Datta, W. Tian, S. Hong, R. Reifenberger, J. I. Henderson and C. P. Kubiak, *Phys. Rev. Lett.* 79, 2530 (1997).
19. J. G. Kushmerick, D. B. Holt, J. C. Yang, J. Naciri, M. H. Moore and R. Shashidhar, *Phys. Rev. Lett.* 89, 086802 (2002).
20. J. Taylor, M. Brandbyge and K. Stokbro, *Phys. Rev. Lett.* 89, 138301 (2002).
21. V. B. Engelkes, J. M. Beebe and C. D. Frisbie, *J. Am. Chem. Soc.* 126, 14287 (2004).
22. J. G. Kushmerick, *Mater. Today* 8, 26 (2005).
23. A. Salomon, D. Cahen, S. Lindsay, J. Tomfohr, V. B. Engelkes and C. D. Frisbie, *Adv. Mater.* 15, 1881 (2003).
24. A. Salomon, T. Böcking, J. J. Gooding and D. Cahen, *Nano Lett.* 6, 2873 (2006).
25. B. Xu and N. J. Tao, *Science* 301, 1221 (2003).
26. S. M. Sze, *Physics of Semiconductor Devices* (Wiley, New York, 1981).
27. T. W. Kim, G. Wang, H. Lee and T. Lee, *Nanotechnology* 18, 315204 (2007).

28. C. Zhou, M. R. Deshpande, M. A. Reed, L. Jones and J. M. Tour, *Appl. Phys. Lett.* 71, 611 (1997).
29. J. Chen, L. C. Calvet, M. A. Reed, D. W. Carr, D. S. Grubisha and D. W. Bennett, *Chem. Phys. Lett.* 313, 741 (1999).
30. M. A. Ratner, B. Davis, M. Kemp, V. Mujica, A. Roitberg and S. Yaliraki, in *Molecular Electronics: Science and Technology*, The Annals of the New York Academy of Sciences, Vol. 852, eds. A. Aviram and M. A. Ratner (The New York Academy of Sciences, New York, 1998).
31. Although the HOMO-LUMO gap of alkyl chain type molecules has been reported (see Ref. 32), there is no experimental data on the HOMO-LUMO gap for Au/alkanethiol SAM/Au system. 8 eV is commonly used as HOMO-LUMO gap of alkanethiol.
32. C. Boulas, J. V. Davidovits, F. Rondelez and D. Vuillaume, *Phys. Rev. Lett.* (1996).
33. J. G. Simmons, *J. Appl. Phys.* 34, 1793 (1963).
34. R. Holmlin, R. Haag, M. L. Chabinyc, R. F. Ismagilov, A. E. Cohen, A. Terfort, M. A. Rampi and G. M. Whitesides, *J. Am. Chem. Soc.* 123, 5075 (2001).
35. C. Joachim and M. Magoga, *Chem. Phys.* 281, 347 (2002).
36. W. Wang, T. Lee and M. A. Reed, *Phys. Rev. B* 68, 035416 (2003).
37. J. G. Simmons, *J. Phys. D* 4, 613 (1971).
38. J. O. Lee, G. Lientschnig, F. Wiertz, M. Struijk, R. A. J. Janssen, R. Egberink, D. N. Reinhoudt, P. Hadley and C. Dekker, *Nano Lett.* 3, 113 (2003).
39. H. Haick, J. Ghabboun and D. Cahen, *Appl. Phys. Lett.* 86, 042113 (2005).
40. T.-W. Kim, G. Wang, H. Song, N.-J. Choi, H. Lee and T. Lee, *J. Nanosci. Nanotechnol.* 6, 3487 (2006).
41. H. Song, H. Lee and T. Lee, *J. Am. Chem. Soc.* 129, 3806 (2007).
42. K. Slowinski, R. V. Chamberlain, R. Bilewicz and M. Majda, *J. Am. Chem. Soc.* 118, 4709 (1996).
43. N. Majumdar, N. Gergel, D. Routenberg, J. C. Bean, L. R. Harriott, B. Li, L. Pu, T. Yao and J. M. Tour, *J. Vac. Sci. Technol. B* 23, 4 (2005).
44. X. Li, J. He, J. Hihath, B. Xu, S. M. Lindsay and N. Tao, *J. Am. Chem. Soc.* 128, 2135 (2006).
45. N. Majumdar, N. Gergel, D. Routenberg, J. C. Bean, L. R. Harriott, B. Li, L. Pu, T. Yao and J. M. Tour, *J. Vac. Sci. Technol. B* 23, 4 (2005).
46. W. Wang, T. Lee, I. Kretzschmar and M. A. Reed, *Nano Lett.* 4, 643 (2004).
47. G. Lewicki and J. Maserjian, *J. Appl. Phys.* 46, 3032 (1975).
48. D. J. Wold and C. D. Frisbie, *J. Am. Chem. Soc.* 123, 5549 (2001).
49. X. D. Cui, X. Zarate, J. Tomfohr, O. F. Sankey, A. Primak, A. L. Moore, T. A. Moore, D. Gust, G. Harris and S. M. Lindsay, *Nanotechnology* 13, 5 (2002).
50. D. J. Wold, R. Haag, M. A. Rampi and C. D. Frisbie, *J. Phys. Chem. B* 106, 2813 (2002).
51. C. Y. Ng, T. P. Chen and C. H. Ang, *Smart Mater. Struct.* 15, S39 (2006).

52. Khairurrijal, W. Mizubayashi, S. Miyazaki and M. Hirose, *Appl. Phys. Lett.* 77, 3580 (2000).
53. M. Städele, F. Sacconi, A. Di Carlo and P. Lugli, *J. Appl. Phys.* 93, 2681 (2003).
54. H. Song, T. Lee, N. J. Choi and H. Lee, *Appl. Phys. Lett.* 91, 253116 (2007).
55. J. M. Beebe, B. Kim, J. W. Gadzuk, C. D. Frisbie and J. G. Kushmerick, *Phys. Rev. Lett.* 97, 026801 (2006).
56. E. E. Polymeropoulos and J. Sagiv, *J. Chem. Phys.* 69, 1836 (1978).
57. L. H. Yu, C. D. Zangmeister and J. G. Kushmerick, *Phys. Rev. Lett.* 98, 206803 (2007).
58. A. E. Hanna and M. Tinkham, *Phys. Rev. B* 44, 5919 (1991).
59. Y. Azuma, M. Kanehara, T. Teranishi and Y. Majima, *Phys. Rev. Lett.* 96, 016108 (2006).
60. M. A. Rampi, O. J. A. Schueller and G. M. Whitesides, *Appl. Phys. Lett.* 72, 1781 (1998).
61. A number of devices failed during the $I(V,T)$ characterization likely due to continuous electric and thermal sweep. For reliability, we only considered the devices where the conduction mechanism is confirmed by the repeated $I(V)$ measurements in a sufficient wide temperature range (300-80 K).
62. V. B. Engelkes, J. M. Beebe and C. D. Frisbie, *J. Phys. Chem. B* 109, 16801 (2005).
63. The distribution of conductance values would result from a variation of junction geometry such as contact area, as the conductance depends linearly on contact area according to Ohm's law.
64. E. Lörtscher, H. B. Weber and H. Riel, *Phys. Rev. Lett.* 98, 176807 (2007).
65. Y. Hu, Y. Zhu, H. Gao and H. Guo, *Phys. Rev. Lett.* 95, 156803 (2005).
66. M. T. González, S. Wu, R. Huber, S. J. Molen, C. Schönenberger and M. Calame, *Nano Lett.* 6, 2238 (2006).
67. S.-Y. Jang, P. Reddy, A. Majumdar and R. A. Segalman, *Nano Lett.* 6, 2362 (2006).
68. G. Wang, T. W. Kim, H. Lee and T. Lee, *Phys. Rev. B* 76, 205320 (2007).
69. N. B. Zhitenev, A. Erbe and Z. Bao, *Phys. Rev. Lett.* 92, 186805 (2004).
70. X. D. Cui, A. Primak, X. Zarate, J. Tomfohr, O. F. Sankey, A. L. Moore, T. A. Moore, D. Gust, G. Harris and S. M. Lindsay, *Science* 294, 571 (2001).
71. C. Joachim and M. Magoga, *Chem. Phys.* 281, 347 (2002).
72. S.-H. Ke, H. U. Baranger and W. Yang, *J. Am. Chem. Soc.* 126, 15897 (2004).

CHAPTER 8

A HYSTERIC CURRENT/VOLTAGE RESPONSE OF REDOX-ACTIVE RUTHENIUM COMPLEX MOLECULES IN SELF-ASSEMBLED MONOLAYERS

Kyoungja Seo, Junghyun Lee, Gyeong Sook Bang and Hyoyoung Lee*

CRI, Center for Smart Molecular Memory, Department of Chemistry, Sungkyunkwan University, 300 Cheoncheon-dong, Jangan-gu, Suwon 440-746, Korea
*E-mail: hyoyoung@skku.edu

We review on the current-voltage response of redox-active ruthenium complex molecules in self-assembled monolayers. In single molecular junctions with STM, the threshold voltage to the high conductance state in the molecular junctions of Ru^{II} complexes was consistent with the electronic energy gap between the Fermi level of the gold substrate and the lowest ligand-centered redox state of the metal complex molecule. As an active redox center leading to conductance switching in the molecule, the lowest ligand-centered redox state of Ru^{II} complexes was suggested to trap an electron injected from the gold substrate. In molecular monolayer devices of Ru^{II} complexes with conducting polymer (PEDOT:PSS), the hydrophilic terminal thiol group of the alkylthiolate-tethered Ru^{II} terpyridine complexes can interact with the hydrophilic sulfonic acid groups of PEDOT:PSS, preventing the penetration of soft and metal top electrodes into the SAMs to result in high yields. As the number of alkyl chains or as alkyl chain length increases, the device yield of molecular monolayer memory will improve and its retention time will increase, guiding a new strategy for the development of volatile to non-volatile memory.

1. Introduction

1.1. *A Redox-Active Ruthenium Complex as a Molecular Switch*

Molecular electronic elements are very attractive for applications of nanodevices as alternatives to the high cost and low integrated silicon devices.[1-3] As elements in functional molecular electronic devices,[4,5] molecular switches have been designed for the achievement of the concept of a molecular electronic change driven by a photon[6,7] or an electron.[8,9] Redox-active π-conjugated molecules and transition metal complexes having at least two redox states (reduction and oxidation) are promising candidates for application to molecular switches.[10-13] If an applied potential can control the redox states of the molecules sandwiched between two metal contacts, the conducting channel induced by a redox fluctuation can open to the Fermi levels of the contacts and charge transport can occur at a redox state close to the Fermi levels.[14,15] Thus, molecular electronic events such as negative differential resonance (NDR) and conducting switching reflect the electrochemical reaction of molecules.[10,11] Discrete redox states of the molecules via oxidation and reduction reactions can be accessible to transfer and store charges.

Charge transport through the molecular junctions depends significantly on the energy levels of molecules relative to the Fermi levels of the contacts[16,17] and the electronic structure of the molecule.[17,18] In the concept of a molecular switch is the idea that the charge can be efficiently trapped in the molecule; a resonant state (a redox state) of molecules as a conducting channel, where that is the highest occupied molecular orbital (HOMO) or the lowest unoccupied molecular orbital (LUMO), may trap charges injected from close Fermi levels of the contacts. Photochemically- or electrochemically-induced metal-centered or ligand-centered chemical reactions should drive charge or electron trapping into the molecular state.

The transition metal MLCT (metal-to-ligand charge transfer) complexes can offer two redox-centered states (the metal-centered HOMO and the ligand-centered LUMO).[19,20] The chemistry of the metal-centered and ligand-centered reactions for a transition metal-electroactive ligand complex is well-understood.[19,20] In particular, Ru^{II}-centered complexes exhibit stability in their chemical and electronic properties during redox reactions. As a voltage-driven molecular switch, thiol-tethered ruthenium(II) terpyridine complexes were used for measurement of charge transport characteristics of the Ru^{II} complex, the molecular structures were designed to have a centered Ru, and mono- or dithiol-substituted terpyridine ligand groups in the molecular junctions between two metal contacts. In addition, Au nanoparticle attachment on the dithiol-tethered Ru^{II} terpyridine complexes incorporated in n-alkanethiol self-assembled monolayers (SAMs) was used as a simplified symmetric molecular junction, Au-NP/Ru^{II} terpyridine/Au substrate, as a model close to the realistic systems of a molecular switch.

1.2. Design of a Monolayer Non-Volatile Memory Device with a Ruthenium Complex

Organic non-volatile memory is a possible substitute for volatile dynamic random access memory (DRAM), which typically requires a data refresh every few milliseconds, and has the advantage of very low power consumption. However, due to limitations of the nanoscaled device fabrication of vapor-deposited organic resistive random access memory (ORRAM)[21,22] and spin-coated polymer resistive random access memory (PORAM),[23] molecular monolayer memory using organic self-assembled monolayers (SAMs) has become an important issue. Several researchers have demonstrated possible voltage-driven molecular memory devices possessing the advantages of fast response time and highly dense circuits over a photo-driven circuit,[24] using a molecular monolayer for metal-molecule-metal (MMM) devices.[25-27] However, the use of molecular monolayers has been limited due to low device yields, which are mainly attributed to electrical shorting.[28,29] This is especially true for voltage-driven devices. As the top metal electrode is deposited onto the molecular monolayer, energetic metal atoms can degrade the SAM molecules,[30] while the metal particles often penetrate the molecular monolayers to form metallic current paths.[31] Thus, to reduce the degree of electrical shorting, the following issues have all been considered: compactness and robustness of the Langmuir-Blodgett (LB) films[32,33] and SAMs;[34] use of bilayer SAMs;[35] a metal electrode with a nanosized surface area;[36-39] reduction of the surface roughness of the metal electrode;[40] a Pd nanowire[41] or single-walled carbon nanotube[42] as a substitute for the top metal electrode; use of a conducting polymer layer (PEDOT:PSS)[43] as a soft portion of the top metal electrode on the molecular monolayer.

For fabrication of molecular monolayer non-volatile memory (MMNVM), the design of redox-active molecular memory SAM materials becomes a critical factor, especially for the development of a voltage-driven MMNVM that requires direct contact measurement of the memory effect via a molecular monolayer between the bottom and top electrodes. Ru^{II} terpyridine complexes can act as a voltage-driven molecular switch in the solid state molecular junction.[44] In the fabrication of a molecular monolayer memory device, however, the direct use of Ru^{II} terpyridine complexes without alkyl chains results in electrical shorting. Thus, in an implement of molecular monolayer memory circuits with a high yield, it is important to reduce electrical shorting by modifying the Ru^{II} terpyridine complexes with thiol-terminated alkyl chains of varying chain lengths on one or both sides.

2. Materials and Surface Chemistry

2.1. *Material Synthesis and Characterization*

Mono- and dialkylthiolate-tethered ruthenium(II) hexafluorophosphate $(Ru^{II}(tpy)(tpy(CH_2)_nSAc)(PF_6)_2$, denoted by $Ru^{II}(tpy)(tpyC_nS)$, and $Ru^{II}(tpy(CH_2)_nSAc)_2(PF_6)_2$ complexes, denoted by $Ru^{II}(tpyC_nS)_2$, $n = 0, 7,$ and 13), as shown in Fig. 1, were designed for the fabrication of a vertical MMM device using only a single monolayer.

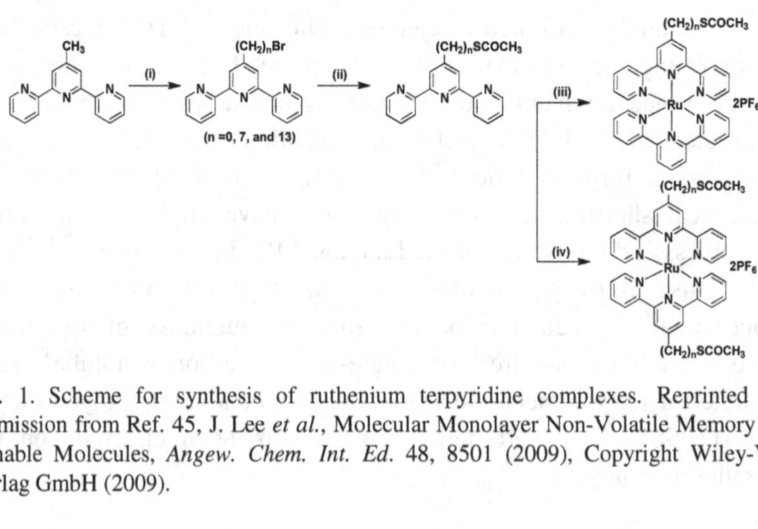

Fig. 1. Scheme for synthesis of ruthenium terpyridine complexes. Reprinted with permission from Ref. 45, J. Lee *et al.*, Molecular Monolayer Non-Volatile Memory with Tunable Molecules, *Angew. Chem. Int. Ed.* 48, 8501 (2009), Copyright Wiley-VCH Verlag GmbH (2009).

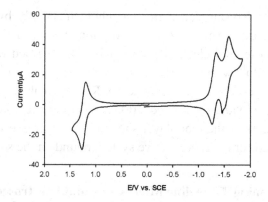

Fig. 2. Cyclic voltammogram for a 3 mM $Ru^{II}(tpy)(tpyC_{13}SAc)$ solution in acetonitrile containing 0.1 M $TBAPF_6$ at 0.1 V s^{-1}. Reprinted with permission from Ref. 44, K. Seo et al., Molecular Conductance Switch-On of Single Ruthenium Complex Molecules, J. Am. Chem. Soc. 130, 2553 (2008), Copyright American Chemical Society (2008).

Discrete redox states due to metal- and ligand-centered reactions of Ru^{II} terpyridine complexes were clearly characterized by cyclic voltammetry (Fig. 2). The redox formal potential respective to the Ru^{III}/Ru^{II} reaction was approximately +1.2 V_{SCE}. The first two redox couples of terpyridine ligand were observed in the negative potential region and the redox formal potential respective to the $[Ru^{II}(tpy)_2]^{2+}/[Ru^{II}(tpy)(tpy)^-]^+$ reaction was approximately −1.2 V_{SCE}. On the other hand, the typical MLCT electronic transition (λ_{max} = 477 nm) of Ru^{II} terpyridine complexes was characterized by the UV-Vis absorption spectra.

2.2. Surface Chemical Analysis

The surfaces of the self-assembled molecular film (3.0 nm thickness) of $Ru^{II}(tpyC_nS)_2$ on Au were characterized using UV-Vis spectroscopy and spectroscopic ellipsometry. The relatively intense and broad absorption band in the visible region, responsible for the dark red color, was due to the spin allowed by a $d \rightarrow \pi^*$ MLCT transition.[46] To further confirm these SAMs, samples were examined using X-ray photoelectron spectroscopy (XPS). The measured $S(2p_{3/2})$ binding energy of 162.0 eV was in excellent agreement with other studies.[47] The Ru^{II} terpyridine complex

SAMs on ITO were also characterized photochemically by the MLCT electronic transition in the UV-Vis adsorption spectra of the solid state and were characterized electrochemically by well-defined redox current peaks of Ru^{III}/Ru^{II} with a surface coverage of 7.4×10^{-11} mol cm^{-2}. Surface coverage for the corresponding Ru was calculated from the charge under the oxidation and reduction peaks of cyclic voltammograms. The peak currents of the monolayer show a linear increase with the scan rate, as expected for the redox-active system bound on the surface.[48]

3. UHV-Scanning Tunneling Microscopy and Spectroscopy

3.1. *Current/Voltage Response of Single Ruthenium Complexes*

To achieve of single molecular junctions, thiol-tethered Ru^{II} terpyridine complexes incorporated in *n*-alkanethiol matrixes (e.g., 1-octanethiol (OT) or 1-dodecanethiol (DDT) SAMs) on Au(111) were prepared. The bias-induced conductance switching of Ru^{II} complexes in the single molecular junctions was probed by using scanning tunneling spectroscopy (STS) in an ultrahigh vacuum. STM images for the Ru^{II}(tpy)(tpyS) SAMs on Au(111) showed molecular globes presumed to be individual molecules or molecular bundles while an ordered structure of Ru^{II} terpyridine complex SAMs was not observed (Fig. 3(A)). To construct single molecular junctions in dielectric barriers, the incorporation of Ru^{II} terpyridine complexes was conducted in *n*-alkanethiol SAMs, and single molecules or molecular bundles of Ru^{II}(tpy)(tpyS) were observed as protrusions at the edge of gold vacancy islands and at the domain boundaries of well-ordered *n*-alkanethiol (Fig. 3(B)).

Most of the bright protrusions were continuously imaged, while some of them rarely displayed stochastic switching as the molecules blinked on and off in the STM images.[49,50] Single molecules or molecular bundles were defined by a cross-sectional analysis of STM images. The size of the Ru terpyridine complex (i.e., Ru^{II}(tpy)$_2$) was estimated to be 11.3 Å according to MM2 calculations in Chem3D. STM images reflect the conductance and physical height/length, which allows the physical difference to be determined if imaged Ru terpyridine complexes are

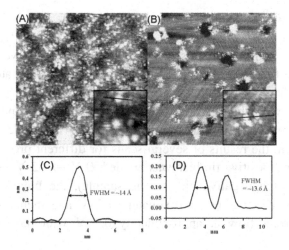

Fig. 3. STM images (88 × 88 nm²) of (A) a pure Ru^{II}(tpy)(tpyS) SAM and (B) a Ru^{II}(tpy)(tpyS) incorporated 1-octanethiol (OT) SAM on Au(111). (C) and (D) Cross-sectional analysis for the inset images (A) and (B), respectively. Reprinted with permission from Ref. 44, K. Seo *et al.*, Molecular Conductance Switch-On of Single Ruthenium Complex Molecules, *J. Am. Chem. Soc.* 130, 2553 (2008), Copyright American Chemical Society (2008).

single molecules or molecular bundles. From the results of a cross-sectional analysis of both STM images for a pure Ru^{II}(tpy)(tpyS) SAM and a Ru^{II}(tpy)(tpyS) incorporated OT SAM, molecular globes or white protrusions that are approximately 13 ~ 14 Å in the FWHM (full width at half maximum) could be defined as a single molecule (Figs. 3(C) and 3(D)) in which larger molecular globes such as those that are >20 Å could be defined as molecular bundles.

After successive STM imaging for the Ru^{II}(tpy)(tpyS) incorporated DDT SAM (Fig. 4(A)), the current-voltage (*I-V*) curves of the current response to the tip-bias voltage were measured from the Ru^{II}(tpy)(tpyS) molecule marked with an arrow (in Fig. 4(A)) without tunneling current feedback. The current values were recorded while the tip-bias voltage was swept from zero to positive or negative directions in a cycle. After finishing the current-voltage measurement, the current feedback was restored and tip-scanning continued. This procedure was performed for all molecular junctions. A representative *I-V* curve for a molecular junction of $Ru^{II}(tpy)(tpyS)$ is shown in Fig. 4(B). The tip-bias voltage was swept in a cycle, 0 V → +2 V → 0 V → −2 V → 0 V. Hysteretic *I-V* curves were obtained in both sweep directions. *I-V* curves were measured two or three times on the top of the same protrusions repeatedly during one scan. The average number of reproducible cycles for hysteretic *I-V* curves was two, which strongly depended on the drift condition of the STM system. Thus, for a statistical analysis, hysteretic *I-V* curves obtained from the results of several scans for different protrusions were used. When a positive tip-bias was applied, 0 V → +2 V → 0 V, (i.e., application of a negative bias to the sample), the tunneling currents suddenly increased after approximately 1.5 V and then returned to a residual current flow at approximately 0.7 V with the reverse scan (Fig. 4(B)).

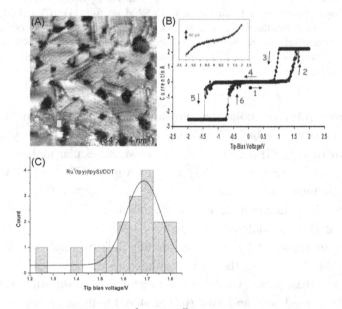

Fig. 4. (A) STM image (34 × 34 nm²) of a $Ru^{II}(tpy)(tpyS)$ incorporated 1-dodecanethiol (DDT) SAM on Au(111). (B) Current-voltage (*I-V*) characteristics measured from a $Ru^{II}(tpy)(tpyS)$ molecule marked with an arrow in (A). The inset is *I-V* curve measured from the DDT molecule. (C) Histograms of the threshold voltage for the current switch-on in the single $Ru^{II}(tpy)(tpyS)$ junctions. Reprinted with permission from Ref. 44, K. Seo et al., Molecular Conductance Switch-On of Single Ruthenium Complex Molecules, J. Am. Chem. Soc. 130, 2553 (2008), Copyright American Chemical Society (2008).

However, conductance switching for the single DDT molecules used for a control experiment was not observed as shown in the inset of Fig. 4(B) and the curves showed symmetrical and sigmoidal *I-V* characteristics. The histograms of the threshold voltage for the current switch-on in molecular junctions of the $Ru^{II}(tpy)(tpyS)$ incorporated DDT SAM suggest that the molecule switched to a high conductance state primarily at 1.70 ± 0.025 V (Fig. 4(C)). On the other hand, in molecular junctions of the $Ru^{II}(tpy)(tpyS)$ incorporated OT SAM, the molecule switched primarily to a high conductance state at approximately 1.75 ± 0.025 V, as shown in the histograms of the threshold voltage for the current switch-on. This type of conductance switching was observed in all molecular junctions with Ru^{II} terpyridine molecules, including $Ru^{II}(tpy)(tpyC_nS)$, $n = 7$ and 13 (not shown).

The molecular junctions with a dithiol-tethered Ru^{II} terpyridine complex (i.e., $Ru^{II}(tpyS)_2$) formed in an OT SAM. Figures 5(A) and 5(B) show STM images of dithiol-tethered Ru^{II} terpyridine complexes incorporated into the OT SAMs and a statistical analysis of threshold voltage for the current switch-on. The molecular junctions consisted of the STM tip/one free thiol group of the $Ru^{II}(tpyS)_2$/Au substrate. Single or bundles of $Ru^{II}(tpyS)_2$ molecules as well as monothiol-tethered Ru^{II} terpyridine

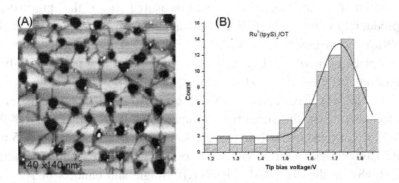

Fig. 5. (A) STM image of a $Ru^{II}(tpyS)_2$ incorporated 1-octanthiol (OT) SAM on Au(111) at a constant tunneling current of 20 pA with a tip-bias of 1.2 V. (B) Histograms of the threshold voltage for the current switch-on in the single $Ru^{II}(tpyS)_2$ junctions. Reprinted with permission from Ref. 44, K. Seo *et al.*, Molecular Conductance Switch-On of Single Ruthenium Complex Molecules, *J. Am. Chem. Soc.* 130, 2553 (2008), Copyright American Chemical Society (2008).

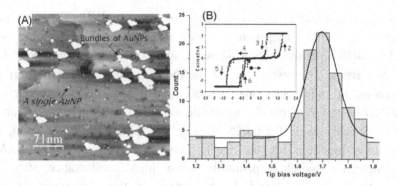

Fig. 6. (A) STM image of a Au-NPs-capped $Ru^{II}(tpyS)_2$ (Au-NP/$Ru^{II}(tpyS)_2$) incorporated OT SAM on Au(111). The image was obtained at the constant tunneling current of 20 pA with a tip-bias of 1.2 V in a vacuum. (B) Histograms of the threshold voltage for the current switch-on in the single Au-NP/$Ru^{II}(tpyS)_2$ junctions. Reprinted with permission from Ref. 44, K. Seo et al., Molecular Conductance Switch-On of Single Ruthenium Complex Molecules, J. Am. Chem. Soc. 130, 2553 (2008), Copyright American Chemical Society (2008).

complexes (i.e., $Ru^{II}(tpy)(tpyS)$) were observed as bright protrusions in the OT SAMs. Some of the protrusions at the edge of the gold vacancy islands displayed stochastic switching. However, the protrusions at the boundaries of the ordered OT domains were imaged stably and the hysteretic I-V characteristics were measured from the protrusions reproducibly in both sweep directions (not shown). In the junction of the tip/$Ru^{II}(tpyS)_2$/substrate, the threshold voltage for the current switch-on takes place primarily at 1.75 ± 0.025 V (Fig. 5(B)) in a positive tip-bias direction of $0 V \rightarrow +2 V \rightarrow 0 V$.

On the other hand, as a model close to realistic systems of a molecular switch, the simplified symmetric molecular junctions of the Au-NP/$Ru^{II}(tpyS)_2$/Au substrate were formed via the attachment of 5 nm gold nanoparticles (Au-NPs) on the top of the inserted dithiol-tethered $Ru^{II}(tpyS)_2$ in the OT SAM (Fig. 6(A)). Single and bundles of Au-NPs were observed at the edge of the gold vacancy islands and at the boundaries of ordered OT domains. I-V curves of single Au-NP junctions showed stable current hysteresis in both bias directions (the inset of Fig. 6(B)), compared to the I-V curves of bundled Au-NP junctions that sometimes show hysteretic I-V characteristics in only one bias direction.

However, current hysteresis in only one bias direction was occasionally observed in molecular junctions of Ru^{II} terpyridine complexes. This may be due to unstable molecular junctions. In this work, this phenomenon was disregarded. On the other hand, in Au-NP junctions, I-V curves can show a charging effect of the nanoparticle,[51,52] which is dependent of the nanoparticle size. According to a previous report,[52] smaller nanoparticles (1.5 nm) had a greater charging effect compared to larger nanoparticles (5.4 nm). The apparent conductance is theoretically smaller by 10% than the actual molecular conductance in the case of a 5.4 nm nanoparticle, which can appear as a wide current-suppressed region in I-V curves.[52] However, magnified I-V curves for the Au-NP/Ru^{II}(tpyS)$_2$ junctions at an approximate zero bias condition were nearly linear without noticeable current-suppressed regions (not shown); slight oscillation was also observed, which can be attributed to the thermal fluctuation.

The molecular conductance switch-on of the Au-NP-capped molecular junctions takes place primarily at 1.70 ± 0.025 V (according to the histograms in Fig. 6(B)) in a positive tip-bias direction of 0 V → +2 V → 0 V. Therefore, the results of conducting switching in different junctions with mono- and dithiol-tethered Ru^{II} terpyridine complexes demonstrate that the bias-induced switching occurs due to the intrinsic nature of the molecules, although the role of the internal conformation change cannot be ignored. The switch-on threshold voltage of Ru^{II} terpyridine complexes in the molecular junctions, Au (or Pt/Ir)/Ru^{II} terpyridine complexes/Au substrate, can be approximated in the range of 1.70 ~ 1.75 V.

3.2. A Proposed Model for Electron Trapping in Ru^{II} Terpyridine Complexes

When the Fermi levels of electrodes align to molecular redox formal potentials, resonant tunneling may take place across reduction-oxidation states which can be expected due to the chemical nature of the charge-trap states. Vacuum levels of the ionization and the electron affinity of molecules on metals can be closely approximated from electrochemical potential scales. The first electrochemical oxidation and reduction potentials should be approximately the first ionization energy and the first electron affinity levels of thin film supported on metals, respectively.[53] To convert an electrochemical potential referenced to a saturated calomel electrode (SCE) to a vacuum level, it is possible to utilize the simplified model offered by Hipps et al.,[11,53,54] V_{abs} (eV) = 4.7 eV + E^0(SCE), in which E^0 is the redox formal potential and 4.7 eV are approximated according to the vacuum level, ~4.5 eV for the NHE (normal hydrogen electrode) and a 0.24 V difference between the SCE and the NHE reference electrode.[55] For reduction processes, this model was in very good agreement with UPS (ultraviolet photoelectron spectroscopy) observations in many cases.[53] However, the polarization stabilization of ions by the surrounding molecules and image charges induced in the metal substrate can lead to the ionization potential of electrochemical reactions greater than that of the gas phase (e.g., it was to be approximately 0.5 to 1.0 eV for a thin film of NiOEP).[53] Thus, for oxidation processes (e.g., the metal-centered oxidation of a transition metal-organic ligand complex), the equation offered by Armstrong et al.[54,56] was used, and the ionization energies V_i = 4.7 eV + $(1.7)E^{ox}(SCE)_{1/2}$, in which $E^{ox}(SCE)_{1/2}$ is the half-wave oxidation potential. Therefore, the redox formal potentials can be converted to comparable solid state potentials in STM using two equations, V_a = 4.7 eV + $E^{red}(SCE)_{1/2}$ and V_i = 4.7 eV + $(1.7)E^{ox}(SCE)_{1/2}$, where $E^{red}(SCE)_{1/2}$ and $E^{ox}(SCE)_{1/2}$ are the half-wave reduction and oxidation potentials, respectively.

A simplified molecular orbital diagram for an octahedral transition metal complex,[19] which consists of two discrete redox states (the metal-centered highest occupied molecular orbital (HOMO) and the ligand-centered lowest unoccupied molecular orbital (LUMO)), can be used in molecular orbital configurations of Ru^{II}(tpy)(tpyS) as depicted in Fig. 7. From the results of electrochemical measurements (Fig. 2), the first oxidation occurs near +1.2 V_{SCE} (i.e., $Ru^{III} - e^- \rightarrow Ru^{II}$) and the first reduction occurs near −1.2 V_{SCE} (i.e., $[Ru^{II}(tpy)_2]^{2+} + e^- \rightarrow [Ru^{II}(tpy)(tpy)^-]^+$). These redox formal potentials can be converted to the vacuum levels using two equations offered by Hipps et al.[53] and Armstrong et al.,[56] and the energy levels of the first metal-centered oxidation and the first ligand-centered reduction are 6.74 and 3.4 V below the vacuum, respectively (V_i = 4.7 eV + (1.7) × 1.2 = 6.74 eV and

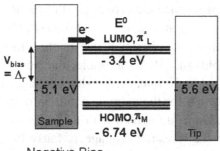

Fig. 7. Proposed charging process into the ligand-centered LUMO of Ru^{II} terpyridine complexes. Reprinted with permission from Ref. 44, K. Seo *et al.*, Molecular Conductance Switch-On of Single Ruthenium Complex Molecules, *J. Am. Chem. Soc.* 130, 2553 (2008), Copyright American Chemical Society (2008).

V_a = 4.7 eV − 1.2 = 3.4 eV). Actual solid state ionization and electron affinity energies of HOMO and LUMO in the monolayer should deviate slightly from those in a vacuum.[53] However, charge transport can be discussed in terms of the relative energy levels between the LUMO and HOMO levels and metal Fermi levels.[10,16] When a negative or positive bias is applied to the sample, the metal Fermi-level of either Pt/Ir or gold should go up toward a molecular orbital level of the Ru complex to allow electron trapping from the metal nearby. The Fermi levels of the STM tip (Pt/Ir) and the substrate (gold) are approximately 5.6 and 5.1 V below the vacuum, respectively.[16] Electron transfer reactions should occur through the LUMO level because a HOMO level is far from the metal Fermi levels. Application of a negative sample-bias drives the Fermi levels of the gold substrate to align to the ligand-centered LUMO of the Ru^{II} complexes (Fig. 7). Thus, approximately 1.7 V (5.1 V − 3.4 V = 1.7 V) of bias is needed to bring the Fermi level of the gold substrate up to the ligand-centered formal potential, which is close to the typically observed threshold voltage of 1.70 ~ 1.75 V.

For an understanding of the hysteretic *I-V* characteristics in a negative tip-bias direction of 0 V → −2 V → 0 V, the threshold voltage of the current switch-on was analyzed in both molecular junctions in the Ru^{II}(tpy)(tpyS)/gold substrate and the Au-NP/Ru^{II}(tpyS)$_2$/gold substrate. When a negative bias is applied to the tip (0 V → −2 V → 0 V), Pt/Ir should go up toward a molecular orbital level of the Ru complex and the threshold voltage of the conductance switch-on should be larger than that obtained in a positive bias direction due to the different metal Fermi levels between gold and Pt/Ir. In molecular junctions of the Pt/Ir tip/Ru^{II}(tpy)(tpyS)/gold substrate, the threshold voltage of -1.95 ± 0.025 V was primarily measured for the current switch-on (Fig. S5A, Supporting Information), which roughly agreed with the expected effect. However, the distribution of the threshold voltage is fairly broad. On the other hand, in simplified symmetric Au-NP/Ru^{II}(tpyS)$_2$ junctions (i.e., Pt/Ir tip/Au-NP/Ru^{II}(tpyS)$_2$/gold substrate), Au nanoparticle capping had a significant effect on the threshold voltage values. Histograms of the threshold voltage revealed that the current switch-on takes place primarily at -1.75 ± 0.025 V, the absolute value of the threshold voltage that is consistent with that determined in a positive tip-bias direction. In the Au-NP/Ru^{II}(tpyS)$_2$/gold substrate junctions, Ru^{II} terpyridine complexes were supported on gold in both contacts in which the absolute threshold voltage for current switch-on may be similar in both bias directions.

In the all molecular junctions of Ru^{II} terpyridine complexes including simplified symmetric junctions of an Au-NP-capped Ru^{II} dithiol-tethered terpyridine complex, the threshold voltage of switch-on was comparable to the first redox formal potential of the terpyridine ligand supported on gold. The proposed model can postulate that trapping the electron on the ligand center of Ru complexes leads to conductance switching.

4. A MMNVM Device of Ru^{II} Terpyridine Complexes

4.1. *Fabrication of a Large-Area Molecular Device*

In molecular monolayer devices of dialkylthiolate-tethered ruthenium(II) terpyridine complexes, it is expected that the hydrophilic terminal thiol group of the alkylthiolate-tethered Ru^{II} terpyridine complexes can interact with the hydrophilic sulfonic acid groups of PEDOT:PSS, preventing the penetration of soft and metal top electrodes into the SAMs to result in high yields.[43] The direct measurement of the hysteretic *I-V* characteristics of the MMNVM, the write-multiple read-erase pulse cycles, and the retention time dependence on alkyl chain lengths was also reported. The processing of a molecular device with diameters of 25 μm is schematically depicted (Fig. 8).

The devices were electronically stable and possessed a thermal endurance up to 300°C, indicating that reduction of electrical shorting through modification of the Ru^{II} terpyridine complexes with thiolate-terminated alkyl chains was a crucial means of improving device yield. For more details, one thiol of the terminal dialkylthiols was introduced to attach the Ru^{II} complexes onto a gold surface with the other thiol of the dialkylthiol. This allowed for hydrophilic interactions with the sulfonic acid of the PEDOT:PSS. This resulted in the prevention of a permeation of PEDOT:PSS onto the SAMs as the monoalkylthiol could not contribute to a hydrophilic interaction due to the presence of a hydrophobic terpyridine end group. In addition, longer alkyl chains effectively prevented penetration of the PEDOT:PSS and top metal electrode, resulting in a higher device yield. In fact, as the number of alkyl chains or alkyl chain length increased, the device yield improved sharply from nearly 0 to 81%. All current-voltage (I-V) measurements were performed in a vacuum to avoid the influence of moisture and oxygen, especially for PEDOT:PSS.

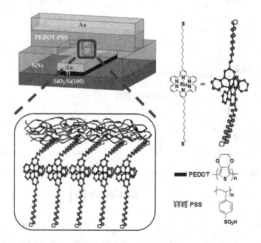

Fig. 8. The schematic cross-section of the device layout of PEDOT:PSS on the ruthenium complexes. Reprinted with permission from Ref. 45, J. Lee *et al.*, Molecular Monolayer Non-Volatile Memory with Tunable Molecules, *Angew. Chem. Int. Ed.* 48, 8501 (2009), Copyright Wiley-VCH Verlag GmbH (2009).

4.2. Current/Voltage Response of a MMNVM Device with Ru^{II} Terpyridine Complexes

Figure 9 shows the *I-V* characteristics of the Au/Ru^{II}(tpyC$_{13}$S)$_2$/Au junction with a diameter of 25 μm. The *I-V* characteristics of the devices were recorded by scanning the applied voltage from 0 to +2 V and then to −2 V, followed by a reverse scan from −2 to +2 V. The positive bias corresponded to positive voltages applied to the top metal pad, whereas negative bias corresponded to negative voltages applied to the top pad. The top inset of Fig. 9 shows a control experiment using only PEDOT:PSS, without ruthenium SAMs between the top and bottom electrodes. This experimental setup showed nearly ohmic behavior with kilo-ohm resistance. The *I-V* curve of the Ru^{II} complex was asymmetric, as depicted in Fig. 9. The negative bias region always indicated a larger current and hysteresis. An enlarged hysteretic *I-V* curve is shown in the bottom inset of Fig. 9. This result resembles the literature diagram reported for the Ru^{II}(bipyridyl)$_2$(triazolopyridyl) complex prepared after spin-coating, with a thickness of 80 nm.[57]

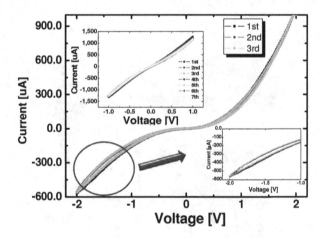

Fig. 9. Hysteretic *I-V* characteristics of the molecular monolayer device (Au/Ru^{II} complex, Ru^{II}(tpyC$_{13}$S)$_2$/PEDOT:PSS/Au). The top inset, a control experiment using PEDOT:PSS without ruthenium SAMs between the top and bottom electrodes. Reprinted with permission from Ref. 45, J. Lee *et al.*, Molecular Monolayer Non-Volatile Memory with Tunable Molecules, *Angew. Chem. Int. Ed.* 48, 8501 (2009), Copyright Wiley-VCH Verlag GmbH (2009).

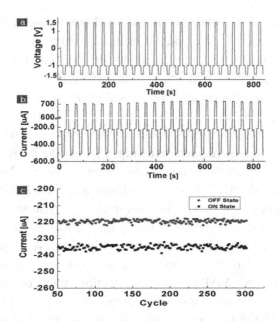

Fig. 10. (a) and (b), Write-multiple read-erase-multiple read (WRER) cycles of a molecular monolayer device containing Ru^{II} complex ($Ru^{II}(tpyC_{13}S)_2$) for rewritable data storage application. The writing (W), reading (R), erasing (E), and reading (R) voltages were −1.5, −1, +1.5, and −1 V, respectively. (c), Currents on the ON state and OFF states as a function of the number of WRER cycles. 3.0×10^2 cycles were tested in inert conditions. Reprinted with permission from Ref. 45, J. Lee *et al.*, Molecular Monolayer Non-Volatile Memory with Tunable Molecules, *Angew. Chem. Int. Ed.* 48, 8501 (2009), Copyright Wiley-VCH Verlag GmbH (2009).

Stable conductivity switching behavior makes it possible for non-volatile molecular memory phenomenon to be tested under a voltage pulse sequence, specifically, a write-read-erase-read (WRER) cycle. In such a cycle, the high (write) and low (erase) conducting states were repeatedly induced and the read states monitored in-between the high and low conducting states. A section of the voltage sequence and corresponding current from the device is shown in Figs. 10(a) and 10(b). As seen in Figs. 10(a) and 10(b), the device can be programmed to a high conductivity state using a −1.5 V pulse and to a low conductivity state using a +1.5 V pulse with multiple current measurements for reading at −1.0 V. This device shows no significant degradation after several hundred write-and-erase cycles. This stable conductivity switching behavior made it possible for metal-SAM-conducting polymer-metal junctions to be used as non-volatile memories with write-multiple read-erase-multiple read operations, as shown in Fig. 10(b). Each multiple reading was measured eight times after each write or erase pulse. The WRER cycles can be repeatedly performed in an excess of 300 cycles, as depicted in Fig. 10(c).

Figure 11 shows a typical result of the retention time carried out under inert conditions. As seen in the Fig. 11, once the device is switched to the ON-state by applying a negative voltage pulse at −2.0 V, the ON state was retained after 390 s with insignificant degradation. When the ON-state was switched back to the OFF-state by a positive voltage pulse applied at an amplitude of +2.0 V, the OFF-state was sustained even after 390 s with little degradation. Conversely, in the case of shorter alkyl chains, $Ru^{II}(tpyC_7S)_2$, the retention time was reduced to near 200 s. For the application of MMNVM, the retention time of Ru complex $Ru^{II}(tpyC_{13}S)_2$ remained sufficiently large in comparison with a few milliseconds of DRAM and furthermore, the retention time of the Ru complex could be extended to several days by utilizing a nanowire as a channel,[58] based on our previous report with $Ru^{II}(tpy)(tpyC_7S)$.[59]

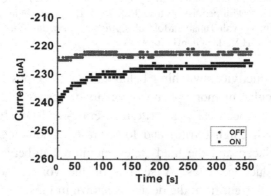

Fig. 11. Retention times of the ON and OFF states of the molecular monolayer device, Au/Ru^{II} complex ($Ru^{II}(tpyC_{13}S)_2$)/PEDOT:PSS/Au, probed with currents under −1.0 V. The ON and OFF states have been induced by −2.0 V (writing) and 2.0 V (erasing), respectively. Reprinted with permission from Ref. 45, J. Lee *et al.*, Molecular Monolayer Non-Volatile Memory with Tunable Molecules, *Angew. Chem. Int. Ed.* 48, 8501 (2009), Copyright Wiley-VCH Verlag GmbH (2009).

In addition, the ratio between the probe currents corresponded to the ON/OFF ratio obtained in the *I-V* characteristics as depicted in Fig. 9. Although the ON/OFF ratio of the molecular monolayer was not large in comparison to the 4-5 orders of the ON/OFF ratio of our organic bulk memory device using the same materials at a 60 nm thickness, a direct observation of the non-volatile memory effect via the molecular monolayer was very meaningful for designing a nanoscaled circuit and architecture that did not require a large ON/OFF ratio under consideration of a faster response time and low power operation, and for fabricating large-array molecular memory devices toward the realization of ultimately miniaturized devices. In order to understand the performance of the present MMNVM, the conduction mechanism for the switching behavior was investigated. It is known that the presence of large mobile counter anions $(PF_6^-)_2$ in the Ru complex and conducting polymer PEDOT:PSS, used as a soft portion of the top electrode, do not play important roles in the conduction mechanism.[44,57] Thus, the conductance switching mechanism of $Ru^{II}(tpy(CH_2)_{7 and 13}S)_2$ with alkyl chains should be similar to that of the $Ru^{II}(tpy)(tpyS)$ without alkyl chains, previously reported by the authors.[44,57] The conductance switching mechanism was at least a two-step tunneling process where the electron first tunnels from the electrodes into the metal-ligand complex center (electron reduction), while maintaining an OFF-state. The electron then tunnels out from the terbipyridine ligand of the ruthenium complex to the electrodes (electron oxidation), while maintaining an ON-state. The hysteresis observed was then the result of the charging-uncharging the self-assembled monolayer that acts as a capacitor.[57] The transport mechanism, with variable low temperature measurements at only the ON or OFF states, was direct tunneling, independent of temperature.

5. Conclusion

The hysteretic *I-V* characteristics in single molecular junctions of mono- and dithiol-tethered Ru^{II} terpyridine complexes were measured by STS in vacuum. The tunneling current initially stays switched off until the sharp threshold voltage is aligned with the electronic energy of the ligand-centered LUMO, which originates with the reduction of the terpyridine ligand. Thus, the threshold voltage can be determined by the electronic energy gap between the Fermi level of the metal contacts and the lowest ligand-centered redox state of the metal complex molecule. For molecular memory applications, these results can provide guidance in design that improves the charge-trapping efficiency of various ligands with different metal substrates in the solid state.

A novel MMNVM was demonstrated using new Ru^{II} terpyridine complexes with thiol-terminated alkyl chains of different chain lengths on one or both sides. The dialkyl chains and longer alkyl chains effectively prevented penetration of the PEDOT:PSS and the top metal electrode into the SAMs, resulting in a higher device yield. As the alkyl chain length increased, retention time increased. Write-multiple read-erase-multiple read pulse cycles were repeatedly operated at low driving voltages with a reasonable ON/OFF ratio. This device showed no significant degradation after several hundred pulse cycles. This is the first known achievement of its kind toward the development of a voltage-driven MMNVM that does not involve the use of state-of-the-art techniques and uses a commercially available microsized metal electrode well. Although retention time and switching cycle remain far from satisfactory compared to current Si technology, our leading-edge research on molecular monolayer memory can provide new insights into the design of voltage-driven functional molecules and the fabrication of MMNVM.

Acknowledgment

This work was supported by the Creative Research Initiatives (project title: Smart Molecular Memory) of MEST/NRF.

References

1. L. A. Bumm, J. J. Arnold, M. T. Cygan, T. D. Dunbar, T. P. Burgin, L. Jones, II, D. L. Allara, J. M. Tour, and P. S. Weiss. *Science* 271, 1705 (1996).
2. C. P. Collier, E. W. Wong, M. Belohradsky, F. M. Raymo, J. F. Stoddart, P. J. Kuekes, R. S. Williams, and J. R. Heath. *Science* 285, 391 (1999).
3. C. P. Collier, G. Mattersteig, E. W. Wong, Y. Luo, K. Beverly, J. Sampaio, F. M. Raymo, J. F. Stoddart, and J. R. Heath. *Science* 289, 1172 (2000).

4. J. M. Tour, M. Kozaki, and J. M. Seminario. *J. Am. Chem. Soc.* 120, 8486 (1998).
5. A. S. Blum, J. G. Kushmerick, D. P. Long, C. H. Patterson, J. C. Yang, J. C. Henderson, Y. Yao, J. M. Tour, R. Shashidhar, and B. R. Ratna. *Nat. Mater.* 4, 167 (2005).
6. S. Sortino, S. Petralia, S. Conoci, and S. Di Bella. *J. Am. Chem. Soc.* 125, 1122 (2003).
7. S. Sortino, S. Petralia, S. Conoci, and S. D. Bella. *J. Mater. Chem.* 14, 811 (2004).
8. D. I. Gittins, D. Bethell, D. J. Schiffrin, and R. J. Nichols. *Nature* 408, 67 (2000).
9. H. X. He, X. L. Li, N. J. Tao, L. A. Nagahara, I. Amlani, and R. Tsui. *Phys. Rev. B* 68, 045302 (2003).
10. R. A. Wassel, G. M. Credo, R. R. Fuierer, D. L. Feldheim, and C. B. Gorman. *J. Am. Chem. Soc.* 126, 295 (2004).
11. J. He, Q. Fu, S. Lindsay, J. W. Ciszek, and J. M. Tour. *J. Am. Chem. Soc.* 127, 11932 (2006).
12. X. Xiao, L. A. Nagahara, A. M. Rawlett, and N. Tao. *J. Am. Chem. Soc.* 127, 9235 (2005).
13. L. Cai, M. A. Cabassi, H. Yoon, O. M. Cabarcos, C. L. McGuiness, A. K. Flatt, D. L. Allara, J. M. Tour, and T. S. Mayer. *Nano Lett.* 5, 2365 (2005).
14. T. Albrecht, A. Guckian, J. Ulstrup, and J. G. Vos. *Nano Lett.* 5, 1451 (2005).
15. R. A. Wassel, G. M. Credo, R. R. Fuierer, D. L. Feldheim, and C. B. Gorman. *J. Am. Chem. Soc.* 126, 295 (2004).
16. K. Kitagawa, T. Morita, and S. Kimura. *Langmuir* 21, 10624 (2005).
17. B. S. Kim, J. M. Beebe, C. Olivier, S. Rigaut, D. Touchard, J. G. Kushmerick, X. Y. Zhu, and C. D. Frisbie. *J. Phys. Chem. C* 111, 7521 (2007).
18. C. Li, W. Fan, D. A. Straus, B. Lei, S. Asano, D. Zhang, J. Han, M. Meyyappan, and C. Zhou. *J. Am. Chem. Soc.* 126, 7750 (2004).
19. V. Balzani, A. Juris, M. Venturi, S. Campagna, and S. Serroni. *Chem. Rev.* 96, 759 (1996).
20. J. P. Sauvage, J. P. Collin, J. C. Chambron, S. Guillerez, C. Coudret, V. Balzani, F. Barigelletti, L. De Cola, and L. Flamigni. *Chem. Rev.* 94, 993 (1994).
21. K. L. D. Tondelier, D. Vuillaume, C. Fery, and G. Haas. *Appl. Phys. Lett.* 85, 5763 (2004).
22. H. S. W. Tang, G. Xu, B. S. Ong, Z. D. Popovic, J. Deng, J. Zhao, and G. Rao. *Adv. Mater.* 17, 2307 (2005).
23. M. A. D. Ma, J. A. Freire, and I. A. Hummelgen. *Adv. Mater.* 12, 1063 (2000).
24. H. B. A. A. J. Kronemeijer, T. Kudernac, B. J. V. Wees, B. L. Feringa, P. W. M. Blom, and B. D. Boer. *Adv. Mater.* 20, 1467 (2008).
25. M. M. F. B. D. Boer, Y. J. Chabal, W. Jiang, E. Garfunkel, and Z. Bao. *Langmuir* 20, 1539 (2004).
26. M. A. R. R. Haag, R. E. Holmin, and G. M. Whiteside. *J. Am. Chem. Soc.* 121, 7895 (1999).
27. P. A. L. R. K. Smith, and P. S. Weiss. *Prog. Surf. Sci.* 75, 1 (2004).

28. S. Y. C. M. D. Austin. *Nano Lett.* 3, 1687 (2003).
29. G. W. T. Kim, H. Lee, and T. Lee. *Nanotechnology* 18, 315204 (2007).
30. A. V. W. G. L. Fisher, A. E. Hooper, T. B. Tighe, K. B. Bahnck, H. T. Skriba, M. D. Reinard, B. C. Haynie, R. L. Opila, N. Winograd, and D. L. Allara. *J. Am. Chem. Soc.* 124, 5528 (2002).
31. M. A. R. J. Chen, A. M. Rawlett, and J. M. Tour. *Science* 286, 1550 (1999).
32. G. M. C. P. Collier, E. W. Wong, Y. Luo, K. Beverly, J. Sampaio, F. M. Raymo, J. F. Stoddart, and J. R. Health. *Science* 289, 1172 (2000).
33. G.-Y. J. Y. Chen, D. A. A. Ohlberg, X. Li, D. R. Stewart, J. O. Jeppesen, K. A. Nielsen, J. F. Stoddart, and R. S. Williams. *Nanotechnology* 14, 462 (2003).
34. L. A. E. J. C. Love, J. K. Kriebel, R. G. Nuzzo, and G. M. Whitesides. *Chem. Rev.* 105, 1103 (2005).
35. H. C. G. S. Bang, J.-R. Koo, T. Lee, R. C. Advincula, and H. Lee. *Small* 4, 1399 (2008).
36. M. R. D. C. Zhou, M. A. Reed, L. Jones, and J. M. Tour. *Appl. Phys. Lett.* 71, 611 (1997).
37. J. C. M. A. Reed, A. M. Rawlett, D. W. Price, and J. M. Tour. *Appl. Phys. Lett.* 78, 3735 (2001).
38. N. G. N. Majumdar, D. Routenberg, J. C. Bean, L. R. Harriott, B. Li, L. Pu, Y. Yao, and J. M. Tour. *J. Vac. Sci. Technol. B* 23, 1417 (2005).
39. T. L. W. Wang and M. A. Reed. *Phy. Rev. B* 68, 035416 (2003).
40. H. L. D.-H. Kim, C.-K. Song, and C. Lee. *J. Nanosci. Nanotechnol.* 6, 3470 (2006).
41. D. Z. C. Li, X. Liu, S. Han, T. Tang, C. Zhou, W. Fan, J. Koehne, J. Han, M. Meyyappan, A. M. Rawlett, D. W. Price, and J. M. Tour. *Appl. Phys. Lett.* 82, 645 (2003).
42. B. C. J. He, A. K. Flatt, J. J. Stephenson, C. D. Doyle, and J. M. Tour. *Nat. Mater.* 5, 63 (2006).
43. P. W. M. B. H. B. Akkerman, D. M. D. Leeuw, and B. D. Boer. *Nature* 441, 69 (2006).
44. A. V. K. K. Seo, J. Lee, G. S. Bang, and H. Lee. *J. Am. Chem. Soc.* 130, 2553 (2008).
45. H. C. J. Lee, S. Kim, G. S. Bang, and H. Lee. *Angew. Chem. Int. Ed.* 48, 8501 (2009).
46. J.-P. C. J.-P. Sauvage, J.-C. Chambron, S. Guillerez, and C. Coudret. *Chem. Rev.* 94, 993 (1994).
47. Y. L. S. Y. V. Zubavichus, M. K. Nazeeruddin, S. M. Zakeeruddin, M. Grätzel, and V. Shklover. *Chem. Mater.* 14, 3556 (2002).
48. A. J. Bard and L. R. Faulkner. *Electrochemical Methods, Fundamentals and Applications* (John Wiley and Sons, New York, 1980).
49. Z. J. Donhauser, B. A. Mantooth, K. F. Kelly, L. A. Bumm, J. D. Monnell, J. J. Stapleton, D. W. Price, Jr., A. M. Rawlett, D. L. Allara, J. M. Tour, and P. S. Weiss. *Science* 292, 2303 (2001).

50. H. Basch, R. Cohen, and M. A. Ratner. *Nano Lett.* 5, 1668 (2005).
51. A. E. Hanna and M. Tinkham. *Phys. Rev. B* 44 (1991).
52. T. Morita and S. Lindsay. *J. Am. Chem. Soc.* 129, 7262 (2007).
53. K. W. Hipps. *Handbook of Applied Solid State Spectroscopy: Scanning Tunneling Spectroscopy*, Vol. 7, ed. D. R. Vij (Springer Verlag, 2006).
54. L. Scudiero, D. E. Barlow, and K. W. Hipps. *J. Phys. Chem. B* 106, 996 (2002).
55. R. Memming. *Comprehensive Treatise of Electrochemistry*, Vol. 7 (Plenum, New York, 1983).
56. A. Schmidt, N. R. Armstrong, C. Goeltner, and K. Muellen. *J. Phys. Chem.* 98, 11780 (1994).
57. S. D. B. Pradhan. *Chem. Mater.* 20, 1209 (2008).
58. J. S. L. M. Jung, W. Song, Y. Kim, S. D. Lee, N. Kim, J. Park, M.-S. Choi, S. Katsumotos, H. Lee, and J. Kim. *Nano Lett.* 8, 3189 (2008).
59. J. L. I. Choi, G. Jo, K. Seo, N.-J. Choi, T. Lee, and H. Lee. *Appl. Phys. Exp.* 2, 015001 (2009).

CHAPTER 9

CHARACTERISTICS OF CHARGE TRANSPORT AND ELECTRIC CONDUCTION IN VIOLOGEN SELF-ASSEMBLED MONOLAYERS

Nam-Suk Lee[1], Dong-Yun Lee[2], Dong-Jin Qian[3], Young-Soo Kwon[2] and Hoon-Kyu Shin[4,*]

[1]*Department of Chemistry, Hanyang University, Seoul, Korea*
[2]*Department of Electrical Engineering, Dong-A University, Korea*
[3]*Department of Chemistry, Fudan University, China*
[4]*National Center for Nanomaterials Technology,
Pohang University of Science and Technology, Pohang, Korea*
E-mail: shinhk@postech.ac.kr

This study performed a study on the characteristics of self-assembled monolayers by fabricating monolayers using a self-assembly method in which viologen molecules were applied to an Au(111) substrate. The forming process of such monolayers was verified through the change in the resonance frequency of quartz crystal microbalance (QCM) and the morphology determined by using a scanning tunneling microscopy (STM). Based on the results of the measurement of the I-V characteristics of self-assembled viologen monolayers at a nanolevel using STM, it can be seen that the currents in its I-V characteristics according to the change in the length of viologen molecules were reduced from 292 nA to 40.6 nA in the region of the forward bias of +1.65 V. In addition, in the electric conduction mechanism at a high electric field region more than 1 V, it was verified that the Fowler-Nordheim tunneling dominated the mechanism.

1. Introduction

A technical level in the field of molecular electronics throughout the inside and outside of the country has been determined as a forming level in basic technologies for developing such molecular electronics and not been presented as a level that develops sophisticated molecular electronics.[1] Thus, it is necessary to perform some preceding tasks from the ultra-thin layers technology of organic materials to the observation, analysis, and operation technology of such ultra-thin layers in order to maximally use the functions presented in these materials. It is very important to fabricate functional organic molecules as ultra-thin layers and study on the electric characteristics of monolayers and single molecules for developing molecular electronics.[2] In particular, the viologen used in this study is a type of organic material and represents an advantage that shows a narrow electrochemical range in oxidation and reduction.[3] Also, it has been used as a medium of electron transport and as an electron acceptor.[4] In addition, it has been applied as various photochemical, photoelectrochemical, and solar energy conversion systems[5,6] based on its property that changes light energies to chemical energies and used as electrochromic displays using the change in its colors caused by their oxidation and reduction reactions.

In a recent study, in the results of the study performed by the research team of Professor David Schiffrin at the University of Liverpool, it presented that a STM tip, which is applied to gold nanoparticles and viologen molecules, plays a role in a molecular switch. The viologen as a functional material has been largely studied since it was first introduced by Michaelis in 1933.[7] That is, it can be expected that viologen derivatives are to be applied in lots of various fields, such as molecular electronics, EL elements, LC, nanoscaled composites, bioengineering element composites, drug deliveries, and other applications. Therefore, this study attempts to investigate the electrical characteristics of such viologen molecules and the possibility of the application of the molecules as molecular elements.

The viologen monolayers represented reversible oxidation and reduction reactions in a specific voltage range in which the results of these reactions were obtained by the electron transport caused by the carrier of ions in electrolytes. Based on the results, this study analyzed the charge transport characteristics using the essential characteristics of viologen molecules in qualitative and quantitative manners. Furthermore, this study analyzed the electric conduction characteristics and its mechanism by measuring the surface structures of viologen monolayers and changes in tunneling currents as a nanolevel using the STM, which has a resolution that is able to observe atoms.

2. Materials and Experiment

2.1. *Materials*

Since the viologen (1,1'-disubstituted-4,4'-bipyridinium) was first introduced by Michaelis in 1933, it has been largely studied in recent years and can be reversibly reacted as two different stages in its oxidation and reduction as illustrated in Fig. 1.[8]

Studies on the early viologen have conducted with respect to oxidation and reduction indicators, artificial photosynthesis, and solar energy storages in the field of biology,[9,10] and some types of viologen derivatives presented a certain activity as a weed-killer. In recent years, there are some applications that use electrode surface modification to viologen,[11,12] and studies on the conducting polymers that introduce viologen have also been conducted.[13,14] The reason that various studies on such viologen are conducted is that the positive ion radical obtained by using the reduction with a single electron satisfies a stable condition in most of solvent conditions that does not include oxygen and catalyzers. The neutral molecular viologen obtained by using the reduction with two electrons represents larger reduction potentials than that of the positive ion radical and that is able to reduce various

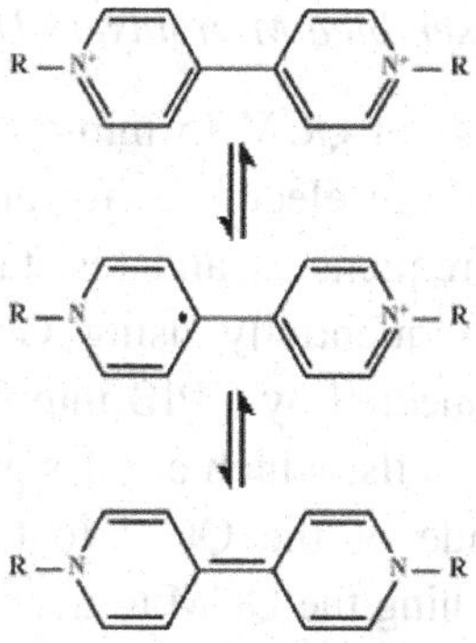

Fig. 1. Three redox state of bypyridinium moiety.

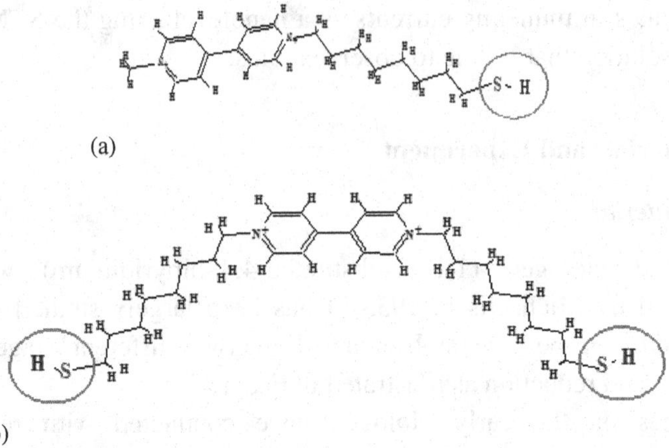

Fig. 2. Chemical structure of viologen used in this study (a) VC_8SH, (b) HSC_8VC_8SH.

materials. Also, the viologen shows relatively low oxidation and reduction voltages and has an advantage that represents almost perfect reversibility in oxidation and reduction reactions.

This study synthesized a thiol group that can be absorbed with Au at the end of viologen and adjusted the number of thiol groups in order to control the structure of molecules. Figure 2 shows the chemical structure of viologen with the adjusted thiol groups.

2.2. Experiment

2.2.1. Analysis of Self-Assembled Monolayers Using QCM and STM

This study used 9 MHz At-cut QCM (5 mm-diameter, Seiko Instrument Inc., Japan) with sputtered Au electrodes for analyzing a self-assembly process. The resonance frequencies and resistances of the QCM were measured and analyzed sequentially using QCA 922 (Seiko EG&G, Japan) and computers connected by GPIB interfaces, respectively. Also, this study applied specific cells, which are designed to perform reactions by contacting a single side of the QCM to the solution and used to analyze the QCM, by attaching the QCM to the cells.[15] Figure 3 illustrates the diagram of the cells used to analyze the QCM.

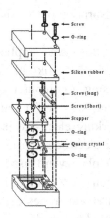

Fig. 3. Liquid cell assembly of quartz crystal for liquid analysis.

In addition, in the viologen derivatives that were self-assembled to an Au substrate for 24 h under the concentration of 1 mM, the self-assembled monolayers were verified using STM. The STM measurement was performed at room temperature, and the scanning in the measurement was processed by a constant current mode. The HOPG surface was observed before scanning the sample surface in order to verify the normal state of the probe. The reason that the HOPG was selected in this process is that it is possible to observe the surface as an atomic level of resolution not only for a high vacuum condition but also in air. The normal state of the probe was verified by investigating the surface of well-arranged graphite single crystal using a 10 nm × 10 nm scan size. For verifying the fact that the STM is accurately operated while it has an atomic level of resolution, this study observed the surfaces of the highly-oriented pyrolytic graphite (HOPG), which has a crystallized surface, and Si wafers. Because the HOPG is stable in air where carbon atoms form certain lattices, it has been largely used to calibrate x and y axes for measuring STM images.

2.2.2. Cyclic Voltammetry Test and Charge Transport Measurement

This study performed a cyclic voltammetry test using the QCM self-assembled by using viologen in which the self-assembly was not implemented by chemical adsorption but washing physically precipitated molecules using a solvent.

The cyclic voltammetry test was applied using the Potentiostat (PerkinElmer, USA) and QCA 922 (Seiko EG&G, Japan), and the results of the test was processed using a computer connected by using GPIB. The QCM in which viologen was self-assembled to a 0.196 cm^2 Au electrode was used as a working electrode in all tests. In addition, for configuring a cell with three electrodes, a 1.5 × 4.5 cm^2 Pt plate and an Ag/AgCl (MW-4130, BAS) plate were used as counter and reference electrodes, respectively. Then, the oxidation and reduction reactions were measured using this three electrodes cell. The test equipments and measuring process are presented in Fig. 4.

This study performed the cyclic voltammetry test in three different electrolytes, such as 0.1 M $NaClO_4$, Na_2SO_4, and Na_3PO_4, in order to verify the influence of the negative ion on the oxidation and reduction of the viologen in which the voltage ranges were configured by 0 to -1 V. All cyclic voltammetry tests were repeated by 10 times, and the electrolyte used in these tests was 18.3 MΩ ultra-pure water, which was applied after fully purifying it in Ar gas for 30 min.

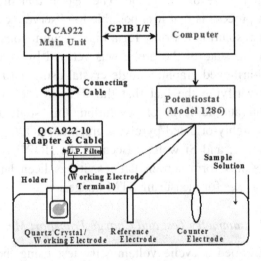

Fig. 4. Schematic diagram of electrochemical detection sensing system used.

2.2.3. Analysis of the Characteristics of Electric Conduction

STM is a microscope that can observe individual atoms in a real space. It has a simple principle in its operation in which a specific voltage level is applied to the space between a conductive sample and a metal probe and represents tunnel current as the space is approached up to 1 nm. The current corresponds to the change in the distance between two materials and is very sensitive to the change. The change is represented as an exponential function. As the change in the distance shows 0.1 nm, the tunnel current is to be changed by the scale of one order.[16]

For producing a video of atoms, the probe is inserted to the surface of the sample using a piezoelement while the tunnel current is maintained as a constant level. If the change in voltages that is applied to the piezoelement is detected and visualized, the surface structure of the sample can be observed as an atomic scale.

For observing the characteristics of electric conduction in self-assembled monolayers, the surface image and voltage-current characteristics were measured using the system by Veeco as shown in Fig. 5.

Because the tunnel current ranged by 1 ~ 10 nA can be detected and controlled as a higher level than 1%, the resolution for the vertical direction is to be determined by about 0.01 nm. Also, in the resolution within the face, the sharper end of the probe shows more higher

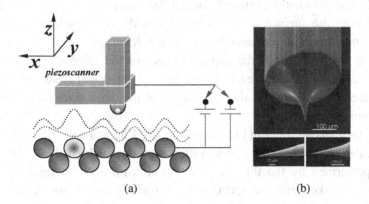

Fig. 5. Principle of STM system and the image of STM tip.

resolution levels than other cases. Based on a model calculation under the conditions that the radius of curvature and distance in the end of the metal probe are determined by R and Z, respectively, the resolution within the face can be determined as Eq. (1):

$$[0.2(R+Z)]^{1/2}. \qquad (1)$$

In the case of the end of the probe that has one atom, the values of R and resolution within the face are determined by 0.1 nm and 0.3 nm, respectively.

3. Analysis of Self-Assembled Monolayers Using STM

STM has a resolution that is able to observe atoms in air. In the case of the distance between a conductive sample and a metal probe that is approached by about 0.01 nm while a specific voltage is applied to this distance, there is a quantum physical tunneling effect that shows some currents in which electrons pass through an energy barrier by applying a proper voltage between these two sides even though these two conductors are separated. If the distance is changed by 0.1 nm, the tunnel current will be varied by a scale of one order.

The obtaining of the images of atoms or molecules using STM, it is possible to observe the surface structure of the sample up to the scale of atom or molecule based on the image processing using the detection of the change in tunneling currents during the injection through moving the probe onto the surface of the sample using the piezoelement while the tunneling current is maintained as a constant level.

This study observed the surface images of the viologen derivatives, which were self-assembled to the surface of an Au(111) substrate, and presented a possibility of the application of the derivatives as molecular elements using STM.

In this study, the Au(111) substrate used as a lower electrode was fabricated using a thermal evaporation system in which the fabrication was performed by the thermal evaporation with 100 nm thickness Au on the prebaked mica at 320°C for 2 h under maintaining the vacuum of 6.5×10^{-7} Torr. The viologen derivatives were self-assembled on the fabricated Au(111) substrate at the concentration of 1 mM for 24 h. The fabricated samples after completing the self-assembly were washed using ethanol and stored these samples in a desiccator after drying them using N_2 gas.

The STM measurement was performed at room temperature using the UHV-STM (UNISOKU, USM-1200) in which a probe made by using a Pt-Ir(80:20) alloy was used. Also, the voltages applied to the space between the STM tip and the sample were ranged from -2.5 V to +2.5 V. The scanning was applied by a constant current mode at the tunneling region of ~0.20 nA, and the change in tunneling currents was verified using STM/STS.

Figures 6(a)-6(d) show the data analyzed by using the ChemDraw for viologen. In addition, an ellipsometer was used to measure the thicknesses of viologen molecules, which were self-assembled on the Au(111) substrate, VC_8SH, $VC_{10}SH$, HSC_8VC_8SH, and $HSC_{10}VC_{10}SH$. In the results of the measurement using the ellipsometer, the heights of VC_8SH, $VC_{10}SH$, HSC_8VC_8SH, and $HSC_{10}VC_{10}SH$ were 0.33 nm, 0.74 nm, 0.84 nm, and 0.87 nm, respectively. Although there were some little differences between these values and the results analyzed by using STM, the results of these two cases showed similar figures within a specific error range because the heights obtained by using STM were determined as the convolution of the electric conductivity and physical height of samples and that caused some differences in heights.

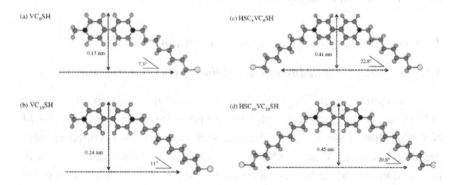

Fig. 6. Chemical Structure of viologen derivatives. (a) VC_8SH, (b) $VC_{10}SH$, (c) HSC_8VC_8SH, and (d) $HSC_{10}VC_{10}SH$.

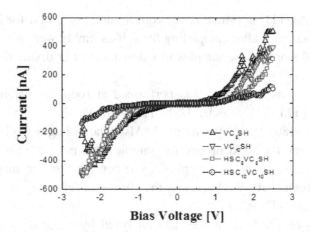

Fig. 7. I-V characteristic curves of viologen derivatives (VC$_8$SH, VC$_{10}$SH, HSC$_8$VC$_8$SH, and HSC$_{10}$VC$_{10}$SH).

The STM is able to measure the spectra measurement of tunnel spectrum (STS). It measures the change in tunnel currents by varying bias voltages while feedbacks are halted under the application of set-point bias voltages after moving the probe to the region that is to be measured. Figure 7 illustrates the I-V characteristics of the viologen monolayers measured by using STS. The I-V characteristics were observed by fixing the STM tip at a specific point after measuring four different viologen monolayers.

4. Cyclic Voltammetry and Charge Transport

4.1. *Oxidation and Reduction According to Changes in Electrolytes*

The previous results reported the relationship between the cyclic voltammetry curve and the peak current versus injection speed in a 0.1 M NaClO$_4$ solution for VC$_8$SH. The peaks in oxidation and reduction were presented at -0.45 V and -0.51 V, respectively, in which the currents in the oxidation and reduction showed the same levels. Also, it can be seen that the oxidation and reduction of the positive radicals, which were caused by the one electron reduction presented by V^{2+} ⇄ V$^+$, were reversibly generated as it is expected that the peak currents of the oxidation and reduction showed the same levels, and the relationship between the peak current and the injection speed was determined as a linear way.[17] Although the oxidation and reduction peaks were generated at a certain constant voltage level regardless of the type of positive ion due to the reaction of $V^{2+} \rightleftarrows V^+$ in viologen monolayers, the scales of peak currents decreased by the order of SO_4^{2-}, ClO_4^-, and PO_4^{3-}. It is considered that there are some differences in the amount of transported charges caused by the limited molar conductivity according to the mobility in each ion.

4.2. Characteristics of Interfacial Charge Transport Caused by the Change in Mass

The previous results reported the change in the resonance frequencies of QCM measured in the cyclic voltammetry test of VC_8SH and HSC_8VC_8SH. As results, the change in resonance frequencies occurred by oxidation and reduction reactions simultaneously and that can be presented as a change in mass using the Sauerbrey equation. According to the reduction reaction in two stages, the change in resonance frequencies increased by two stages, and the decrease in resonance frequencies occurred by two stages according to the oxidation. It shows a process that reassociates the ClO_4^- ion, which is associated to the N^+ ion that is an electrochemical active place of the viologen monolayers, is separated by its reduction according to the oxidation. The entire changes in the resonance frequencies of VC_8SH and HSC_8VC_8SH were 18.1 Hz and 8.4 Hz, respectively, and that can be converted by the change in masses as 19.36 ng and 8.98 ng, respectively. The numbers of associated and dissociated ions obtained from the mass of ClO_4^- ion for VC_8SH and HSC_8VC_8SH were 2.33×10^{13} and 1.08×10^{13}. It can be considered that the results were affected by the electrical double layer that formed the confronted layers of charges, which were located at both the electrode and solution in which the Au electrode where the viologen monolayers were unself-assembled represented electricity at the boundary between the metal surface and the solution, because the viologen monolayers were not self-assembled on the surface of the Au electrode of the QCM.[18]

In the case of the 0.1 M Na_2SO_4 electrolyte solution, the changes in the resonance frequencies were 19.3 Hz and 9.5 Hz and that can be converted by the change in masses as 20.65 ng and 19.0 ng. Also, the numbers of associated and dissociated Cl^- ions obtained from the change in masses were 2.49×10^{13} and 1.22×10^{13}, respectively. In the case of the 0.1 M Na_3PO_4 electrolyte solution, the changes in the resonance frequencies were 16.0 Hz and 7.1 Hz and that can be converted by the change in masses as 17.16 ng and 7.60 ng. Also, the numbers of associated and dissociated Cl^- ions obtained from the change in masses were 2.06×10^{13}, 9.1×10^{12}, respectively.

5. Electric Conductive Characteristics of Viologen Derivatives

In general, the charge insertion at the interface between metal and organic materials can be explained by the thermionic emission (dominated in a low bias level) and Fowler-Nordheim tunneling (dominated in a high bias level). That is, as a strong electric field is applied to the interface between metals and conductors, the barrier is decreased due to the Schottky effect and that shows a decrease in the width, which is forward to the same electric field in a Fermi level. Then, tunneling occurs at this position.[19] If the energy of the electron that is approached to the interface within metal shows the distribution of Fermi-Dirac, the tunnel current, I, can be obtained using Eqs. (2) and (3):

$$I = V^2 \exp\left(\frac{-x}{V}\right) \quad (2)$$

$$x = \frac{8\pi(2m)^{1/2}\Phi^{3/2}}{3qh} \quad (3)$$

where m is electron effective mass, and h is Plank constant. The I-V relationship in Eqs. (2) and (3) shows that it is possible to obtain a line, which has a negative value as $\log(I/V^2) \propto -1/V$. Figure 8 shows the Fowler-Nordheim plot obtained from the I-V characteristics shown in Fig. 7 for each viologen molecule, VC_8SH, $VC_{10}SH$, HSC_8VC_8SH, and

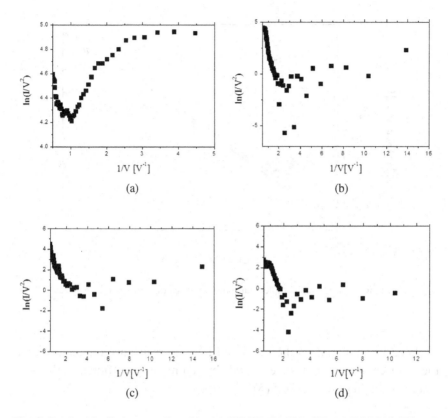

Fig. 8. Fowler-Nordheim tunneling plot. (a) VC_8SH, (b) $VC_{10}SH$, (c) HSC_8VC_8SH, and (d) $HSC_{10}VC_{10}SH$, respectively.

$HSC_{10}VC_{10}SH$. A linear relation that has a negative gradient in a high voltage region determined by ~1 V was obtained. Then, it is verified that the Fowler-Nordheim tunneling dominated the conduction mechanism in a high electric field region for the viologen molecules, VC_8SH, $VC_{10}SH$, HSC_8VC_8SH, and $HSC_{10}VC_{10}SH$.[20,21]

Figure 9 shows the resistance values for the lengths of molecules at 1.63 ~ 1.67 V positive bias regions in the viologen derivatives measured in self-assembled monolayers. In the results, it was considered that the resistance values increased according to the increase in the lengths of molecules and that caused decreases in tunneling currents.

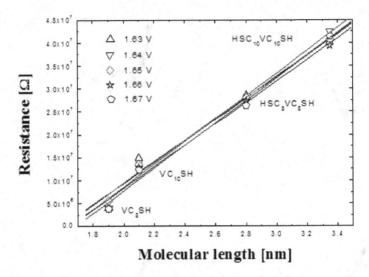

Fig. 9. Resistance versus molecular length plots of the viologen derivatives.

In addition, based on the measured I-V and resistance characteristics, the barrier height (φ) determined by damping coefficient (β) was calculated using Eqs. (4) and (5)[22]:

$$R = R_0 \exp(\beta d) \quad (4)$$

$$\beta = \frac{4\pi}{h}(2m\Phi)^{1/2}. \quad (5)$$

Figure 10 illustrates the differential conductance (dI/dV-V) for each molecule in VC_8SH, $VC_{10}SH$, HSC_8VC_8SH, and $HSC_{10}VC_{10}SH$. In the characteristics of dI/dV-V, there is symmetric or asymmetric characteristic according to the lengths of the molecules of viologen derivatives, i.e., the scale of the conductivity in molecules, because the dI/dV-V represents such symmetric or asymmetric characteristic according to the lengths of viologen molecules and positions of the STM tip.[23]

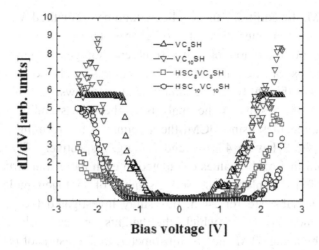

Fig. 10. dI/dV-V characteristics of the viologen derivatives.

In this study, it was possible to verify the energy gap of viologen molecules through the dI/dV-V characteristics. The differential conductance (dI/dV-V) corresponds to the state density in molecular systems. As the lock-in is detected for reducing the influence of noises, a stable spectrum will be obtained. Because the dI/dV-V represents an empty state that does not include electrons as a positive bias voltage and a filled state that includes electrons as a negative bias voltage, the difference in the edge region between positive and negative biases is to be determined as the energy gap. It was also possible to verify the decrease in tunneling currents due to the increase in tunnel barrier heights caused by the decrease in the hole injection barrier height and the increase in energy gaps according to the increase in the lengths of viologen molecules.

6. Conclusion

This study observed the change in the structure of viologen molecules in a nanoscale and analyzed the electric conduction characteristics. Also, this study analyzed the forming of the monolayers of viologen molecules using QCM and the morphology of self-assembled viologen monolayers using STM. In addition, the cyclic voltammetry method was used to analyze the characteristics of the charge transport caused by the oxidation and reduction of self-assembled viologen monolayers. The characteristics of the electric conduction was analyzed using STM. The results of this study can be summarized as follows:

First, in the results of the analysis of the self-assembly process of viologen molecules using QCM, the amounts of the forming of VC_8SH and HSC_8VC_8SH were 436 ng and 217 ng, respectively. These values agreed with that of the values calculated in each molecular structure and weight. After completing the self-assembly of viologen molecules on an Au(111) substrate for 24 h, this study investigated the forming of viologen monolayers in which the heights of the molecules were observed by using STM and an ellipsometer. In the results of the analysis of the morphology of the viologen molecules, which were formed on the Au(111) substrate using STM, the heights of viologen derivatives, VC_8SH, $VC_{10}SH$, and HSC_8VC_8SH were 0.13 nm, 0.24 nm, 0.41 nm, and 0.45 nm, respectively. The results showed the same as the values measured by using the ellipsometer.

Second, in the results of the simultaneous analyses of the I-V characteristics and changes in the resonance frequencies of QCM through cyclic voltammetry tests, it was verified that there were reversible oxidation and reduction reactions. Also, it was possible to verify the current characteristics according to the transport of electrolytes.

Third, in the results of the I-V characteristics of self-assembled viologen monolayers using STM at a nanolevel, it was investigated that the I-V characteristics according to the change in the lengths of viologen molecules represented a decrease in currents from 292 nA to 40.6 nA in the positive bias region of +1.65 V. In addition, in the conduction mechanism at a high electric field ranged by ~1 V, it was verified that the Fowler- Nordheim tunneling dominated the mechanism.

Acknowledgment

This research was supported by Strategic Technology under the Ministry of Knowledge Economy (MKE).

References

1. A. Aviram, M. A. Ratner, *Chem. Phys. Lett.*, Vol. 29, p. 277 (1974).
2. A. Ulman, "An Introduction Organic Films", Academic Press, Boston, MA, 1991.
3. M. C. Petty, "An Introduction to Molecular Electronics", Edward Arnold, London, 1996, p. 131.
4. A. Ulman, *Chem. Rev.*, Vol. 96, p. 1533 (1996).
5. R. C. Young, T. J. Meyer, D. G. Whitten, *J. Am. Chem. Soc.*, Vol. 97, p. 4781 (1975).
6. M. D. Ward, J. R. White, A. J. Bard, *J. Am. Chem. Soc.*, Vol. 105, p. 27 (1983).
7. L. Michaelis, E. S. Hill, *J. Am. Chem. Soc.*, Vol. 55, p. 1481 (1933).
8. L. A. Summers, "The Bipyridinium Herbicides", Academic Press, New York, 1980.
9. J. H. Fendler, *Phys. Chem.*, Vol. 89, p. 2730 (1985).
10. T. J. Meyer, *Acc. Chem. Res.*, Vol. 22, No. 5, p. 163 (1989).
11. G. S. Ostrom, D. A. Buttry, *J. Phys. Chem.*, Vol. 99, p. 15236 (1995).
12. S. Cosnier, C. Innocent, Y. Jouanneau, *Anal. Chem.*, Vol. 66, p. 3198 (1994).
13. G. Inzelt, "Charge Transport in Polymer-Modified Electrodes: In Electroanalytical Chemistry", Vol. 18, Marcel Dekker, New York, 1994.
14. R. H. Terrill, J. E. Hutchison, R. W. Murray, *J. Phys. Chem. B*, Vol. 101, p. 1535 (1997).
15. H. Muramatsu, J. M. Dick, E. Tamiya, I. Karube, *Anal. Chem.*, Vol. 59, p. 2760 (1987).
16. G. Binnig, C. F. Quate, Ch. Gerber, *Phys. Rev. Lett.*, Vol. 56, No. 9, p. 930 (1986).
17. J. Y. Ock, H. K. Shin, D. J. Qian, J. Miyake, Y. S. Kwon, *Jpn. J. Appl. Phys.*, Vol. 43, No. 4B, p. 2376 (2004).
18. D. Y. Lee, A. K. M. Kafi, S. H. Park, Y. S. Kwon, *J. Nanosci. Nanotechnol.*, Vol. 6, p. 3657 (2006).
19. M. A. Reed, J. Chen, A. M. Rawlett, D. W. Price, J. M. Tour, *Appl. Phys. Lett.*, Vol. 78, p. 3735 (2001).
20. J. M. Beebe, B. S. Kim, J. W. Gadzuk, C. D. Frisbie, J. G. Kushmerick, *Phys. Rev. Lett.*, Vol. 97, p. 026801 (2006).
21. Y. Noguchi, T. Nagase, R. Ueda, T. Kamikado, T. Kubota, S. Mashiko, *Jpn. J. Appl. Phys.*, Vol. 46, No. 4B, p. 2683 (2007).
22. T. Ishida, W. Mizutani, Y. Aya, H. Ogiso, S. Sasaki, H. Tokumoto, *J. Phys. Chem. B*, Vol. 106, p. 5886 (2002).
23. C. Arena, B. Kleinsorge, J. Robertson, W. I. Milne, M. E. Welland, *J. Appl. Phys.*, Vol. 85, p. 1609 (1999).

CHAPTER 10

TIME-AVERAGED DEUTERIUM NMR STUDIES OF THE DYNAMIC PROPERTIES FOR A LOW MOLAR MASS NEMATIC

Akihiko Sugimura[1] and Geoffrey R. Luckhurst[2]

[1]*Department of Information Systems Engineering, Osaka Sangyo University, 3-1-1 Nakagaito, Daito, Osaka 574-8530, Japan*
E-mail: sugimura@ise.osaka-sandai.ac.jp
[2]*School of Chemistry, University of Southampton, Highfield, Southampton, SO17 1BJ, United Kingdom*
E-mail: G.R.Luckhurst@soton.ac.uk

There have been many investigations of the alignment of nematic liquid crystals by either a magnetic and/or an electric field. The basic features of these hydrodynamic processes have been characterized for the systems in their equilibrium and non-equilibrium states. To complement the experiments theoretical models, based on continuum theory, have been developed which successfully describe the static and dynamic phenomena. This technique has proved to be especially important for the investigation of liquid crystals. In this chapter, we describe some of our studies of the field-induced oscillatory behaviour of the nematic director for low molar mass nematics using deuterium NMR but with a sinusoidal electric as well as a static magnetic field to align the director. The director orientations have been observed from the time-resolved NMR experiments but we have found that they are also available, more directly, from the time-averaged NMR spectrum. The maximum and minimum angles made by the director with the magnetic field were determined, as a function of frequency, from the NMR spectrum averaged over many thousand cycles of the oscillations. These NMR experiments can give dynamic as well as static information concerning the nematic phase and could be used to provide a new route to certain material properties of the nematic.

1. Introduction

Nuclear magnetic resonance (NMR) spectroscopy is widely used in the study of liquid crystals. Deuterium NMR has proved to be especially important for the investigation of liquid crystals because the spectra of specifically or fully deuteriated materials are rather simple compared to the corresponding proton NMR spectra.[1-5] The quadrupolar splitting for deuterons observed in the liquid crystal phase is related, primarily, to the second rank orientational order parameter for the C-D bond direction and so deuterium NMR has been widely used for the study of the orientational order of liquid crystals as well as their phase transitions.[2] The quadrupolar splitting is also determined by the angle between the director and the magnetic field. In consequence, deuterium NMR spectroscopy has proved to be a powerful method with which to investigate the director orientation and its distribution, as well as the director dynamics in nematic[6-25] and smectic liquid crystals.[26-29] In more recent years deuterium NMR spectroscopy, combined with continuum theory, has been applied successfully to investigate the static director distribution in thin nematic liquid crystal cells, with different film thicknesses and different surface anchoring strengths, subject to both magnetic and electric fields.[11-13] Deuterium NMR spectroscopy has also been employed to investigate the dynamic director alignment process in a thin nematic film following the application or removal of an electric field.[14-25] This technique has the added advantage that the presence of the magnetic field of the spectrometer ensures that during the electric field-induced alignment the director rotates as a monodomain,[13] which facilitates the analysis of the results, except for the case when the angle made by the magnetic field with electric field is close to a right angle.

The rotational viscosity coefficient, γ_1, for the director is one of the central quantities in determining the response times of liquid crystal display devices based on nematics.[30] A particularly apposite method with which to determine γ_1 is to generate a non-equilibrium state by applying a field at some orientation to the director and then to monitor this as the system returns to equilibrium.[31] For example, for the alignment of a nematic with a positive diamagnetic anisotropy, $\Delta\tilde{\chi}$, by a magnetic field, the director orientation, θ, with respect to the field, is predicted[32] to vary with time according to

$$\tan\theta(t) = \tan\theta_0 \exp\left(-\frac{t}{\tau_M}\right), \qquad (1)$$

in the absence of other external constraints. Here, θ_0 is the angle made by the director with the field at time zero and the magnetic relaxation time is given by

$$\tau_M = \frac{\gamma_1 \mu_0}{\Delta\tilde{\chi} B^2}, \qquad (2)$$

where μ_0 is the magnetic constant and B is the magnetic flux density. One powerful method by which the director orientation can be monitored is deuterium NMR spectroscopy and in experiments using this technique the magnetic field causing the director alignment is provided by the spectrometer.[9] Such experiments have been extended by using an electric field both to create the original non-equilibrium state of the system and to drive the director alignment.[14] The dynamics of the changes in the director orientation are given, within the framework of the Leslie-Ericksen hydrodynamic theory, by the torque-balance equation[14]

$$\gamma_1 \frac{d\theta}{dt} = -\frac{\Delta\tilde{\chi}}{2\mu_0} B^2 \sin 2\theta + \frac{\varepsilon_0 \Delta\varepsilon}{2} E^2 \sin 2(\alpha - \theta). \qquad (3)$$

Here α is the angle between the magnetic and electric fields, $\Delta\tilde{\varepsilon}$ is the dielectric anisotropy and ε_0 is the vacuum permittivity. This equation ignores the elastic terms which is justified because in our experiments the director remains uniformly aligned; in addition, the inertial term is ignored. The director orientation varies according to Eq. (1) when the electric field is switched off; when the electric field is switched on and provided $\Delta\tilde{\varepsilon}$ is positive the orientation now changes according to

$$\tan(\theta(t) - \theta_\infty) = \tan(\theta_0 - \theta_\infty)\exp\left(-\frac{t}{\tau_C}\right). \qquad (4)$$

Here, θ_∞ is the limiting value of the director orientation as t goes to infinity and the composite relaxation time, τ_C, is given by

$$\tau_C = (\tau_M^{-2} + 2\tau_M^{-1}\tau_E^{-1}\cos 2\alpha + \tau_E^{-2})^{-1/2}, \qquad (5)$$

where τ_E is the relaxation time for the pure electric field-induced rotation of the director

$$\tau_E = \frac{\gamma_1}{\varepsilon_0 \Delta \tilde{\varepsilon} E^2}. \tag{6}$$

The experiments combining both electric and magnetic fields have the advantage that the ratio of anisotropic susceptibilities, $\Delta\tilde{\varepsilon}/\Delta\tilde{\chi}$, is readily measured and so if $\Delta\tilde{\varepsilon}$ is available independently then the rotational viscosity can be determined from the relaxation times.[5,14]

In the experimental investigations it is usual to employ a high frequency electric field, typically several kHz, to ensure that disruption of the director orientation by conductive motion of ions is eliminated.[14] However, in the torque-balance equation (see Eq. (3)), the electric field is taken to be constant. Clearly the periodic variation in the square of the electric field will influence the director dynamics, thus as the electric field increases so the relaxation time, given by Eq. (5), will decrease caused by a reduction in τ_E and as the field decreases so τ_C will increase caused by an increase in τ_E. In addition, the direction of the director alignment will be reversed when the electric field starts to decrease; however, the precise moment when this occurs will depend on the frequency of the electric field in comparison with the composite relaxation rate, τ_C^{-1}. The importance of these variations to the validity of the analysis, based on the assumption that the magnitude of the electric field is constant, will also depend on the frequency of the electric field and the composite relaxation rate for the director. Here, we consider this question both theoretically and experimentally. As we shall see such considerations have lead us to a novel variation in the NMR method used to study field-induced director dynamics and hence determine certain nematic properties.

The layout of this chapter is the following. In the next section we give the theoretical background to the dynamic alignment of the nematic director subject to a static magnetic field and a sinusoidal electric field; from this we calculate the time dependence of the director orientation, in particular as a function of frequency. In Sec. 3 we describe the NMR experiments that have been performed. The results are presented in the following section where their significance is discussed and comparison is made with an analogous experiment in which the nematic is spun about an axis orthogonal to the magnetic field. The outcomes of this study are given in Sec. 5 together with our conclusions.

2. Theory

With a sinusoidal electric field the E^2 term in the torque-balance equation must be replaced by

$$E^2 = E_0^2 \sin^2 2\pi ft, \tag{7}$$

where f is the frequency of the sinusoidal field. In addition, the dielectric anisotropy will also depend on f but for the frequency range and low molar mass materials with which we are concerned it is a good approximation to ignore this dependence.[33] The torque-balance equation can then be written in terms of generalised or scaled coordinates as

$$\frac{d\theta}{dt^*} = -\frac{1}{2}\sin 2\theta + \frac{\rho}{2}\sin^2 2\pi f^* t^* \sin 2(\alpha - \theta). \tag{8}$$

The generalised time coordinate is given by

$$t^* = \frac{t}{\tau_M}, \tag{9}$$

where the magnetic relaxation time, τ_M, is given by Eq. (2), the generalised frequency is

$$f^* = \frac{f}{\tau_M^{-1}}, \tag{10}$$

and the parameter ρ is the ratio of electric to magnetic anisotropic energies

$$\rho = \mu_0 \varepsilon_0 \left(\frac{\Delta\tilde{\varepsilon} E_0^2}{\Delta\tilde{\chi} B^2} \right). \tag{11}$$

It depends, therefore, on the material property, $\Delta\tilde{\varepsilon}/\Delta\tilde{\chi}$, and the experimental parameter, E_0/B. For a given value of α, the angle between the two fields, the time dependence of the director orientation will be determined by ρ and the scaled frequency, f^*. To obtain this

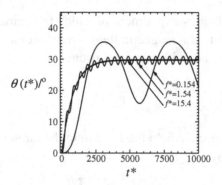

Fig. 1. The scaled time dependence of the angle made by the director with the magnetic field, $\theta(t^*)$, calculated with $\alpha = 44.7°$ and $\rho = 3.47$ for f^* equal to 15.4, 1.54 and 0.154.[20]

dependence we have solved the generalised torque-balance equation numerically using a finite difference method with the time step chosen according to the procedure described by Killian and Hess.[34]

The results of these calculations are shown in Fig. 1 for α equal to 44.7°, the angle used in the majority of our experiments, and which removes the degeneracy in the alignment pathway for the director found for $\alpha = 90°$,[3,23] that is the director moves uniformly and the angle made with the aligning field decreases. The ratio ρ was given the value of 3.47 which is typical of 4-pentyl-4'-cyanobiphenyl (5CB) and the field strengths used in our experiments. The director was taken to be parallel to the magnetic field prior to the application of the electric field. The highest scaled frequency used in the calculations was 15.4 and to place this in context we note that when τ_M is of the order of ms the frequency corresponding to this value of f^* is about 15 kHz. As we can see, for this value of f^* the angle θ increases from zero and then levels off, apart from an extremely weak, barely discernible, high frequency oscillation superimposed on this curve. The time dependence of the director orientation is indistinguishable from that calculated from Eq. (4) with θ_0 set equal to zero and the limiting director orientation, θ_∞, given, in terms of the variable, ρ, as[14]

$$\cos 2\theta_\infty = \frac{2 + \rho \cos 2\alpha}{(4 + 4\rho \cos 2\alpha + \rho^2)^{1/2}}. \quad (12)$$

However, when f^* is reduced by a factor of ten to 1.54 the magnitude of the oscillations grows significantly and these are now clearly visible. We find that the frequency of the oscillations is equal to that of the square of the electric field although the director orientation exhibits a phase lag with the maxima and minima being reached after the square of the electric field has passed through its respective maxima and minima. Of more interest is the fact that the director is predicted to rotate through a greater angle than that expected at high frequencies, when the director experiences the root mean square field. This occurs because the slower rate of change in the electric field allows the director a greater opportunity to respond to the higher electric field, E_0. Reduction of the scaled frequency by a further factor of ten to 0.154 makes such effects clearly apparent, as we can see in Fig. 1. Thus the frequency of the oscillations has manifestly decreased and the maximum director orientation has increased significantly although the reduction in the minimum orientation is even larger. What is happening is that as the rate of change in E^2 is slowed so the director is able to follow the electric field more closely and will approach the limiting value of θ_{max} given by

$$\cos 2\theta^0_{max} = \frac{1+\rho\cos 2\alpha}{(1+2\rho\cos 2\alpha + \rho^2)^{1/2}}, \qquad (13)$$

where the superscript zero indicates the maximum value at zero frequency, by analogy the minimum angle at zero frequency should be denoted by θ^0_{min} and here this is zero. At the other extreme when the electric field vanishes the director should be aligned parallel to the magnetic field corresponding to θ of zero. This clearly does not occur because the appropriate field-induced relaxation time is too long to enable the electric field to be followed instantaneously. We would expect, therefore, that as the frequency is reduced still further so the minimum angle made by the director with the magnetic field will tend to zero whereas there will be little change in the value of the maximum angle which is already close to its limiting value. This non-symmetric behaviour is certainly what is observed, as the results in Fig. 2 for the time dependence of $\theta(t^*)$ predicted for f^* after it has been reduced by a further factor of ten to 0.0154 clearly show. We see that θ_{max} has increased slightly and has reached its limiting value, θ^0_{max}, of 36.5°; in

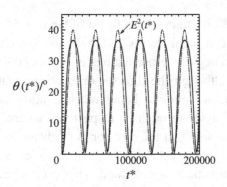

Fig. 2. The scaled time dependence of $\theta(t^*)$ calculated with $\alpha = 44.7°$, $\rho = 3.47$ and $f^* = 0.0154$ for long scaled times, t^*; also shown is the time dependence of the square of the electric field.[20]

marked contrast θ_{min} has decreased significantly to 1°, which is essentially the limiting value, θ^0_{min}, of zero. The results shown in this figure, with its longer time range, predict that the oscillations in the director orientation continue unattenuated indefinitely. We can also see that at this low scaled frequency the director is able to follow almost exactly the square of the electric field as it increases and decreases. That is the phase lag between the director and the field has almost vanished although it is more apparent at the minima than the maxima in the director orientation.

A particularly intriguing feature that has emerged from these calculations is that after the system has reached its quasi-equilibrium or stationary state, that is when θ_{max} and θ_{min} no longer vary with time, the difference between them changes with the scaled frequency. Since this frequency depends on the magnetic relaxation time the values of θ_{max} and θ_{min} also contain information concerning the field-induced director dynamics. The dependence of the maximum and minimum angles made by the director with the magnetic field on the scaled frequency is shown in Fig. 3. As is apparent, at high frequencies the two angles are identical ($\equiv \theta_\infty$) and given by Eq. (12) but as the frequency is lowered a difference between them gradually emerges, grows rapidly and then reaches its maximum limiting value, $\theta^0_{max} - \theta^0_{min}$, as f^* tends to zero. In between these two limiting values for the frequency the value of f^*,

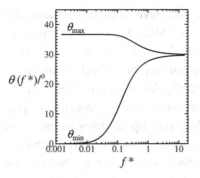

Fig. 3. The scaled frequency dependence, shown on a logarithmic scale, of the maximum, θ_{max}, and the minimum, θ_{min}, angle made by the director with the magnetic field predicted for a system with $\alpha = 44.7°$ and $\rho = 3.47$.[20]

corresponding to a particular θ_{max} and θ_{min}, will be determined by the director dynamics as well as by ρ and α. Similarly, at low frequencies the time dependence of the director orientation does not provide information about the dynamic properties of the nematic since the variation in $\theta(t^*)$ simply follows that in $E^2(t^*)$. At high frequencies θ_{max} and θ_{min} are identical and so the director orientation is necessarily independent of time. Strictly it is the effective field, formed by combining E and B, rather than the electric field alone, that controls the director alignment; we shall return to this aspect of the director oscillations later.

In the following section we describe experiments designed to test some of the predictions resulting from our numerical analysis of the torque-balance equation including the time dependence of the electric field. Here the technique used to determine the director orientation is deuterium NMR spectroscopy; this has the significant advantage of being able to monitor the deviation of the director from a state of uniform alignment across the nematic slab.[13] This enables us to check one of the basic assumptions of our analysis, namely that the nematic is aligned as a monodomain.

3. The Experiments

The nematogen chosen for our studies was 4-pentyl-4'-cyanobiphenyl (5CB) because its physical properties and behaviour, in particular the field-induced alignment, have been well-characterized.[14,31,35] To obtain the simplest deuterium NMR spectrum the sample was deuteriated in the α-position of the pentyl chain.[13] As we shall see the deuterium NMR spectrum of a monodomain of 5CB-d_2 contains a single quadrupolar doublet because the dipolar coupling between the two deuterons is negligible in comparison with the linewidth. In principle, this could be reduced by decoupling the dipolar interactions to neighbouring protons which broaden the lines inhomogeneously.[36] The sample of 5CB-d_2 was contained in a cell 56.4 μm thick composed of two glass slides each coated with indium oxide to form the electrodes, although some measurements were also made with a cell 97 μm thick. The surfaces of the electrodes were not treated in any way and so the surface anchoring strength was of the order of 10^{-7} Jm^{-2}, corresponding to weak anchoring. The sinusoidal electric field was produced using a function generator, Wave Factory WF1943, NF Electronic Instruments and a high power amplifier, model 4005 NF Electronic Instruments.

The deuterium NMR spectra were measured using a JEOL Lambda 300 spectrometer which has a magnetic flux density of 7.05T. A quadrupolar echo sequence, sketched in Fig. 4, was used to record the free induction decay (FID) which was then Fourier transformed to obtain the conventional frequency domain spectrum; also shown in this figure is the sinusoidal electric field. This field was applied for a total time, t_a, corresponding to an integer number of cycles and finishing after the FID acquisition was complete, the dead time of the receiver coil, t_D, was about 10 μs, the time between pulses was 38 μs, each pulse was 7.5 μs long, the acquisition time for the FID was 2.56 ms and finally the delay time, PD, was given values in the range 50 to 200 ms with the longer times being used at the lower frequencies. This delay allows the director to be realigned parallel to the magnetic field following the removal of the electric field and, incidentally, allows the deuterium spins to relax back to their equilibrium distribution. The experiment was performed by activating a trigger to switch on the electric field, at the same time the trigger initiates the sequence which leads to the measurement of a single FID. The time at which the measurement takes place, following the application of the field, is controlled by the interval PI3 (see Fig. 4) although the total time to the acquisition of the FID is the sum of PI3, t_D,

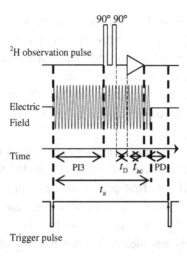

Fig. 4. The quadrupolar-echo pulse sequence used to obtain the FID following the application of a sinusoidal electric field. The horizontal triangular symbol in the ^{2}H observation pulse sequence denotes the data acquisition. The time during which the electric field is applied is t_a, the dead time for the receiver coil is t_D and the time taken to acquire the FID is t_{ac}. PI3 is the time between the trigger pulse and the start of the pulse sequence and PD is a delay to allow the director to be aligned parallel to the magnetic field and the nuclear spins to relax.[20]

and the time between pulses. The second trigger shown in the figure initiates another complete cycle allowing the FIDs to be averaged before Fourier transformation to give the spectrum. To obtain spectra with a good signal-to-noise ratio typically 10,000 FIDs were averaged for the 56.4 μm thick cell with necessarily fewer for the thicker cell.

As we have seen in the previous section, for certain frequencies the maximum and minimum director orientations also contain information about the field-induced director dynamics. It is clearly possible to determine these limiting director orientations from the spectra measured in a time-resolved experiment but there is a simpler method. This is to average the NMR spectra recorded over many cycles of the electric field for the resultant sum of spectra is found, as we shall see in the following section, to be dominated by peaks associated with the extreme director orientations. To obtain such a time-averaged spectrum the sinusoidal electric field is applied continuously, the FIDs are measured at regular intervals and summed prior to Fourier transformation. The time between

FIDs is equal to the sum of PD, the acquisition time, the pulse lengths and the delay between pulses; the important feature of this time is that it is chosen so that the acquisitions do not occur at simple multiples of the periodicity of the director oscillations. This is to ensure that the large number of FIDS summed, typically 40,000 for the thicker cell, is representative of all orientations reached by the director as it is rotated by the oscillating electric field.

In the experiments the sample cell was positioned in the probe head so that the electric field made an angle of about 45° with the magnetic field. To do this the cell was first placed in the goniometer of the probe head so that the angle between the cell normal and the vertical direction, that is B, is approximately 45° as judged by eye. The cell was then oriented more accurately by measuring the quadrupolar splitting when the nematic sample was subject to an applied voltage of $60V_{RMS}$ which aligns the director essentially parallel to the electric field. The angle, θ, made by the director with the magnetic field can be determined from the quadrupolar splitting, $\Delta\tilde{v}(\theta)$, by using

$$\Delta\tilde{v}(\theta) = \Delta\tilde{v}_0 P_2(\cos\theta), \tag{14}$$

where $\Delta\tilde{v}_0$ is the splitting when the director is parallel to the magnetic field and $P_2(\cos\theta)$ is the second Legendre function. At the desired angle of 45° the quadrupolar splitting should be one quarter of that in zero electric field; the accuracy (± 0.2 kHz) with which the quadrupolar splitting can be determined means that the angle α can be measured to $\pm 0.15°$. In our experiments α was found to be 44.7° from the voltage dependence of the quadrupolar splitting. Most of the measurements, which we describe here, were made at a temperature of 295 K although some were also made at the slightly lower temperature of 288 K, that is in the supercooled nematic. The applied voltage was 70.7 V for the thinner cell; this corresponds to the field strength, E_0, of 1.26 MVm^{-1} and for the thicker cell the potential was increased to give E_0 of 1.41 MVm^{-1}.

4. Results and Discussion

We have already reported results for the time dependence of the director orientation obtained using a sinusoidal electric field with a frequency of 10 kHz and these do not exhibit any oscillations in $\theta(t)$.[14,19] This behaviour is in accord with our calculations since at this frequency the magnetic relaxation time, τ_M, for 5CB[19] is 1.54 ms at 295 K which gives a scaled frequency of 15.4. The other parameter controlling the oscillatory behaviour of the director orientation is the ratio, ρ; this is calculated from $\Delta\tilde{\varepsilon}/\Delta\tilde{\chi}$ which for 5CB-d_2 is 0.97×10^7 at 295 K[19] and from the field strengths, 7.05 T and 1.26 MVm^{-1}, used in our experiments ρ is found to be 3.45. The calculations using f^* of 15.4 and ρ equal to 3.47, close to that relevant for our experiments, show only very weak high frequency oscillations in $\theta(t^*)$ (see Fig. 1). To observe such oscillations experimentally we need to reduce the frequency of the applied electric field and based on our calculations, described in the previous section, they should certainly be apparent when f^* is 0.154, corresponding to a frequency of 100 Hz. We have recorded the NMR spectra of 5CB-d_2 for this frequency over a time span of 10 ms which corresponds to one cycle of the electric field. A selection of these spectra is shown in Fig. 5. Initially ($t = 0.1$ ms) the director is parallel to the magnetic field and the spectrum consists of a simple quadrupolar doublet with a splitting of 53.2 kHz, corresponding to $\Delta\tilde{\nu}_0$. After 1 ms the quadrupolar splitting has decreased, but only slightly, as the director is moved from being parallel to the magnetic field. In addition to the main spectral feature, that is the quadrupolar doublet, some small spectral oscillations are apparent just inside the doublet. These result from the variation in the director orientation during the time over which the FID is acquired[37] and should not be confused with the oscillations in the director orientation. The weakness of the spectral oscillations means that the angle made by the director with the magnetic field can still be determined from the quadrupolar splitting obtained from the spectrum via Eq. (14). As the time increases so the quadrupolar splitting is seen to decrease corresponding to an increase in the angle made by the director with the magnetic field; the splitting reaches a minimum after about 3 ms. It is of interest that at 3 ms the oscillations have vanished because at the maximum in $\theta(t)$ the director orientation does not change so rapidly with time. As a consequence the quadrupolar splitting is essentially constant during the time taken to measure the FID. Then at 4 ms the quadrupolar splitting is seen to have increased slightly corresponding to a decrease in

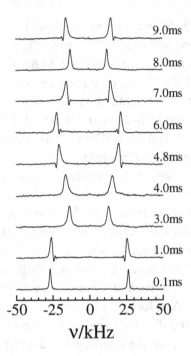

Fig. 5. The time dependence of the deuterium NMR spectrum of 5CB-d_2 recorded at 295 K with a sinusoidal electric field of 100 Hz, for different values of the time following the application of the electric field of strength 1.26 MVm^{-1}.[20]

the angle made by the director with the magnetic field. At 4.8 ms the splitting has increased still further as the reduced electric field results in the director being aligned more by the magnetic field. The beginning of a weak oscillation is apparent in the spectrum but now on the outside of the quadrupolar doublet which shows that the director motion is to increase the quadrupolar splitting.[37] After 6 ms the splitting in the quadrupolar doublet has barely changed since, after 4.8 ms, the director orientation has passed through a minimum and is now increasing again. This change in the direction of the director movement is immediately apparent because the weak spectral oscillations now occur on the inside of the quadrupolar doublet whereas at 4.8 ms they were on the outside.[37] At 7 ms the quadrupolar splitting has decreased corresponding to the continuing increase in the angle made by the director with the magnetic field. The oscillations still occur on the inside of the quadrupolar doublet but are not so pronounced, showing that the quadrupolar splitting continues to increase with time but not so rapidly. After 8 ms no oscillations are apparent and the quadrupolar splitting has continued to decrease; this suggests that the splitting is close to its minimum value. In the final spectrum, recorded at 9 ms, the quadrupolar splitting has increased as the director is aligned by the magnetic field. The weak oscillations occur on the outside of the doublet which indicates that the director orientation is continuing to decrease as the minimum in $\theta(t)$ is approached.

The results for the time dependence of the angle made by the director with the magnetic field extracted from the spectra shown in Fig. 5 and many others obtained in this experiment are given in Fig. 6. It is immediately apparent that following the application of the sinusoidal electric field the angle, $\theta(t)$, increases, passes through a maximum, decreases and passes through a minimum before increasing to a maximum again. It is also apparent that the maximum angle has grown slightly with time, in agreement with our calculations (see Fig. 1). To obtain a more detailed comparison with theory we have calculated the time dependence of the director orientation using parameters previously

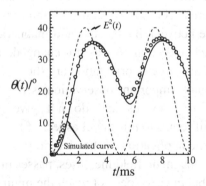

Fig. 6. The time dependence of the angle, $\theta(t)$, made by the director with the magnetic field measured for 5CB-d_2 at 295 K for a 100 Hz sinusoidal electric field with E_0 of 1.26 MVm^{-1}. The time dependence of the square of the electric field is shown as the dashed line. The solid line shows the predicted time dependence using parameters $\rho = 3.45$ and $\tau_M = 1.54$ ms previously determined for 5CB and with the angle, α, between the electric and magnetic fields of 44.7°.[20]

determined for 5CB-d_2 at 295 K, namely τ_M equal to 1.54 ms[19] and ρ equal to 3.45. In the calculations the angle between the electric and magnetic fields was set equal to the experimental value of 44.7°. The overall agreement found with this parameter set is very good, as is clearly apparent from the results shown in Fig. 5. At the initial stage there are some small deviations and these are caused by the large error in estimating the angle from the quadrupolar splitting when $\theta(t)$ is close to zero (see Eq. (14)). However, this cannot account for the more apparent but still small differences found at longer times, especially at the start of the second oscillation where the angles are relatively large in comparison with theory. We have tried to improve the agreement by varying the parameters but have found that those determined in previous experiments[19] provide the best fit. We are unaware as to the source of this modest disagreement between experiment and the hydrodynamic theory. We also show in Fig. 6 the time variation of the square of the electric field and this illustrates quite clearly that the director is unable to remain in phase with the electric field. It is under such conditions that the measurements are able to provide information about the field-induced director dynamics.

In order to investigate the oscillations in the director orientation for significantly longer times, we have studied the same sample cell but under slightly different conditions. Thus the temperature was lower at 288 K and the angle between the electric and magnetic fields was larger at 55.0°; the electric field strength was not changed. The results for the time dependence of the director orientation are shown in Fig. 7. The most obvious and potentially important observation is that the oscillations in $\theta(t)$ extend over many cycles and do not give any indication of attenuation. We shall return to this novel aspect of the director behaviour shortly. Following the application of the electric field the angle made by the director with the magnetic field increases, passes through a maximum and then decreases but clearly does not reach the minimum value parallel to the magnetic field. The maximum angle reached by the director grows during the first few cycles as expected from theory (see Fig. 1) and then levels off corresponding to the stationary state of the system. Similarly the minimum values of $\theta(t)$ also increase slightly after the first cycle but then are constant. The values of θ_{max} and θ_{min} are larger than for the

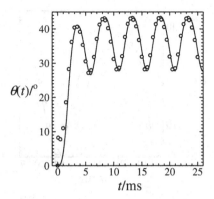

Fig. 7. The time dependence of the angle, $\theta(t)$, made by the director with the magnetic field for 5CB-d_2 at 288 K subject to a sinusoidal electric field with a frequency of 100 Hz measured over many cycles. The solid line shows the predicted variation in $\theta(t)$ calculated with $\rho = 3.45$ and $\tau_M = 2.60$ ms for the angle between the fields equal to 55°.[20]

sample at 295 K (see Fig. 6) but this difference is not caused by the change in temperature but rather the increase in the angle between the two fields from 44.7° to 55.0°. This increase in α causes the limiting director orientation, θ_∞, about which the director fluctuates in the stationary state to increase (see Eq. (12)). We have predicted the time dependence of the oscillations in $\theta(t)$ using the hydrodynamic theory and in these calculations the parameter ρ was assigned the same value as for the higher temperature since $\Delta\tilde{\varepsilon}/\Delta\tilde{\chi}$ is found to be independent of temperature.[19] The magnetic relaxation time has not been determined at 288 K and so was treated as a fitting parameter; the value giving the best fit to the results in Fig. 7 was 2.60 ms. This result is consistent with earlier measurements of τ_M for 5CB[19] as well as with measurements of γ_1[38] both made at higher temperatures. The agreement between the predicted oscillations and those found experimentally is clearly very good, again providing strong support for the validity of the analysis, the choice of parameters and the experiment.

The director oscillations are predicted to increase in magnitude with decreasing frequency (see Figs. 1 and 2) and to test this aspect of the theory we have made time-resolved NMR measurements at 10 Hz using the same 5CB-d_2 cell and under the original conditions, that is E_0 equal to

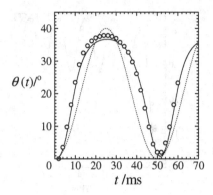

Fig. 8. The time dependence of the director orientation with respect to the magnetic field, $\theta(t)$, measured for 5CB-d_2 at 295 K following the application of a 10 Hz sinusoidal electric field. The dashed line shows the time dependence of the angle between the effective field and the magnetic field direction. The solid line shows the predicted variation with $\rho = 3.45$ and $\tau_M = 1.54$ ms obtained from previous measurements and with α equal to 44.7°.[20]

1.26 MVm^{-1}, α of 44.7° and at 295 K. The quality of the spectra was found to differ slightly from those shown in Fig. 5 for 100 Hz because the minor spectral oscillations are reduced in intensity. This occurs because at the lower frequency for the electric field the director moves more slowly and so its orientation does not change during the acquisition of the FID.[37] The time dependence of the director orientation until just beyond the first cycle is shown in Fig. 8. The striking feature of these results is the large angle made by the director at the maximum and the fact that at the minimum the director returns to be essentially parallel to the magnetic field. In addition, the form of the oscillations is not symmetric being significantly broader at the maximum than at the minimum, this is in qualitative agreement with the theoretical predictions shown in Fig. 2. It can be understood in the following way. At such low frequencies the director moves faster than the electric field and so the director should remain parallel not to the field itself but to the effective field resulting from the combination of the electric and magnetic fields. The angle, θ_∞, made by this effective field with the magnetic field direction is given by Eq. (12) but now with ρ multiplied by a time dependence originating from that of the electric field (see Eq. (7));

this gives

$$\cos 2\theta_\infty = \frac{1+2\rho\sin^2(2\pi ft)\cos 2\alpha}{(1+4\rho\sin^2(2\pi ft)\cos 2\alpha + 4\sin^4(2\pi ft)\rho^2)^{1/2}}. \qquad (15)$$

The time dependence of this angle, calculated for $\rho = 3.45$, $f = 10$ Hz and $\alpha = 44.7°$, is shown as the dashed line in Fig. 8 and we see immediately that the director does indeed remain parallel to the effective field except in the vicinity of the minimum where there is a small phase lag. This is well-accounted for by the hydrodynamic theory using the same parameters as in the calculation of θ_∞ and with $\tau_M = 1.54$ ms. The result is shown as the solid line in Fig. 8 and clearly predicts the small phase lag. Overall the agreement between theory and experiment is good but not quite perfect; thus the maximum angle is predicted to be about 2° smaller than that observed. We have tried to improve the agreement by modifying the parameters ρ and τ_M but we could not obtain any significant improvement over the good agreement that we had already found. We also note that the calculations were, in fact, relatively insensitive to the magnetic relaxation time since the director is moving at almost the same rate as the effective field. The important point to note is again the quality of the agreement with theory at this significantly lower frequency for the electric field.

The oscillatory director motion that we have observed seems to have elements in common with that predicted and observed for a nematic subject to a static magnetic field and spun about an axis orthogonal to the field.[39,40] The behaviour of the director, both static and dynamic, in this experiment can also be obtained from the torque-balance equation[39]

$$\gamma_1 \left(\frac{d\theta}{dt} - \Omega \right) = -\frac{\Delta\tilde{\chi}B^2}{2\mu_0}\sin 2\theta, \qquad (16)$$

again the elastic and inertial contributions to the torque-balance equation are neglected here. At equilibrium, when $d\theta/dt$ is zero, the solution to the torque-balance equation gives the director orientation with respect to the magnetic field as

$$\sin 2\theta = \frac{\Omega}{\Omega_C}, \qquad (17)$$

where Ω_C is given by

$$\Omega_C = \frac{\Delta\tilde{\chi}B^2}{2\mu_0\gamma_1}. \tag{18}$$

It is of interest that this critical angular velocity, Ω_C, is just one half the inverse of the magnetic relaxation time, τ_M (see Eq. (2)). The theory predicts, therefore, that as the director is spun at an angular velocity, Ω, so the director orientation moves away from being parallel to the magnetic field until when Ω is equal to Ω_C the director is at 45° to the field. Beyond this point the theory predicts that the director orientation will now vary with time according to

$$\tan[(\Omega^{*2}-1)^{1/2}(t^*-t_0^*)] = \left[\frac{\Omega^*+1}{\Omega^*-1}\right]^{1/2} \tan\left(\theta - \frac{\pi}{4}\right), \tag{19}$$

obtained by integration of Eq. (16). Here the scaled variables are $\Omega^* = \Omega/\Omega_C$, $t^* = t\Omega_C$ and t_0^* is an arbitrary time origin. To see the effect of this dynamic process for the director on the spectrum we show the time dependence of $P_2(\cos\theta)$ which is proportional to the quadrupolar splitting (see Eq. (14)) in Fig. 9 for different scaled angular velocities. Here the results are plotted as a function of the scaled time starting at zero where the initial director orientation is at 45° to the magnetic field corresponding to $P_2(\cos\theta)$ of 1/4. For a scaled angular

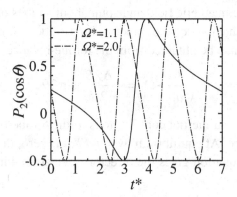

Fig. 9. The time dependence of $P_2(\cos\theta)$ for the scaled angular velocities, Ω^*, of 1.1 (solid line) and 2.0 (dot-dashed line) calculated from Eq. (19).

velocity of 1.1, that is just above the critical velocity, the value of $P_2(\cos\theta)$ decreases slowly because the magnetic torque restricts the spinning motion of the director. However, this restraining torque decreases as θ tends to 90° and so $P_2(\cos\theta)$ changes more rapidly. After the director passes 90° when t^* is about 2.9 the magnetic torque changes sign and now assists in the rotation of the director, as is apparent from the rapid change in $P_2(\cos\theta)$ from -1/2 to 1. When $P_2(\cos\theta)$ reaches the value of 1 corresponding to θ of 180° the magnetic torque again changes sign and restricts the rotation of the director. This pattern of behaviour is repeated and so the director rotation can be envisaged as being slow as it moves from parallel to orthogonal to the magnetic field to being extremely rapid as it moves from being orthogonal to parallel to it. The periodicity in the director rotation is then π and not $\pi/4$. The extremes in the director orientation during the sample rotation are unlike those predicted for the oscillating electric field experiment which depend on the angle between the two fields, the frequency of the electric field and the ratio of the electric and magnetic energies. To reach these extreme director orientations α would need to be essentially 90°, the electric energy would have to be greater that the magnetic and the frequency would need to be low. It seems that these conditions have been approached but not quite reached.[23] Just as the dynamic behaviour in the oscillating field experiment depends on the frequency of the field so that in the spinning experiment depends on the angular velocity of the sample. This is clearly apparent from the results, in Fig. 9, obtained with a larger scaled angular velocity of 2.0. Now the scaled time taken for one cycle, from 45° to 225°, has decreased from about 6.8 when Ω^* is 1.1 to 1.8 when Ω^* is increased to 2.0. In addition, the director rotation from 45° to 90° is still slower than for that from 90° to 180° but the difference is clearly not as great as for the slower sample rotation. This difference in the nature of the director rotation induced by sample spinning clearly results because the larger angular velocity makes the viscous torque greater in comparison with the magnetic torque. Such a difference is also reflected by the scaled average angular velocity, ω^*, which theory predicts to be given by

$$\omega^* = (\Omega^{*2} - 1)^{1/2}, \tag{20}$$

where the director does not rotate as fast as the sample because of an effective friction resulting from the magnetic torque. In the limit that $\Omega^* \gg 1$ the magnetic friction is small and so the director should rotate with the angular velocity of the sample. These various predictions have been studied using ESR spectroscopy in a similar manner to our NMR experiments.[39,40] One technically important difference between the two methods is the much smaller magnetic field strength (~0.3 T) used in ESR relative to NMR spectroscopy (~7.0 T) which means that the critical spinning speed is about 550 times smaller in ESR than NMR, this reduction has clear practical advantages. ESR experiments[39,41] have revealed that immediately the sample is spun above Ω_C the director also starts to spin but with the slightly smaller average angular velocity predicted. However, the oscillations in the observed spectral intensity, caused by the director rotation, were rapidly attenuated and after approximately 30 cycles the spectrum adopts a two-dimensional powder pattern. This results from a random distribution of the director in the plane orthogonal to the spinning axis. It has been postulated[42] that this total loss of director coherence during sample spinning could result from small fluctuations in the spinning speed of the sample. Although there are clear differences in the two experiments it is certainly of interest to ask why the director orientation remains coherent in the oscillating electric field experiment but not in the spinning experiment. If the postulate[42] concerning the spinning experiment is applicable then the coherence might be attributed to the high stability of the electric field, that is in E_0 and f. To explore this possibility we have repeated the oscillating electric field experiment at a frequency of 100 Hz but with white noise superimposed on this; the relative amount of noise was set at 10%. The time-resolved NMR spectra were then measured and although there was a slight broadening of the spectral lines no loss of director coherence was observed at least over eleven cycles. This is apparent from the time dependence of the director orientation obtained for the final seven cycles and shown in Fig. 10. It would appear that the stability of the oscillating electric field employed in our experiments cannot be responsible for the retention of the director coherence found over many oscillations in the director orientation. Another explanation is that in the oscillating electric field experiment the director is subject to a

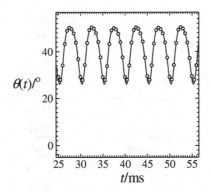

Fig. 10. The time dependence of the director orientation with respect to the magnetic field, $\theta(t)$, measured for 5CB-d_2 at 295 K after the application of a 100 Hz sinusoidal electric field with superimposed 10% white noise. The solid line shows the dependence predicted by the hydrodynamic theory with $\rho = 3.45$, $\tau_M = 1.54$ ms and $\alpha = 55.6°$.[20]

homogeneous field which is responsible for its motion. This is, implicitly, the situation envisaged in the theory for the rotating director[39] whose motion is taken to be spatially uniform. However, in the ESR experiment the director is rotated by spinning the sample tube and so now the torque on the director is not uniform and has to be transmitted from the surface of the spinning tube through elastic interactions. In addition, this may lead to the generation of line or wall defects which together with their migration and annihilation could disrupt the uniformity of the director.[43]

We turn now to the results obtained for the time-averaged NMR spectra and these are shown in Fig. 11, for 5CB-d_2 at a temperature of 288 K, for a range of frequencies of the sinusoidal electric field. It is apparent that instead of a single quadrupolar doublet expected for a monodomain nematic there are two doublets with additional spectral intensity between them indicating that the director adopts a range of orientations during the application of the electric field with the limiting values determining the extent of the two spectral halves. This does not mean that the director orientation is randomised, at any given instant, in the sample but rather that the individual FIDs being averaged correspond to different director orientations of the monodomain, unlike the situation for the spinning experiment. For the lowest frequency of the electric field

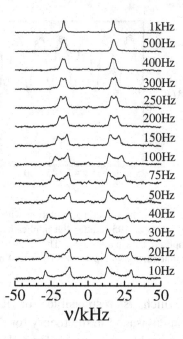

Fig. 11 The time-averaged NMR spectra for 5CB-d_2 measured for values of the frequency, between 10 Hz and 1 kHz, of the sinusoidal electric field with E_0 of 1.41 MVm^{-1} at 288 K.[20]

the spread in the quadrupolar splitting observed in the average spectrum is significant, corresponding to the large oscillations in the angle made by the director with the magnetic field. As the frequency of the field is increased so the oscillations in the director orientation are reduced in magnitude and consequently the spread in the quadrupolar splittings. At higher frequencies the difference in the extreme quadrupolar splittings decreases still further until the two doublets collapse to a single doublet indicating that the director adopts a unique orientation, in accord with the analysis given in Sec. 2. The values of the two extreme quadrupolar splittings can be estimated directly from the spectrum and used to determine θ_{max} and θ_{min} via Eq. (14). These angles are shown in Fig. 12(a) as a function of frequency and, as theory suggests, at low frequencies, when the director is able to follow the oscillations in the effective field faithfully, the two angles are independent of frequency. However, at about 10 Hz the difference in the two angles starts to decrease relatively rapidly with θ_{min} exhibiting greater changes than θ_{max}, as we had predicted (see Fig. 3). At frequencies above 300 Hz the two limiting director orientations are essentially the same.

It is clear from a comparison of the experimental data shown in Fig. 12(a) and the theoretical predictions given in Fig. 3 that, at a qualitative level, the agreement between theory and experiment for the frequency dependence of θ_{max} and θ_{min} is good. However, to make a quantitative comparison we need to know the values of ρ, τ_M and the angle between the two fields, α. As we have seen already ρ takes the value 4.37 and τ_M is measured to be 2.60 ms at this temperature; the angle between the two fields is 44.7°. The frequency dependence of θ_{max} and θ_{min} predicted with these parameters is shown as the solid lines in Fig. 12(a) and are clearly in good quantitative agreement with experiment, except for θ_{min} at low frequencies where the experimental values are slightly larger than those predicted. This could result, at least in part, from the large errors in the angles extracted from the quadrupolar splittings when the director is almost parallel to the magnetic field (see Eq. (14)). To test this possibility we have compared the primary experimental data, namely the quadrupolar splittings, with theory.

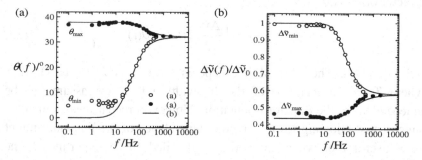

Fig. 12. The frequency dependence, on a logarithmic scale, of (a) the limiting director orientations, θ_{max} and θ_{min}, for 5CB-d$_2$ at 288 K determined from the time-averaged spectra and (b) the limiting quadrupolar splittings, $\Delta\tilde{v}_{max}(\equiv\theta_{min})$ and $\Delta\tilde{v}_{min}(\equiv\theta_{max})$ measured from the time-averaged spectra and scaled with $\Delta\tilde{v}_0$. The predicted variation of both angles and quadrupolar splittings, calculated with the parameters $\rho = 4.37$ and $\tau_M = 2.60$ ms determined previously for 5CB and with an angle, α, between the electric and magnetic fields of 44.7°, are shown as the solid lines.[20]

In Fig. 12(b) we show the frequency dependence of the limiting splittings, $\Delta\tilde{v}_{max}$ and $\Delta\tilde{v}_{min}$; in labelling these splittings we have chosen to use the subscripts max and min employed for the director orientations. However, because of the relationship between the director orientation and the splitting this leads to the counter-intuitive situation that $\Delta\tilde{v}_{max}$ is smaller than $\Delta\tilde{v}_{min}$. The frequency dependence of these limiting splittings, predicted by theory using the same parameter set as for the director orientation, is shown as the solid line in Fig. 12(b). The prediction of the splittings is clearly in far better agreement with theory than for the director orientation which shows that the discrepancy found for the director orientation originates from the error in determining this from the quadrupolar splittings.

An additional test of our theoretical model, employed to understand the oscillatory motion of the director, is to simulate the time-averaged spectrum from the predicted time dependence of the director orientation. The theoretical spectrum is given by the time average

$$I(v) = \oint L(v, \tilde{v}_{\pm}(\theta(t)), T_2^{-1}) dt, \tag{21}$$

where v is the frequency, $\tilde{v}_{\pm}(\theta(t))$ denotes the resonance frequencies for the two components of the quadrupolar doublet and these depend on the director orientation via

$$\tilde{v}_{\pm}(\theta(t)) = v_0 \pm \left(\frac{\Delta\tilde{v}_0}{2}\right) P_2(\cos\theta(t)), \tag{22}$$

(cf. Eq. (14)). The spectral lineshape is written as $L(v, \tilde{v}_{\pm}(\theta(t)), T_2^{-1})$ where T_2^{-1} is a measure of the linewidth which we assume to be independent of the director orientation. However, recent measurements[44] suggest that allowing for the angular dependence of the linewidth might be a useful tool when attempting a detailed fit to experiment. The lineshape is taken to be a Lorentzian, that is

$$L(v, \tilde{v}_{\pm}(\theta(t)), T_2^{-1}) = \frac{T_2/\pi}{1 + T_2^2(v - \tilde{v}_{\pm}(\theta(t)))^2}. \tag{23}$$

Since the director orientation is calculated for discrete values of time it is convenient to replace the integral in Eq. (21) with a summation over a large number of times, uniformly distributed over one cycle for the variation in the director orientation. The high quality of the agreement between the simulated and the experimental spectra is illustrated, in Fig. 13, for the time-averaged spectrum measured for 5CB-d_2 at 288 K with a frequency of 10 Hz for the electric field. To simulate this spectrum the time dependence of the director orientation in the stationary state was calculated using the same parameters employed to determine θ_{max} and θ_{min}. The quadrupolar splitting, $\Delta \tilde{\nu}_0$, and linewidth parameter, T_2^{-1}, were both measured from the NMR spectrum recorded in zero electric field; they were found to be 59.3 kHz and 1.0 kHz, respectively. The simulated time-averaged spectrum was obtained by summing 10,000 spectra and is shown in Fig. 13. It is clear that there is good quantitative agreement with experiment; of course, the simulation does not include the small central peak seen in the experimental spectrum and thought to originate either from an isotropic component of the sample or from electrical interference by the wires connecting the cell. The agreement provides further support for the validity of the theory used to evaluate the time dependence of the director orientation, the parameters determined for 5CB from more conventional experiments and the experimental methodology. It also suggests that the director oscillations continue

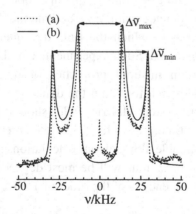

Fig. 13. The time-averaged NMR spectrum for 5CB-d_2 at 288 K with a 10 Hz sinusoidal electric field and strength 1.41 MVm^{-1} (a) experimental and (b) simulated using the parameters $\rho = 4.37$ and $\tau_M = 2.60$ ms obtained from previous measurements and with α equal to 44.7°.[20]

unattenuated for the several thousand oscillations involved in measuring the time-averaged spectra.

5. Conclusions

We have found, using deuterium NMR spectroscopy, that the application of a static magnetic field and a sinusoidal electric field to a nematic, with positive dielectric and diamagnetic anisotropies, results in an oscillatory motion of the director between the two fields and in the plane containing them. The oscillations are observed to take place without loss of director coherence, at any given time, that is, the director is uniformly aligned, at least for the large number of oscillations we have observed directly and for the many thousand used to obtain the time-averaged spectra. The director oscillations and their magnitude are strongly dependent on the frequency of the electric field with the difference in the limiting director orientations being a maximum at low frequencies and tending to zero at high frequencies. These limiting orientations have been observed directly from the time-resolved NMR experiments but we have found that they are also available, more directly, from the time-averaged NMR spectrum. The results that we have obtained for the time dependence of the director orientation and the limiting values of this, both as a function of frequency, are well-explained using a hydrodynamic theory based on the torque-balance equation in which the time dependence of the electric field is specifically included. Such experiments could be used to provide a new route to certain material properties of the nematic, namely, $\Delta\tilde{\varepsilon}/\Delta\tilde{\chi}$ and $\gamma_1/\Delta\tilde{\chi}$, although to obtain the dynamic quantity, $\gamma_1/\Delta\tilde{\chi}$, the frequency of the sinusoidal electric field should be related to the magnetic relaxation time, τ_M. However, the precise form of this relationship is not available although our calculations (see Fig. 3) suggest that the experimental behaviour will be most dependent on the director dynamics when the frequency is of the order of $1/10\,\tau_M$.

Acknowledgments

This work was supported by a Scientific Grant-in Aid from the Japan Society for the Promotion of Science (JSPS) and the HRC project of the

Ministry of Education, Culture, Sports, Science and Technology of Japan, and was carried out as an Anglo-Japanese joint research project of the International Exchange program supported by The Royal Society and JSPS.

References

1. J. C. Rowell, W. D. Phillips, L. R. Melby and M. Panar, J. Chem. Phys., **43**, 3442 (1965).
2. G. R. Luckhurst, J. Chem. Soc., Faraday Trans. 2, **84**, 961 (1988); V. Domenica, M. Geppi and C. A. Veracini, Proc. Nucl. Magn. Reson. Spect., **50**, 1 (2007).
3. A. F. Martins, P. Esnault and F. Volino, Phys. Rev. Lett., **57**, 1745 (1986).
4. M. Winkler, D. Geschke and P. Holstein, Liq. Cryst., **17**, 283 (1994).
5. M. Bender, P. Holstein and D. Geschke, J. Chem. Phys., **113**, 2430 (2000).
6. R. Stannarius, G. P. Crawford, L. C. Chien and J. W. Doane, J. Appl. Phys., **70**, 135 (1991).
7. S. M. Fan, G. R. Luckhurst and S. J. Picken, J. Chem. Phys., **101**, 3255 (1994).
8. J. R. Hughes, G. Kothe, G. R. Luckhurst, J. Malthête, M. E. Neubert, I. Shenouda, B. A. Timimi and M. Tittelbach, J. Chem. Phys., **107**, 9252 (1997).
9. E. Ciampi, J. W. Emsley, G. R. Luckhurst and B. A. Timimi, J. Chem. Phys., **107**, 5907 (1997).
10. P. Esnault, J. P. Casquilho, F. Volino, A. F. Martins and A. Blumstein, Liq. Cryst., **7**, 607 (1990).
11. A. Sugimura, K. Nakamura, T. Miyamoto, P. J. Le Masurier, B. A. Timimi, T. H. Payne and G. R. Luckhurst, Proc. of the 4th Int. Display Workshops, 65 (1997).
12. G. R. Luckhurst, A. Sugimura and B. A. Timimi, Mol. Cryst. Liq. Cryst., **347**, 297 (2000).
13. G. R. Luckhurst, T. Miyamoto, A. Sugimura, T. Takashiro and B. A. Timimi, J. Chem. Phys., **114**, 10493 (2001).
14. C. J. Dunn, G. R. Luckhurst, T. Miyamoto, A. Sugimura and B. A. Timimi, Mol. Cryst. Liq. Cryst., **347**, 167 (2000).
15. G. R. Luckhurst, T. Miyamoto, A. Sugimura and B. A. Timimi, Thin Solid Films, **393**, 399 (2001).
16. G. R. Luckhurst, T. Miyamoto, A. Sugimura and B. A. Timimi, Mol. Cryst. Liq. Cryst., **347**, 167 (2000).
17. G. R. Luckhurst, T. Miyamoto, A. Sugimura and B. A. Timimi, Mol. Cryst. Liq. Cryst., **402**, 103 (2003).
18. G. R. Luckhurst, A. Sugimura and B. A. Timimi, J. Chem. Phys., **116**, 5099 (2002).
19. G. R. Luckhurst, T. Miyamoto, A. Sugimura and B. A. Timimi, J. Chem. Phys., **117**, 5899 (2002).
20. G. R. Luckhurst, T. Miyamoto, A. Sugimura, B. A. Timimi and H. Zimmermann, J. Chem. Phys., **121**, 1928 (2004).

21. D. Kamada, G. R. Luckhurst, K. Okumoto, A. Sugimura, B. A. Timimi and H. Zimmermann, Mol. Cryst. Liq. Cryst., **441**, 131 (2005).
22. G. R. Luckhurst, A. Sugimura, B. A. Timimi and H. Zimmermann, Liq. Cryst., **32**, 1389 (2005).
23. G. R. Luckhurst, A. Sugimura and B. A. Timimi, Liq. Cryst., **32**, 1449 (2005).
24. A. Sugimura and G. R. Luckhurst, in *Nanotechnology and Nano-Interface Controlled Electronic Devices*, eds. M. Iwamoto, K. Kaneto and S. Mashiko, Elsevier Science B, 2003, Chap. 16.
25. A. Sugimura and G. R. Luckhurst, in *Nuclear Magnetic Resonance Spectroscopy of Liquid Crystals*, ed. R. Dong, World Scientific and Imperial College Press, 2009, Chap. 10.
26. J. W. Emsley, J. E. Long, G. R. Luckhurst and P. Pedrielli, Phys. Rev. E **60**, 1831 (1999); J. W. Emsley, G. R. Luckhurst and P. Pedrielli, Chem. Phys. Lett., **320**, 255 (2000).
27. G. R. Luckhurst, Mol. Cryst. Liq. Cryst., **347**, 121 (2000).
28. G. R. Luckhurst, T. Miyamoto, A. Sugimura and B. A. Timimi, Mol. Cryst. Liq. Cryst., **347**, 147 (2000).
29. G. R. Luckhurst, T. Miyamoto, A. Sugimura and B. A. Timimi, Mol. Cryst. Liq. Cryst., **394**, 74 (2003).
30. M. Schadt, in *Topics in Physical Chemistry*, Vol 3: Liquid Crystals, ed. H. Stegemeyer, Steinkopff, Damstadt, 1994, Chap. 6.
31. J. K. Moscicki, in *The Physical Properties of Liquid Crystals: Nematics*, eds. D. A. Dunmur, A. Fukuda, G. R. Luckhurst, INSPEC, London, 2001, Chap. 8.2.
32. R. A. Wise, A. Olah and J. W. Doane, J. Phys. (Paris) **36** (C1) 117 (1975).
33. H. Kresse, in *The Physical Properties of Liquid Crystals: Nematics*, eds. D. A. Dunmur, A. Fukuda, G. R. Luckhurst, INSPEC, London, 2001, Chap. 6.2.
34. A. Killian and S. Hess, Z. Naturforsch., **44a**, 693 (1989).
35. M. Bender, P. Holstein and D. Geschke, J. Chem. Phys., **113**, 2430 (2000).
36. J. W. Emsley, G. R. Luckhurst and C. P. Stockley, Mol. Phys., **44**, 565 (1981).
37. A. M. Kantola, G. R. Luckhurst, A. Sugimura and B. A. Timimi, Mol. Cryst. Liq. Cryst., **402**, 117 (2003); A. M. Kantola, G. R. Luckhurst, A. Sugimura and B. A. Timimi, J. Chem. Phys. to be submitted (2009).
38. K. Skarp, S. T. Lagerwall and B. Stebler, Mol. Cryst. Liq. Cryst., **60**, 215 (1980).
39. F. M. Leslie, G. R. Luckhurst and H. J. Smith, Chem. Phys. Lett., **13**, 368 (1972).
40. S. G. Carr, G. R. Luckhurst, R. Poupko and H. J. Smith, Chem. Phys., **7**, 278 (1975).
41. S. K. Khoo and G. R. Luckhurst, Liq. Cryst., **15**, 729 (1993); M. P. Eastman, B. A. Freiha, B. Kessel and C. Allen Chang, Liq. Cryst., **1**, 147 (1986).
42. L. Orian, G. Feio, A. Veron, A. Polimeno and A. F. Martins, 19th Int. Liq. Cryst. Conf. Extended Abstract, P653 (Edinburgh, 2002).
43. P. G. de Gennes and J. Prost, *The Physics of Liquid Crystals*, Second Edition, Clarendon Press, Oxford, 1993, p. 223.
44. D. Hamasuna, G. R. Luckhurst, A. Sugimura, B. A. Timimi, K. Usami and H. Zimmermann, Thin Solid Films, **517**, 1394 (2008).

CHAPTER 11

TRAINING AND FATIGUE OF CONDUCTING POLYMER ARTIFICIAL MUSCLES

Keiichi Kaneto

Graduate School of Life Science and Systems Engineering, Kyushu Institute of Technology, 2-4 Hibikino, Wakamatsu, Kitakyushu 808-0196, Japan
E-mail: kaneto@life.kyutech.ac.jp

Artificial muscles (soft actuators) based on electrochemomechanical deformation of conducting polymer films, e.g. polypyrrole and polyaniline are mentioned in their behavior of strain, stress, time response and training under tensile loads. The typical strain and stress of polypyrrole actuators are 20-30% and 22 MPa, respectively, which are larger than those of skeletal muscles. Under higher tensile loads, the artificial muscle shows creeping, which is serious problem for the practical application. The mechanism of creeping has been studied and found that the artificial muscles are strengthened by creeping, namely, the training or experience of high tensile loads. Fatigue of artificial muscle by electrochemical cycles is also found to originate from loose of conductivity due to stress.

1. Introduction

Various kinds of robots are being developed, not only for automatic manufacturing systems in factories but also welfare robots for handicapped and old persons. Presently robots are mostly driven by electrical motors, which are easily controlled, however, heavy and noisy. Light weight, quiet and human friendly driving machines, which could be artificial muscles or soft actuators, are demanded to drive robots. Actuators are materials that deform and change the shape by themselves.

Table 1. Basic type of deformation in actuators.

	Deformations		Mechanism	Materials and Actuators
(a)	Expansion and Contraction		Constant Volume and Shape Change	Muscles, Piezoelectric Actuators, Shape Memory Alloys, Dielectric Elastomers
(b)	Swell and Shrink		In and Out of Solutions and Ions	Conducting Polymers, Gels
(c)	Bending		Electrophoresis of Ions and Molecules Bimorph Structure	Ionic Polymer and Metal Composite (IPMC), Hydro Gels, Polymer Gels, Conducting Polymers
(d)	Twisting		Partial Rotation	Shape Memory Alloys, Composite Materials

The basic types of deformations[1,2] are (a) elongation and contraction, (b) swelling and shrinking, (c) bending and (d) twisting as shown in Table 1. By the combination of these deformations, any sort of motions could be generated.

There are hard actuators based on ceramics and metals, and also soft actuators based on polymers and gels.[2,3] In hard actuators piezoelectric actuators and shape memory actuators have been developed into commercial stage. Piezoactuators can be precisely controlled and currently used for autofocusing camera, electronic weight meters, scanning probe micrographs and ink jet printers. However, their strains are too small to drive robots.

In soft actuators, polymers are suitable as their materials, because of their mechanical toughness, softness and flexibility. The driving forces of the actuators are thermal energy, electrical energy and solution (gas) permeability. Among them, driving by electronic power is quick and easy to control. In Table 2[1-9] characteristics of various kinds of soft actuators driven by electrical energy, except for skeletal muscle for comparison, are shown for their types of deformation, typical operating voltages, strains and stress including response times and cycle lives. In

Table 2. Characteristics of various soft actuators.

Materials	Deformation	Voltage (V)	Strain (%)	Stress (MPa)	Response Time (s)	Cycle Stability	Circumstance
Skeletal Muscle	Expansion and Contraction	-	25	0.4	0.1	◎	Wet
Ion Exchange Membrene (IPMC)[3,4]	Bending	2-3	>3	30	0.1	○	Wet (Water, Ionic Liquids)
Conducting Polymers[1,5]	Swell and Shrink, Bending	1-2	39	22	1	△	Wet (Water Organic Solution)
Dielectric Elastmers[4]	Expansion and Contraction	5,000-6,000	380	8	0.2-1	◎	Dry
Polymer Gels[7]	Bending	500	20	-	0.1	◎	Dry (Organic Liquid)
Hydro Gel[8]	Bending	2-3	-	-	×	×	Wet (Water)
Carbon Nanotube[9]	Bending	3-4	0.9	0.1	5	△	Wet (Ionic Liquids)

the type of deformation conducting polymers are similar to the skeletal muscle. Ionic polymer and metallic composite (IPMC) actuators also demonstrate interesting applications.[10,11]

2. Electrochemomechanical Deformation of Conducting Polymers

Conducting polymers are characterized with π-conjugated polymers and bond alternation along the polymer backbone. The π-electrons are moderately bounded to the polymer backbone, hence, are easily removed and added by the electrochemical oxidation and reduction, respectively. The electrochemical oxidation takes place simultaneously by insertion of same amount of anion as shown in Fig. 1. Upon oxidation the polycations (polaron and bipolaron) are induced in the polymer backbone. The polycations are accommodated by reducing the bond alternation and straighten the conformation of polymer backbone as shown in Fig. 2. Eventually polycations contribute to increase the electrical conductivity up to the metallic regime. Insertion of anions and conformation change of polymer backbone result in expansion, swelling and stiffen of the film, which is named as electrochemomechanical deformation (ECMD). The ECMD reversibly takes place by the electrochemical oxidation and reduction cycle.

In case of polypyrrole (PPy) film prepared electrochemically from dodecylbenzensulfonic acid (DBS), the young modules (E) of the as grown (or oxidized) film was 0.14 GPa, on the other hand E of the reduced film was 0.07 GPa. The result indicates that the film stiffness depends remarkably on the state of oxidation. It is interesting to note that expansion of polyaniline (PANi) films is found to be due to the conformation change of polymer backbone by approximately 1%.[12]

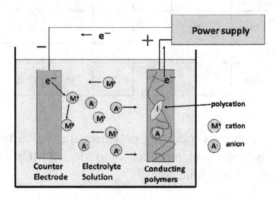

Fig. 1. Schematic drawing of electrochemical oxidation in conducting polymers.

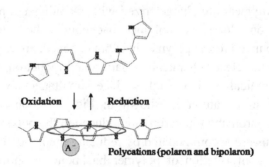

Fig. 2. Polycations (polaron, bipolaron) generated by oxidation.

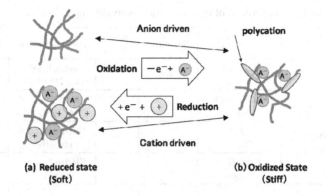

Fig. 3. Mechanisms of ECMD in (a) anion driven and (b) cation driven films.

Characteristics of ECMD strictly depend on a size of anion in electrolyte solution during the electrochemical deposition.[1] As shown in Fig. 3, if the employed anion is small like Cl⁻ and ClO_4^-, the electrochemical reaction takes place by the movement of anions (anion driven ECMD). On the other hand, in case of large anion like DBS, long alkyl chain of DBS entangled and immobilized in conducting polymer network, hence, cations move in and out from the film (anion driven ECMD). In cation driven ECMD, the film shrinks and stiffens upon oxidation, being similar behavior to natural muscles.

In Table 3, characteristics of ECMD in PPy, PANi and polythiophene (PAT) are shown as well as their chemical structures. Strains of the ECMDs are mainly generated by the volume of inserted anions and solvated solutions,[13-15] besides contribution from the conformation change of polymers.[12] Therefore, when larger anions are employed for the anion driven ECMD, the larger strain is obtained.

Morphology of PPy film is controlled by choosing the electrolyte and solution of electrochemical polymerization.[16] For example as shown in Table 4, PPy films prepared in TBATFSI/MB solution showed spongy gel[17] and the largest strain of 39% in ECMD was achieved by the operation in LiTSFI/propylene carbonate solution. On the other hand, PPy films prepared from $TBACF_3SO_3$/MB showed high dense morphology. Although the strain of ECMD was 12.6%, the large stress of 22 MPa was obtained by operation in $NaPF_6/H_2O$.[18]

Table 3. Characteristics of ECMD in various conducting polymers.[16-18,20-24]

Conducting Polymers	Electrolyte	Strain (%)	Stess (MPa)	Film Preparation	Reference
Polypyrrole (PPy)	TBATFSI/H$_2$O Anion	26.5	6.7	Electrochemical BATFSI/MB	13, 16, 17
	NaPF$_6$/H$_2$O Anion	12.4	22	Electrochemical TBACF$_3$SO$_3$/MB	18
	LiCl/H$_2$O Cation	4.9	5	Electrochemical DBS/H$_2$O	20, 21
	NaCl/H$_2$O Cation	0.4	-	Electrochemical PSS/H$_2$O	22
Polyaniline (PANi)	NaCl/H$_2$O Anion	6.7	-	Chemical	23
Polyalkylthiophene (PAT)	TBABF$_4$/CH$_3$CN Anion	3.5 R = C$_6$H$_{13}$	-	Chemical	24
		1.7 R = C$_{12}$H$_{25}$	-		

TBA: Tetra-n-butylammonium, TFSI: bis (trifluoromethansulfonyl)imid, MB: Methyl Benzoate, PSS: Polystylene sulfonic acid

Table 4. Morphology and properties of typical films of PPy/TBATFSI[17] and PPy/TBACF$_3$SO$_3$.[18]

Electrolyte	PPy/TBATFSI	PPy/TBACF$_3$SO$_3$
Morphology	Gel, Sponge	Smooth, High Dense
Density (g/cm^3)	0.63	1.41
Conductivity (S/cm)	80-140	80-140
Tensile Strength (MPa)	20-30	80-100
Strain Before Break (%)	40	60-100

PPy films prepared from aqueous DBS solution exhibited tough and dense film.[19] Although the strain was not large because of cation movement, the ECMD was stable against repeat cycles within comfortable potential range of the operation. The magnitude of strains in the ECMD of PPy/DBS film depended on electrolyte cations in the manner of Li$^+$ > Na$^+$ > K$^+$ as shown in Fig. 4, indicating the cations solvated with some water molecules.[1]

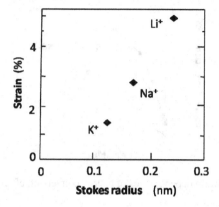

Fig. 4. Relationship between stokes radius of cations and strain in cation driven ECMD of PPy/DBS films.

3. Stress-Strain Characteristics of Polymer Actuators

Tensile stress dependence of ECMD strain of PPy[1] is shown in Fig. 5. The strain shows the maximum at stress free and linearly decreases with increasing the stress until it crosses to the zero of strain. The stress obtained at the interception is a blocking force, which relates to the Young's modulus. As evident from Fig. 5, the performance of PPy/CF$_3$SO$_3$ film is better than that of PPy/DBS, because the size of anion is larger than that of cation.

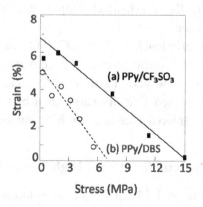

Fig. 5. Stress-strain curves in (a) anion (CF$_3$SO$_3^-$) driven and (b) cation (Li$^+$) driven films.

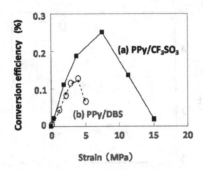

Fig. 6. Strain dependencies of energy conversion efficiencies.

4. Energy Conversion Efficiency

Soft actuator is an energy transducer from electrical energy to mechanical energy. The electrical input energy is obtained from a relation of $E_E = \int ivdt$, where i, v and t are current, voltage and time during contraction of the film, respectively. Mechanical output energy is $E_M = mg\Delta l$, where m, g and Δl are mass of weight, the gravity and the lifting distance of weight, respectively. The E_M is estimated from the area of strain-stress curve in Fig. 5. Energy conversion efficiencies (E_M/E_E in %) obtained from Fig. 5 are shown in Fig. 6 for anion and cation driven ECMDs. It has been found that the electrical input energy is scarcely depend on tensile loads, since the electrochemical reaction is mostly charging and discharging of a secondary battery of conducting polymers. However, the mechanical output strictly depends on the weight and lifting distance.

The conversion efficiencies of PPy/CF$_3$SO$_3$ and PPy/DBS films are 0.25% and 0.13%,[1] respectively. The conversion efficiencies are unexpectedly small, however, most of electrical input energy is considered to be harvested like discharge of a battery. The maximum output energy for the anion driven actuator is estimated to be 0.14 J/g.

5. Training of PPy Actuators Under High Tensile Loads

CV of PPy/DBS film in 1 M LiCl aqueous solution at various tensile stresses is shown in Fig. 7(a). With increasing the tensile stresses,

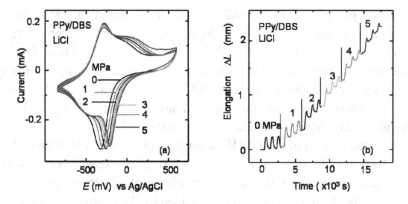

Fig. 7. (a) CV curves of PPy/DBS film in aqueous 1 M LiCl at the scan rate of 3 mV/s and (b) elongation L of the film under applied tensile stresses.

shoulders at around 0 V in the oxidation cycle increased and shifted to higher potential as seen in Fig. 7(a). Similarly, the peak at around -400 mV in the reduction cycle also shifted to higher potential side. After removal of 5 MPa tensile stresses, the CV at 0 MPa traced almost the same to that of 5 MPa.[22]

During the electrochemical cycles the film was gradually elongated about 20% by creeping as shown by curve in Fig. 7(b), where the elongation ΔL was the incremental increased length of the film. Also the film periodically elongated and contracted at the tensile stress up to 5 MPa due to the ECMD. The creeping resulted from the anisotropic deformation of films (conformation change), slipping and breaking of polymer chains under larger tensile stresses.

The shifting of CV peaks under tensile stress is considered to result from the conformation change (stretching) of polymer backbone due to the creeping. Namely, the stretching of polymer chains reduces the ionization potential to positive side. During the electrochemical cycles, however, complicated and dynamical change of rheological properties is taking place[26] in polymer matrices.

Curves in Fig. 8 shows the injected electrical charge (Q) dependence of elongation (Δl) for each cycle in the PPy/DBS film,[27] derived from Figs. 7(a) and 7(b). The Δl contains the creeping to some extent. In the oxidation process at lower tensile stresses as shown in Fig. 8(a), the Δl was proportional to the Q. At higher stresses and higher charges, however, the film clearly showed saturation in the contraction. For the reduction process, at lower tensile stresses the film approximately elongated linearly with increasing Q as seen in Fig. 8(b). At higher tensile stresses, however, the film did not elongate linearly at lower charges, then steeply elongated at larger charges.

The mechanism of the ECMD strain (Δl) in PPy/DBS film under low and high tensile stresses is proposed. At lower tensile stresses for oxidation (contraction), the linear dependence of Δl versus Q shown by curves 0 and 1 MPa in Fig. 8(a) is explained by the isotropic shrinking of the film. At higher tensile stress of 3-5 MPa the film contracts in isotropic manner for smaller charges. At larger Q the saturation of contraction is accounted by the anisotropic shrinking of the film, namely, shrinking to the thickness direction by squeezing out of cations keeping the film length constant. The contraction force of film may be equilibrium with the tensile stress, due to crosslinking of the film. The ECMD behavior during reduction (elongation) shown in Fig. 8(b) indicates the reversed mechanisms of the oxidation. At lower tensile stress the film expanded by isotropic ways. At high tensile stress the film swollen to the thickness direction at low Q, followed by elongation along the tensile direction at larger Q.

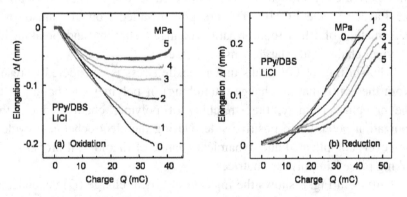

Fig. 8. Electronic charge (Q) dependence of ECMS, Δl in the PPy/DBS film for (a) oxidation (contraction) and (b) reduction (elongation).

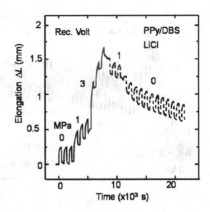

Fig. 9. Cycle dependence ΔL in PPy/DBS film by cycling of rectangular voltages of -850 mV and 600 mV for each 450 s under tensile stresses.

Figure 9 shows the cycle dependence of the ΔL by the application of rectangular voltages under tensile stresses up to 3 MPa. Creeping of the film was remarkably observed at higher tensile stresses. It is noted that after removal of the tensile stress of 3 MPa, the creeping was recovered as seen by the dashed curves in Fig. 9 as reported in our previous paper for NaCl electrolyte.[25] The creeping results from that the tensile stress is larger than the elasticity of the film. During the electrochemical reaction, it is supposed that polymer chains dynamically fluctuate and should be aligned along the external stress. By the completion of electrochemical oxidation, however, the anisotropic deformation (largely expanded state of the polymer along the tensile direction) should be fixed or frozen. This is named as the memory of deformation. After the release of tensile stress, the frozen and expanded state should be relaxed to the original state. Detailed mechanisms of the memory effect have been discussed in our paper.[27]

6. Training of PANi Film

Figure 10 shows the incremental elongation of the ECMD for a PANi film with the original length (L_0) of 10 mm operated by a triangle voltage under tensile loads at the scan rate of 2 mV/s. The electrochemical cycle

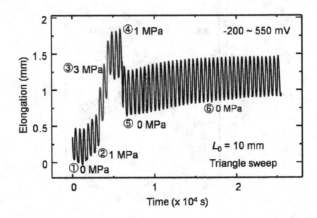

Fig. 10. ECMD of PANi film operated by a triangle sweep at -200 ~ 550 mV Ag/AgCl with the period of 900 s under tensile loads up to 3 MPa, numbers in circles indicate the sequence of measurement.

was repeated three times at the same tensile load as shown by ① ~ ④ in Fig. 10. The film showed cyclic strokes of expansion and contraction (ΔL) being 0.4 ~ 0.6 mm (corresponding strain of 4 ~ 6%). The ΔL is induced by insertion and ejection of bulky anions as well as creeping by high tensile loads. The creeping was remarkable for lager tensile loads. It is interesting to note that by release of tensile loads down to 0 MPa, the creeping was recovered more clearly than the case of PPy. It should be also noted that the ΔL as shown by ⑤ and ④ in Fig. 10 after 3 MPa was substantially larger than ① and ② in Fig. 11 for 0 and 1 MPa, respectively, indicating the large training effect.

Curves in Fig. 11 show the CV obtained in PANi film shown in Fig. 10. By the application of tensile loads, the oxidation peaks shifted to lower potential (easy oxidation) and the reduction peaks also shifted to lower potentials as shown by an arrow. By the removal of tensile loads, the CV traced the nearly same curve of the highest tensile load with a slightly broadening of peaks. The result suggests some permanent change of polymer conformation.

Curves in Fig. 12 show the injected electrical charge (Q) dependences of the elongation (Δl) in PANi film. The Δl was obtained at the 2nd cycle of each tensile stress. The Q was calculated from the CV in Fig. 11. Δl

versus Q in PANi films showed approximately linear dependence, which can be compared with the case of nonlinear behavior[27] of PPy shown in Fig. 8. In the oxidation process as shown in Fig. 12(a), the elongation Δl at large tensile loads was due to creeping to some extent. In the reduction process shown in Fig. 12(b), the negative elongation was the contraction stroke. It should be noted that Δl apparently increased by the experience of high tensile stresses and the Q at ④ ~ ⑥ was apparently larger than ① and ② in Fig. 12(b), indicating the pronounced training effect.

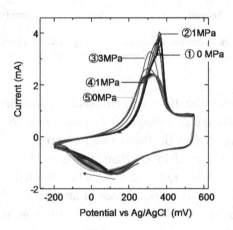

Fig. 11. CV curves of PANi film under tensile stress shown in Fig. 10.

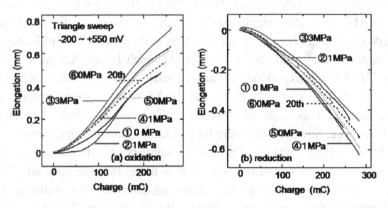

Fig. 12. Elongations of PANi film against charges during (a) oxidation and (b) reduction under tensile loads.

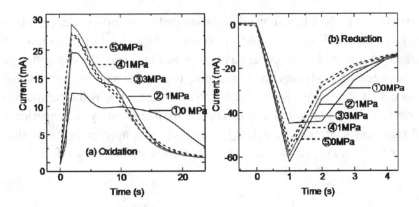

Fig. 13. Time responses of current waveform by the application of rectangular voltage of -200 ~ 550 mV, each 450 s for oxidation and reduction cycles.

Figure 13 shows the time responses of current waveform by the application of rectangular voltage. It is noted that the oxidation took longer time compared with the reduction process. The results are consistent with the case of PPy,[27] namely, the oxidation starts from the film with low conductive reduced state, while the reduction starts from high conductive state. By the experience of large tensile loads the current response became faster, indicating larger diffusion of ions and the increased electrochemical activity of the film.

Table 5 shows the summary of ECMD including the energy conversion efficiency in PANi films. The contraction length (stroke, Δl) of ECMD enhanced significantly by the experience of high tensile stresses (training effect). The contraction strokes of ECMD indicated by the strokes of ④ and ⑤ in Table 4 are much larger than those of ① and ② after the experience of larger tensile loads. It is also interesting to note that training effect in the PANi film is more pronounced compared with the case of PPy film.[27] The reason is not clear at the present stage, however, this may be due to the longer conjugation length of PANi polymer chains. The energy conversion efficiency E_M/E_E in PANi film also is shown in Table 5. It is noted that the injected charge (Q) and the input energy were also enhanced by the training. The maximum conversion efficiency was 0.15% at around 3 MPa, being similar to 0.15% in polyaniline actuators by Smela et al.[28]

Table 5. Summary of ECMD during the contraction (reduction) in a PANi film under tensile loads.

Tensile Stress (MPa)	Contraction Stroke Δl (μm)	Charge Q (mC)	Input Energy E_E (mJ)	Output Work E_M (mJ)	Energy Conversion Efficiency (%)
① 0	490	245	49	0.0048	0.0098
② 1	453	259	51.8	0.0271	0.0523
③ 3	468	282	56.5	0.0843	0.149
④ 1	630	282	56.4	0.0377	0.0668
⑤ 0	635	284	56.7	0.0062	0.0109

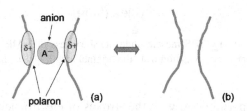

Fig. 14. Schematic drowing of (a) creation ionic crosslink by oxidation and (b) removal of crosslink by reduction in conducting polymers.

During electrochemical reaction, it is considered that polymer chains are dynamically fluctuated by flow of anions and should be easily creped. By completion of the electrochemical oxidation, however, the anisotropic deformation should be fixed or frozen by ionic crosslink as shown in Fig. 14(a). The anion may interact with polycations nearby and crosslink adjacent polymer chains. This is a memory of deformation. After the release of tensile stress and reduction, the frozen and expanded state should be released to the original conformation (recovery of creeping) upon electrochemical cycles. The recovery of creeping is driven by the elasticity or thermal vibration of polymer chains.

7. Fatigue of Artificial Muscles

Figure 15 shows the cycle life or aging of PPy/DBS muscle under the tensile stress of 3 MPa.[25] In each cycle, the elongation and contraction of

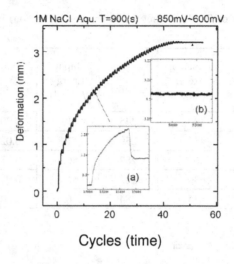

Fig. 15. Cycle life of PPy/DBS muscle under tensile stress of 3 MPa. The deformation (ΔL) was the increment of elongation from the original length of 10 mm.

the film was superimposed with the strokes of electrochemical cycle and creeping as shown by the inset (a) of Fig. 15. The elongation due to creeping came to saturate around 30 cycles and the muscle did not respond to the electric stimuli and died after 52 cycles as shown by the inset (b) of Fig. 15. This mainly results from loss of the conductivity. In facts, the conductivity of 41 S/cm before cycling dropped to 0.1 S/cm after 60 cycles at 3 MPa. Another result was that the conductivity of 28 S/cm before training decreased to 2 S/cm by 18 cycles under the tensile loads of 1, 2, 3 to 5 MPa for the each three cycles. However, without tensile stress the conductivity of 28 S/cm was unchanged by 26 cycles. During the cycling test, the films kept the shape and did not mechanically breakdown. From these results, electrochemical cycle without stress is fairly stable, however, under large tensile stresses the degradation of the π-electron systems severely occurs.

It is interesting to note that as shown in Fig. 15 the maximum creeping was about 30%, which nearly coincided with the tensile strength (breakdown strain).[25] The result indicates that the PPy/DBS film was highly cross-liked. Under high tensile stress, polymer chains were aligned and straighten along the direction of tensile. The energy state of π-electron system could be lowered by straighten of polymer chains, resulting in the shift of oxidation and reduction potential to lower energy. The remarkable shift of reduction peak by the stress in CV of Fig. 7(a) may be related to this energy shift. The oxidation attacks the lowered π-electrons and degrades the electrochemical activity of the films. The detailed mechanism of degradation by electrochemical cycle under tensile stress is not known in detail, however, the thermodynamical behavior of ion diffusion is also taken into account.

8. Prospect of Soft Actuators

Soft actuators based on conducting polymers are being developed with respect to the strain, contraction force and response time, which are more than those of skeletal muscle. However, there are still some problems, such as cycle life and stability as well as using liquid electrolyte. The soft actuators are a new candidate for human friendly machine to drive robots. Mostly present conducting polymers being studied for soft actuators are existing materials, and new materials aimed for soft actuators are needed. Also novel applications of soft actuators has to be considered.

References

1. K. Kaneto, Oyo Butsuri, **76** (12), 1356 (2007) [In Japanese].
2. Ed. Y. Osada, Frontier of Soft Actuators, N.T.S. (2004) [In Japanese].
3. T. Mirfakhrai, J. D. W. Madden and R. H. Baughman, Materials Today, **10**, 30 (2007).
4. K. Oguro, Kagaku to Kogyou, **72**, 162 (1998) [In Japanese].
5. K. Kaneto, M. Kaneko, Y.-G. Min and A. G. MacDiarmid, Synthetic Metals, **71**, 2211 (1995).
6. Q. Pei, M. Rosethal, S. Stanford, H. Prahlad and R. Pelrine, Smart Mater. Struct., **13**, N86-N92 (2004).
7. Toshiro, Zairyou Kagaku, **32**, 59 (1995) [In Japanese].
8. Y. Osada, H. Okuzaki and H. Hori, Nature, **355**, 242 (1992).
9. T. Fukushima, K. Asaka, A. Kosaka and T. Aida, Agrew Chem. Int. Ed., **44**, 2410 (2005).
10. Y. Nakabo, T. Mukai, K. Asaka and K. Ogawa, WW-EAP Newsletter, **7**, 22 (2005).
11. W. Yim, J. Lee and K. J. Kim, Bioinspiration and Biomimetics, **2**, 531 (2007).
12. K. Kaneto and M. Kaneko, Applied Biochemistry and Biotechnology, **96**, 13 (2001).

13. S. Hara, T. Zama, A. Ametani, W. Takashima and K. Kaneto, J. Mater. Chem., **14**, 1516 (2004).
14. S. Hara, T. Zama, W. Takashima and K. Kaneto, Synth. Met., **146**, 199 (2004).
15. K. Kaneto, Oyo Butsuri, **75**, 318 (2006).
16. S. Hara, T. Zama, W. Takashima and K. Kaneto, Polym. J., **36**, 933 (2004).
17. S. Hara, T. Zama, W. Takashima and K. Kaneto, Smart Mater. Struct., **14**, 1501 (2005).
18. S. Hara, T. Zama, W. Takashima and K. Kaneto, Polym. J., **36**, 151 (2005).
19. K. Kaneto, H. Fujisue, M. Kunifusa and W. Takashima, J. Smart Mat. and Struct., **16**, S250 (2006).
20. K. Kaneto, H. Fujisue, K. Yamato and W. Takashima, Thin Solid Film, **516**, 2808 (2008).
21. H. Fujisue, T. Sendai, K. Yamato, W. Takashima and K. Kaneto, Bioinsp. Biomim., **2**, S1 (2007).
22. K. Kaneto, H. Suematsu and K. Yamato, Bioinsp. Biomim., **3**, 035005 (2008) (6pp).
23. W. Takashima, M. Nakashima, S. S. Pandey and K. Kaneto, Electrochimica Acta, **49**, 4239 (2004).
24. M. Fuchiwaki, W. Takashima and K. Kaneto, Mol. Crys. Liq. Crys., **374**, 523 (2002).
25. K. Kaneto, H. Suematsu and K. Yamato, Advances in Science and Technology, **61**, 122 (2008).
26. T. F. Otero, Advances in Science and Technology, **61**, 112 (2008).
27. T. Sendai, H. Suematsu and K. Kaneto, Jpn. J. Appl. Phys., **48**, 051506 (2009) (4p).
28. E. Smela, W. Lu and B. R. Mattes, Synthetic Metals, **151**, 25 (2005).

Part 3
Polymer Electronics

CHAPTER 12

SURFACE PLASMON EXCITATIONS AND EMISSION LIGHTS IN NANOSTRUCTURED ORGANIC FILMS

Keizo Kato

Graduate School of Science and Technology, Niigata University,
2-8050 Ikarashi, Nishi-ku, Niigata 950-2181, Japan
E-mail: keikato@eng.niigata-u.ac.jp

The research is reviewed on the attenuated total reflection (ATR) method utilizing surface plasmon (SP) excitations at the interface between metal and dielectric ultrathin films, that is, the surface plasmon resonance (SPR) method. The ATR method is quite useful for evaluation of the dielectric ultrathin films and applications to sensors and plasmonic devices. Emission lights due to SP excitations are observed through the prism in the Kretschmann configuration for the ATR method, when metal films on the prism or organic films with metal films on the prism are directly irradiated from air by a laser beam, that is, reverse irradiation. The emission lights depend on the surface roughness of the films and are also related to the photoluminescence of organic molecules. The properties and applications of the SP emission lights are described. Among the SP excitation methods, one of the most well known techniques is a prism coupling technique based on the ATR configuration, while a grating coupling technique can be widely used for the sensing or plasmonic devices. The grating coupling excitation method is also described.

1. Introduction

The investigation of properties in organic ultrathin films is very important and many studies on the electrical and optical devices have been carried out. For the development of organic ultrathin film devices with high efficiency, it is quite important particularly to evaluate the structure and optical functions of the ultrathin films. The attenuated total reflection (ATR) method utilizing surface plasmon (SP) excitation at the interface between metal and dielectric ultrathin films is one of very useful techniques for evaluation of dielectric properties of the ultrathin films and sensing.[1,2] SPs are a coupling mode of free electrons and light, and they can be resonantly excited on metal surfaces in the Kretschmann and the Otto configurations by electromagnetic waves due to the total reflection of a p-polarized laser beam.[1] SPs are two-dimensional optical waves and propagate along the surfaces with strong electromagnetic waves, that is, evanescent waves, which decay exponentially away from the surfaces. SPs are also utilized in near-field optics where electromagnetic waves with light frequency are localized in structures smaller than the light wavelength.[3] It is considered that SPs are very important and useful for the application to optical nano-devices.[3]

Recently, emission lights have been observed through the prism in the Kretschmann configuration for the ATR method, when metal ultrathin films on the prism or organic ultrathin films with metal ultrathin films on the prism were directly irradiated from air by a laser beam, that is, reverse irradiation.[3-6] The emission lights depend on the surface roughness of the films.[6] The intensities and the spectra of the emission lights strongly depend on the emission angles and the emission lights are also related to the photoluminescence of organic molecules.[7-9] The relation between the peak wavelengths of the spectra and the emission angles corresponded to the dispersion property of the resonant excitations of SPs in the ATR configuration.[7] From the dispersion properties, it was estimated that the light emission was caused by multiple excitations of SPs.[7-9] The SP emission light property also depended on the nanostructures and the strongest emission was observed at some separation of the dye molecule from the metal surface.[8] The intensity of SP emission light was found to depend on the amount of dyes and nano-spacing between the metal and dyes.[8,9] The SP emission light also depended on the polarization direction of excitation light, which was due to the molecular shape of the dyes.[10]

Among the SP excitation methods, one of the most well-known techniques is the prism coupling technique based on the ATR configuration. The grating coupling technique has not been widely used for SPR sensors or plasmonic devices.[11-14] One major advantage of using the grating-coupling excitation method is the fact that a prism is not necessary; hence, inexpensive and disposable plastics can be used as the substrates, allowing for more flexible configurations.[15,16]

2. SP Excitation and ATR Method

2.1. *SP Excitation*

Figure 1(a) shows the reflection of the p-polarized light at the interface at the interface between a prism and air. The condition of the total reflection of the incident light is given by

$$\theta_i > \theta_c = \sin^{-1}(n_1/n_2) \qquad (1)$$

where θ_i is the incident angle of the incident light, θ_c is the critical angle of the total reflection, and n_1 and n_2 indicate the reflective indices of the prism and air, respectively. Figure 1(b) shows the electric field pattern of the p-polarized incident light at the interface and the evanescent wave produced at the interface.

Figure 2 shows the SP excitation at the interfaces of metal thin films. The electric fields of p-polarized incident light generate vibration of free electrons of the metal thin film and excite the SP. The SP excitations at the interfaces between the metal thin film and air and between the metal thin film and organic ultrathin film are shown in Figs. 2(a) and 2(b),

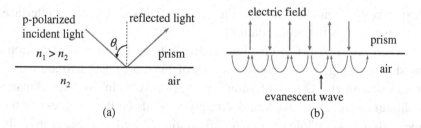

Fig. 1. Reflection of p-polarized incident light and the evanescent wave produced at the interface between prism and air. (a) indicates the total reflection at the interface and (b) indicates the electric field pattern of the p-polarized incident light at the interface and the produced evanescent wave.

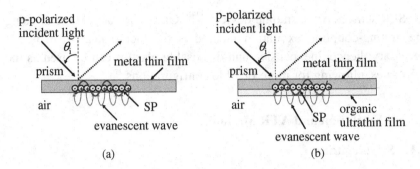

Fig. 2. SP excitation at the interface between metal thin film and air (a) and at the interface between metal thin film and organic ultrathin film (b).

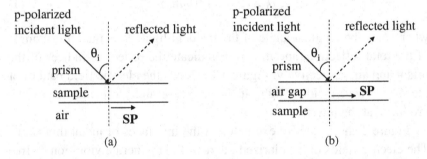

Fig. 3. Prism coupling configurations for the SP excitations. (a) and (b) represent Kretschmann and Otto configurations, respectively. The sample is metal film or metal film with dielectric film.

respectively. You can find that the electric fields of the metal interface are enhanced by the SP excitation.

There are two prism coupling configurations for the SP excitations as shown in Fig. 3. The configurations of Figs. 3(a) and 3(b) are called Kretschmann and Otto configurations, respectively. In the Kretschmann configuration, the SP excitation strongly depends on the thickness of the metal thin film. While the Otto configuration, the SP excitation strongly depends on the thickness of the air gap. The Kretchmann configuration is useful in the visible and ultraviolet region, while the Otto configuration is useful in the infrared and far infrared region.[1]

2.2. Method of ATR, Scattered Light and Emission Light Measurements Utilizing SP Excitations

The sample configurations for the measurements of ATR, scattered lights and emission lights utilizing SP excitations are shown in Fig. 4. Figure 4(a) indicates the ATR and scattered light measurements and Fig. 4(b) indicates the emission light measurement. The glass substrates with organic ultrathin films on metal thin films are brought into optical contact with half-cylindrical prisms using an index matching fluid. The sample configurations are Kretschmann configuration shown in Fig. 3(a).

The experimental system for the measurements utilizing SP excitations is shown in Fig. 5. The samples shown in Fig. 4 are mounted on a computer-controlled goniostage and the incident angle θ_i is changed by rotating the goniostage. For the ATR measurement, the p-polarized light is directed on to the back surface for the metal thin films through the prism. The intensity of the reflected light is detected as a function of θ_i. The reflectivity, that is, the ATR value is obtained from the ratio of the intensities of the reflected light to the incident light. Thus you can obtain the ATR curves as a function of θ_i. For the scattered light measurement, the intensity of the scattered light is observed as a function of the scattered angle θ_s when the incident angle θ_i is set as the resonant angle of the ATR curve. For the emission light measurement, the intensity of the emission light is observed as a function of the emission

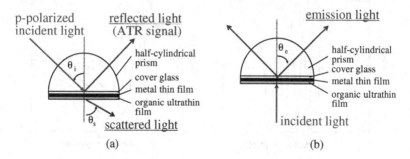

Fig. 4. Sample configurations for the measurements utilizing SP excitations. (a) indicates the sample configuration for ATR and scattered right measurements and (b) indicates that for the emission light measurement by the reverse irradiation.

angle θ_e when the organic ultrathin film with metal ultrathin film on the prism is directly irradiated from air by the incident light, that is, reverse irradiation.[3-6]

Figure 6 shows the calculated ATR curves for dielectric films with the thickness of d nm on metal thin films. The wavelength of the incident

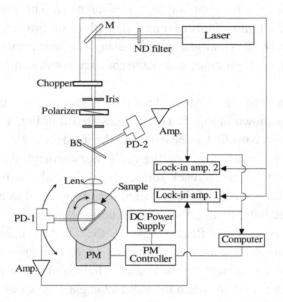

Fig. 5. Experimental system for the measurements utilizing SP excitations.

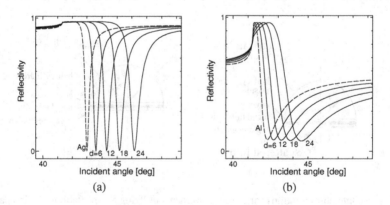

Fig. 6. Caluculated ATR curves for dielectric ultrathin films with the thickness of d [nm] on Ag thin film (a) and on Al thin film (b).

light is assumed to be 632.8 nm in the calculation. For the prism, the complex dielectric constant of 2.295 + i·0, that is, the refractive index of 1.51 for BK-7 glass is used. For the dielectric films, the complex dielectric constant of 2.3 + i·0, that is, the refractive index of 1.50 is used. The dielectric constants and thicknesses of -16.5 + i·0.7 and 50 nm for Ag and -45 + i·17.5 and 15 nm for Al are used. It is though that the ATR curves for the dielectric ultrathin films on Al thin films have a wider half-width than those for dielectric ultrathin films on Ag thin films.

3. Evaluation of Organic Ultrathin Films on Metal Thin Films by ATR Method

3.1. *Evaluation of Surface Roughness of Organic Ultrathin Films by Scattered Light Measurement*

The surface roughness of organic ultrathin films on metal thin films has been evaluated from the scattered light measurement utilizing SP excitation.[17-19] Figure 7 shows the sample configuration used for the scattered light measurement of arachidic acid (C20) LB ultrathin films and a model of surface roughness of the film. I_i is the p-polarized incident light intensity at the incident angle θ_i, and I_r is the reflected light intensities. The scattered light intensity dI_s at the scattered angle θ_s per solid angle $d\Omega$ normalized to be p-polarized incident light can be expressed by[1,20]

$$\frac{dI_s}{I_i d\Omega} = \frac{4\varepsilon_p^{1/2}}{\cos\theta_i} \frac{\pi^4}{\lambda^4} |F|^2 |s(\Delta k)|^2 \tag{2}$$

where $\Delta k = (2\pi/\lambda)(\varepsilon_p^{1/2}\sin\theta_i - \sin\theta_s)$. Here, ε_p is the dielectric constant of the prism, and λ is optical wavelength of the incident light, and $|s(\Delta k)|^2$ is a roughness spectrum. F is a function of Fresnel's reflection transmission coefficients and also depends on θ_i and θ_s. Assuming a Gaussian distribution as an autocorrelation function with the surface corrugation depth δ and the transverse correlation length σ, $|s(\Delta k)|^2$ is given by[1,20]

$$|s(\Delta k)|^2 = \frac{\delta^2 \sigma^2}{4\pi} \exp\left[-\frac{\sigma^2}{4}(\Delta k)^2\right] \tag{3}$$

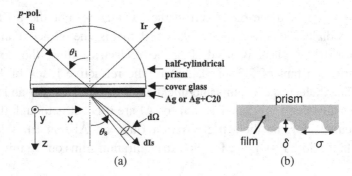

Fig. 7. Sample configuration used for the scattered light measurements (a) and a model of surface roughness (b). θ_i and θ_s are the incident and scattered angles, respectively. I_i and I_r are the incident and reflected light intensities, respectively. dI_s is the scattered light intensity per solid angle $d\Omega$. δ and σ are the surface corrugation depth and the transverse correlation length of an autocorrelation function, respectively.

Equation (1) can be extended for the multi-layered rough surfaces. Assuming that the scattered light intensity is described by a linear superposition of the roughness spectra, several pairs of the roughness parameters can be obtained.[21] Therefore, the scattered light spectra for the C20 LB ultrathin films on the Ag thin films can be described by a linear superposition of the roughness spectra emitted from the interfaces between the metal and LB films and between the LB films and air.

The angular distributions of the scattered light intensities from the Ag thin film and C20 LB ultrathin films on the Ag thin film are shown in Fig. 8(a). The theoretical scattered lights from C20 LB ultrathin films on the Ag thin films were calculated from Eq. (1) using the complex dielectric constants and thicknesses obtained from the ATR measurements. Four pairs of σ/δ for the Ag thin film were used as the parameters of the interface between the Ag and LB films. Since the scattered light in the range of $90° < \theta_s < 0°$ were mostly caused by the roughness of Ag film surfaces, the value of δ after the LB film deposition were modified by fitting the theoretical curves to the experimental ones in the range. The theoretical scattered light from the LB surface was obtained by subtracting the calculated curves of Ag from the experimental data of C20 LB films on Ag.

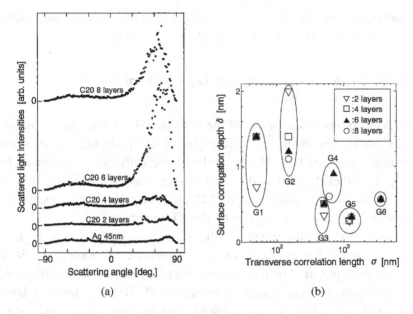

Fig. 8. Scattered light measurements for the arachidic acid (C20) ultrathin films on Ag thin films utilizing SP excitation. (a) shows the angular distributions of the scattered light intensities for the Ag thin film and C20 LB films on the Ag thin film. (b) shows the calculated roughness parameters of C20 LB films with different number of monolayers on Ag thin films.

In the evaluation, the light was considered to be scattered from the interfaces between Ag and LB films and between LB film and air. The calculated roughness parameters of C20 LB films with different number of monolayers on Ag thin films are shown in Fig. 8(b). The δ increased with the number of monolayers. The large δ with small σ in the surface roughness was observed prominently for the C20 LB film with 6 monolayers. From the AFM images of the C20 LB films, the surface roughness with large σ is observed corresponding to the roughness at the interfaces between the cover glass and Ag film and/or between the Ag and LB films. Very large objects with 200-400 nm in diameter were observed with increasing the number of monolayers. The roughness with σ smaller than 100 nm appeared in the LB film with 6 monolayers. The roughness averaged in the measured region increased with the number of monolayers from 2 to 6. The result of the scattered light measurements qualitatively agreed with that of the AFM measurements.

3.2. *Evaluation of Orientations of Liquid Crystal Molecules in a Cell by ATR Method*

The ATR measurement is one of the most useful methods for evaluation of LC molecules in-situ.[22,24] Tilt angles of LC molecules close to the aligning layers and bulk LC molecules in the cells can be estimated by analyzing SP excitations[1] and guided wave excitation modes (GWEM).[23] The photo-induced alignment of LC molecules using photo-sensitive molecules has been investigated by the ATR method.[25]

Figure 9 shows the sample configuration for the ATR measurement of photo-induced in-plane alignments of nematic LC molecules, 5CB. The LC cells prepared with alternate Direct Red 80 (DR80: azo dye) and poly(diallyldimethylammonium chloride) (PDADMAC) layer-by-layer self-assembled films on Au.[25] DR80 contains azo groups exhibiting photo-isomerization and has been used for photo-induced alignment of LC molecules.[25] PDADMAC was used to prepare self-assembled films of well-defined thickness and order. Half-cylindrical prisms (refractive index $n = 1.85$) were used. The self-assembled films had 10 pair-layers on the Au-coated prism and the slide glass. In-plane alignment of the LC molecules on the self-assembled films could also be controlled by the polarization direction of irradiated visible light on the LC cell.

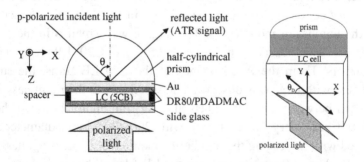

Fig. 9. Sample configuration for the ATR measurement of the LC cell using polycation (PDADMAC) and azo dye (DR80).

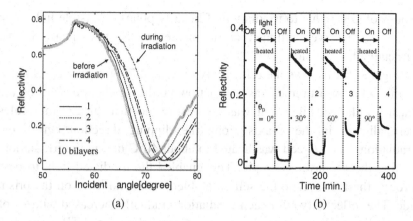

Fig. 10. Evaluation of LC cell using ATR method. (a) shows the ATR properties during irradiation of the polarized light. The directions of polarized light: curve 1 at $\theta_D = 0°$, curve 2 at $\theta_D = 30°$, curve 3 at $\theta_D = 60°$, curve 4 at $\theta_D = 90°$. (b) shows the reflectivity change as a function of the time during irradiation of the polarized light at 70.9°.

Figure 10(a) shows the experimental ATR curves in the region of the resonant angles of the SP excitations for the LC cell at room temperature except the curve during irradiation.[26] Experimental curves were measured before irradiation of the linearly polarized visible light. Curves 1-4 were each measured sequentially after 1-hour irradiation of the visible light, where the angles between the X-axis of the cell, and the direction, that is, θ_D were set to be 0°, 30°, 60° and 90° for the curves from 1 to 4, respectively. The ATR properties for only the aligning layers on the gold film did not exhibit such shifts in the incident angles that increased with the resonant angles. The results showed that the ATR measurements were very sensitive to the irradiation of the polarized light, which induced alignment of the LC molecules in the cell. The resonant angles of the SP excitations, that is, the angles at the minimum reflectivities, increased with θ_D from 0 to 90° in the X-Y plane. From the theoretical ATR curves, the results in Fig. 10(a) indicate that the LC molecules within the penetration length of the SP evanescent fields are aligned perpendicular to the polarized direction of the irradiation light. It was thought that the trans-phase decreased in the polarized direction by cis–trans transitions or trans–cis–trans photo-isomerization of the azo groups of DR80 due to irradiation of linearly polarized visible light. The LC molecules were aligned in the plane by the trans-phase in the perpendicular direction.

Figure 10(b) shows the reflectivity change at 70.9° of the fixed angle in the ATR measurement as a function of the irradiation time of the linearly polarized light.[26] It was also estimated that the LC molecules were aligned in the Y-axis along the dipping direction before the irradiation. The LC cell was heated to about 40°C during the irradiation, making the LC layer isotropic. This enabled the irradiated light to pass through the LC layer to the self-assembled command layer on the prism side. The reflectivity after each irradiation gradually increased with θ_D of the irradiation light. The change corresponded to the ATR properties in Fig. 10(a). This result clearly shows that the alignments of the LC molecules were controlled by the polarized direction of the irradiation light.

Figure 11 shows a model of the preferred alignments of the LC molecules. It can be derived that the in-plane alignments of the LC molecules within the penetration length of the SP evanescent fields were aligned perpendicular to the polarized directions of the irradiation light and that the tilt angle was almost 0°.

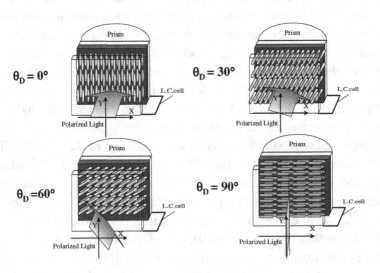

Fig. 11. A model of the in-plane alignments of the LC molecules estimated by ATR mesurements.

3.3. Application of SP Excitations to Organic Photoelectric Cell

In the ATR configurations utilizing SP excitations, remarkably strong optical absorption is expected for organic ultrathin films in photoelectric cells.[27-30] Figure 12(a) shows the layer structure of the photoelectric cell,[29] which enables the SP excitations on the Al surface at the interface between MgF_2 and Al films due to the Otto configuration and on the Ag surface at the interface between Ag film and air due to the Kretschmann configuration, respectively. The MgF_2 film is transparent and it has no optical absorption. Merocyanine (MC) LB film with 10 monolayers is used for the photoelectric cell and it exhibits p-type conduction. Therefore, the Schottky contact is obtained at the interface between MC LB and aluminum films and the Ohmic one is obtained at the interface between MC LB and Ag films. Photoelectric effects have been reported for the Schottky diode using MC dyes.[31]

Figure 12(b) shows the ATR properties of the photoelectric cell measured at 594.1 nm.[29] The theoretical ATR curve was calculated for the structure of prism/MgF_2/Al/Al_2O_3/MC LB film/Ag and fitted the experimental result to obtain the thickness of the MgF_2 layer. Large dip at around 74° and small ones at around 43° were caused by the SP excitations at the interface between MgF_2 and Al films and at the

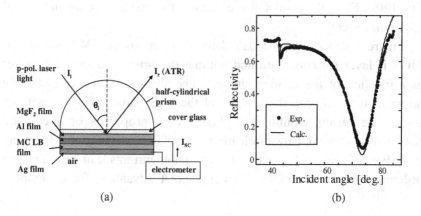

Fig. 12. Structures of photoelectric cell utilizing SP excitations (a) and ATR properties measured at 594.1 nm (b).

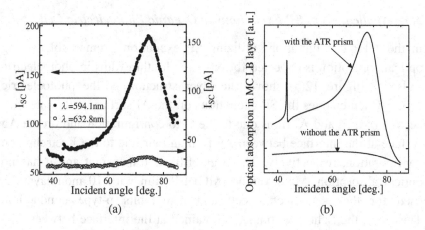

Fig. 13. Short-circuit current I_{SC} at 594.1 nm and 632.8 nm (a), and calculated absorptions in the MC LB layer with and without the prism at 594.1 nm (b).

interface between Ag film and air, respectively. The strong excitation was obtained at the interface between MgF_2 and Al films because of the strong evanescent fields from the prism.

Figure 13(a) shows the short-circuit photocurrent I_{SC} curves as a function of the incident angle of two laser beams at 594.1 nm and at 632.8 nm.[29] The angles of the I_{SC} peaks almost corresponded to the resonant angles of the ATR curves due to the SP excitations shown in Fig. 12(b). From the results, it is estimated that the SP excitations caused the I_{SC} in the cell.

Figure 13(b) shows the calculated absorptions at 594.1 nm in the MC LB layer for cells with and without the prism and the MgF_2 layer as a function of the incident angle.[29] The absorption was considerably larger in the cell with the prism and the MgF_2 layer than one without them. The results exhibit that photoelectric properties of a cell are remarkably improved by fabricating the ATR configuration generating the SP excitations. It is expected the photocurrents can be enhanced utilizing the SP excitations. This is a great advantage for developing solar cell etc.

4. SP Emission Lights

4.1. *Emission Lights from Organic Ultrathin Films*

Figure 14 shows the Kretschmann configuration and a system for detecting emission lights through the prism when the sample is excited by reverse irradiation in the configuration. The organic ultrathin film sample is irradiated at the vertical incidence angle by a p-polarized Ar^+ laser beam at 488.0 nm. The emission lights are observed through the prism in the configuration. For the emission light measurement, merocyanine (MC) dye was used, and the MC LB films and arachidic acid (C20) LB films were prepared as the organic ultrathin films.[8] The samples have the structures of prism/Ag/C20/MC/C20 or prism/Ag/C20/MC/C20. The spectra of the emission lights were measured using a sharp-cut filter below about 520 nm at various emission angles θ_e, where the light was observed.

Figure 15(a) shows emission light from prism/Ag/C20/MC/C20 samples.[8] Arabic numerals 10L/4L/6L etc. in Fig. 15 represent the number of layers for the C20 on Ag or Al film, MC and C20 LB films, respectively. The emission properties depended upon the position of the MC LB films in the hetero films, and the intensities increased with the

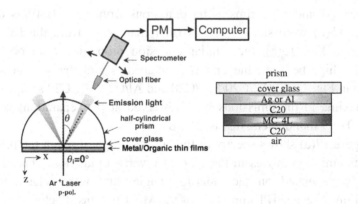

Fig. 14. Sample configuration in the reverse irradiation method for emission light measurements of MC LB film samples.

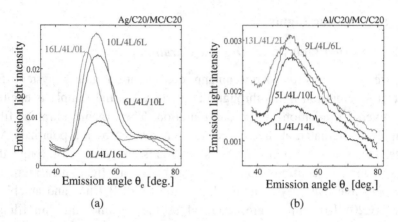

Fig. 15. Emission light properties from MC samples with various separations between metal and MC films for prism/Ag/C20/MC/C20 samples (a) and prism/Al/C20/MC/C20 samples (b).

separation between the MC and Ag films and showed a maximum at the separation of C20 with 10 monolayers. Emission properties from prism/Al/C20/MC/C20 samples with a constant thickness of 19 LB monolayers were also investigated depending upon the separation between Al thin film and the MC LB films. Figure 15(b) shows emission light from prism/Al/C20/MC/C20 samples with various separations between Al and MC films.[8] The emissions from the LB films on Al in Fig. 15(b) were smaller and broader than those from the LB films on Ag in Fig. 15(a), but similar emission properties were observed. Relationships between the emission intensities and the separation are shown in Fig. 16 for Ag/C20/MC/C20 and Al/C20/MC/C20 samples, and the maxima in the emissions were observed at the separations of C20 with 10 or 9 monolayers, that is, 25-28 nm.[8]

Figure 17(a) shows spectra of the emission light through the prism at various emission angles in the case of reverse irradiation.[8] The spectra strongly depended on the emission angles and were related to the photoluminescence (PL) property of the MC LB films. Each spectrum in Fig. 17(a) almost corresponded to a part of the PL spectrum of the MC LB films showing a peak at about 600 nm. The emission spectra were investigated from the dispersion property of SP in the Kretschmann configuration.

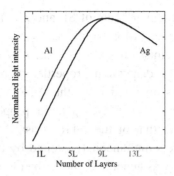

Fig. 16. Relation between emission intensities and separations of MC LB layers from metals.

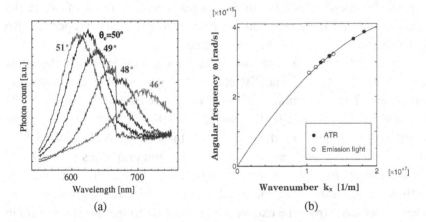

Fig. 17. Spectra of emission lights (a) and dispersion properties (b). Emission lights were measured for Ag/C20(2 layers)/MC(16 layers) sample at various emission angles. The dispersion properties were calculated from ATR and emission light properties.

The dispersion property of SP represents a relation between the wavenumber k_x of SP in the x-direction as shown in Fig. 14 and the angular frequency ω of the wavelength λ. k_x and ω are given by the following equations:

$$k_x = n_p k \sin\theta = n_p(2\pi/\lambda)\sin\theta \quad (4)$$

$$\omega = kc = (2\pi/\lambda)c \quad (5)$$

where n_P is the refractive index of the prism, k ($= 2\pi/\lambda$) is the wavenumber of light in air and c is the light velocity in air. For the ATR properties, θ is the resonant angle of SP and λ is the wavelength of the incident light. For the emission light measurement, θ is the angle of the peak intensity of emission spectrum and λ is the peak wavelength. Figure 17(b) shows the dispersion properties. Closed and open circles indicate the properties obtained from the ATR and the emission light measurements, respectively. The dispersion property of the emission lights agreed well with that of the ATR measurements. It was thought that multiple SPs were simultaneously excited by the reverse irradiation and the emission light was generated due to the dispersion property of SP in the Kretschmann ATR configuration of the prism/Ag/C20/MC LB film. Similar emission lights through the prism are reported for the Ag/Rhodamine-B (RB) LB film by the resonant excitations of SPs in the conventional ATR method and the calculated dispersion properties also coincide with one of the ATR measurements.[7]

Figure 18 shows the emission light for luminescent molecular films due to SP excitation. The SP emission lights involve the following processes. The excitation of SPs directly occurs by the excited MC molecules and/or through the surface roughness, and then the light emission is coupled with the excited SPs through the ATR prism, where the SP emission is influenced largely by the film properties with the MC molecules and/or by the metal surface roughness. It is tentatively estimated that small emissions are due to non-radiative energy transfer or charge transfer[32] from the excited MC molecules to the metal thin film in the small separations and due to smaller induced charges on the metal thin film in the large separations. The electromagnetic modes localized at

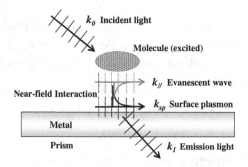

Fig. 18. Emission light for luminescent molecular films due to SP excitation.

the surface are induced by dipoles with electromagnetic waves and have suggested that the dipole-surface interaction will be very important for device applications utilizing near-field optics. It is thought that the phenomenon is very useful for nanostructured device applications.

4.2. SP Emission Lights due to Molecular Luminescence and Interaction

SP emission lights have been investigated for rhodamine-B (RB) LB films. RB is a photosensitizing organic dye showing strong photoluminescence. Figure 19 shows the emission lights as a function of emission angle for the prism/Ag/C20/RB LB films with various separations between the Ag and the RB LB films under the reverse irradiation with two polarized laser beams.[9] The SP emission lights in Figs. 19(a) and 19(b) strongly depend upon the thickness of the separations and show the maxima with the separation of 8 layers. The properties depending upon the separation were similar to that of the photoluminescence. The emission peaks by the p-polarized laser at 0° with the electric fields parallel to the observation plane were much larger than those by the s-polarized one at 90° and the peaks by the s-polarized beam were approximately 15% of those by the p-polarized one.

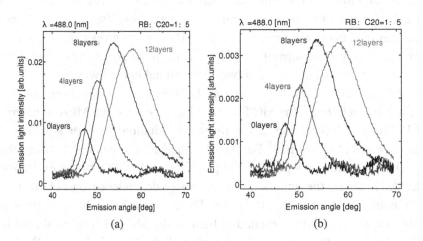

Fig. 19. Emission light for the RB LB film using p-polarized light (a) and s-polarized light (b) at 488.0 nm under reverse irradiation.

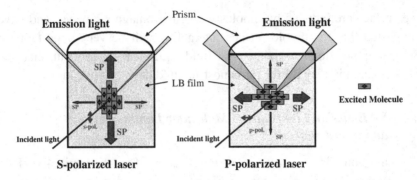

Fig. 20. SP emission due to p- and s- polarized laser beams in the reverse irradiation.

Since orientations of the RB dyes were thought to be almost uniform in the LB film plane, it is tentatively estimated that some RB dyes having the long axis parallel to the polarized plane of the laser were mainly excited by the reverse irradiation, and these excited dyes induce anisotropic SPs which propagate along the Ag surface, which can be observed mainly in the observation plane of the half-cylindrical prism. Figure 20 shows the SP emission due to p- and s- polarized laser beams in the reverse irradiation. Anisotropic molecules are excited mainly by the polarized laser beam with the electric fields parallel to the molecular transition moment. The emission phenomenon includes conversions from three- to two-dimensional optical waves and those from two to three-dimensional ones. It is thought that the SP emission properties depending upon the very short separation of the RB from the Ag are very useful for conversion between three- and two-dimensional optical waves.

Figure 21 shows the emission light curves as functions of the emission angle θ_e for the MC LB film (16 layers) and the MC/CV hetero-LB film (16 layers) due to the reverse irradiation of p- or s-polarized laser beams at 488.0 nm.[10] For the MC and MC/CV hetero-LB films, the peak angles did not depend on the polarization direction of the excitation light. However, the emission light intensity dependence on polarization direction differed between the films. The SP emission light intensity was high for the p-polarized excitation light of the MC LB film, as shown in Fig. 21(a). For the emission light curves of the MC/CV hetero-LB film, the light intensity did not depend on the polarization direction of the

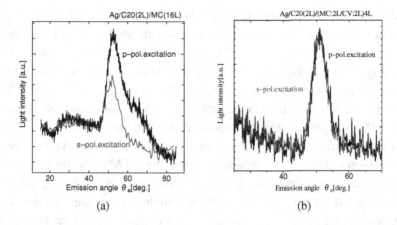

Fig. 21. Emission light intensity curves of MC LB film (a) and MC/CV hetero-LB films (b) observed using p- and s-polarized excitation lights.

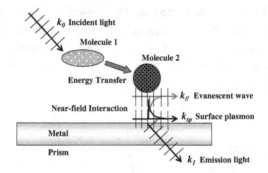

Fig. 22. Emission light due to SP excitations via energy transfer between molecules.

excitation light. The spectra of SP emission lights strongly depend on θ_e. Figure 22 shows the emission light due to SP excitations via energy transfer between molecules.

4.3. *Application of SP Emission Lights to Organic Devices*

Figure 23 shows the emission lights measured at 488.0 nm from the prism/Ag thin film/ Aluminum phthalocyanine chloride (AlClPc) thin film structure.[33] The intensity of the emission light is normalized as the peak intensity of the Ag thin film is unity. The peak angles of the emission light curves correspond to resonant angles of ATR curves, which suggests that the emitted lights were due to SP excitations. The intensity of SP emission light through the prism strongly depends on the nanostructures such as surface roughness of the organic and metal thin films in the Kretschmann configuration. Figures 24(a) and 24(b) show the AFM images of the as-deposited AlClPc thin film on Ag thin film and the film treated with ethanol vapor for 10 min, respectively. Themean roughnesses R_a evaluated from the AFM measurement were 1.6 nm and 9.7 nm for the as-deposited and ethanol-vapor-treated AlClPc thin films, respectively. The surface roughness of the AlClPc thin film was found to increase markedly with ethanol-vapor treatment. The increase of 8.1 nm in the surface roughness of the film due to the ethanol-vapor treatment is extremely large against the thickness of about 10 nm for the as-deposited film. Therefore, the result of the SP emission lights in Fig. 23 was considered to be caused by the large surface roughness of the AlClPc thin film due to the treatment with ethanol vapor. The SP emission light due to molecular luminescence is also applied to the sensing devices using poly(vinyl alcohol) films with fluorescent microsphers.[34]

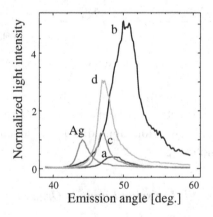

Fig. 23. SP emission lights measured at 488.0 nm from the prism/Ag thin film/aluminum phthalocyanine chloride (AlClPc) thin film structure. a: as-deposited, b: ethanol-vapor-treated, c: H_2O-vapor-treated, d: H_2O-vapor-treated after ethanol-vapor-treated.

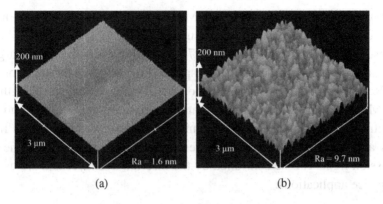

Fig. 24. AFM images of AlClPc thin films on Ag thin films before and after the treatment with ethanol vapor. (a) and (b) show the results for the asdeposited and ethanol-vapor-treated AlClPc thin films, respectively.

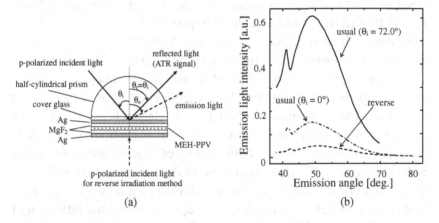

Fig. 25. Sample configuration of ATR and SP emission light measurements for the Ag/MgF$_2$/MEH-PPV/MgF$_2$/Ag structure (a) and the emission light properties for the Ag/MgF$_2$/MEH-PPV/MgF$_2$/Ag sample measured at 488.0 nm by the usual and reverse irradiations (b).

The ATR and SP emission lights have been investigated in the prism/Ag/MgF$_2$/organic dye/MgF$_2$/Ag structure. Poly(2-methoxy-5-(20-ethyl-hexyloxy)-1,4-phenylenevinylene) (MEH-PPV), which is a photoluminescent dye, was used as the organic dye. Figure 25(a) shows the sample configuration.[35,36] The emission light properties for the prism/Ag(25 nm)/MgF$_2$(100 nm)/MEH-PPV(40 nm)/ MgF$_2$(60 nm)/ Ag(40 nm) structure are shown in Fig. 25(b).[35] The emission light measured at the incident angle θ_i of 72.0° of SP excitation had the largest intensity. The emission angle corresponded to the resonant angle of the ATR curve measured at around the photoluminescence wavelength of the MEH-PPV dye. It was found that the emission light intensity and the spectra strongly depended on the emission angles and the emission light was also related to the photoluminescence of the organic dye. The SP emission lights can be controlled by the inserted dye layer and are useful to device applications.

4.4. *Electrochemical SP Excitations and Emission Lights*

SP excitation and the emission light properties are investigated in the doped and dedoped states of regioregular poly(3-hexylthiophene-2,5diyl) (P3HT) thin films.[37] Electrochemical experiments were performed in a conventional three-electrode cell with the Au/SFL11 glass substrate as the working electrode, a platinum wire as counter electrode, and an Ag/Ag$^+$ nonaqueous reference electrode in acetonitrile solution with 0.1 M tetrabutylammonium hexafluorophosphate (TBAPF$_6$) as a supporting electrolyte. The experimental setup for the electrochemical ATR and emission light measurements utilizing the SP excitations is shown in Fig. 26.[37] A Kretschmann configuration of highly refractive prism (SFL11)/Cr/Au/P3HT thin film/acetonitrile solution containing an electrolyte (TBAPF$_6$) structure was used for the measurements. SPs were excited at the metal–dielectric interface, upon total internal reflection of polarized light from a He–Ne laser with the wavelength of 632.8 nm. The optical/electrochemical processes at the Au thin film were detected by monitoring the reflectivity as a function of the incident angle. The SP emission light was obtained by the irradiation of Ar$^+$ laser beam with thewavelength of 488.0 nm. The P3HT thin filmwas luminous upon the light irradiation, and excited SP emission light was measured. A sharp-cut filter that eliminates incident light was used for the measurement of SP emission light intensity. The SCF enables the observation of only SP emission light due to excited fluorescent organic molecules. A photomultiplier tube was used for detecting the emission light.

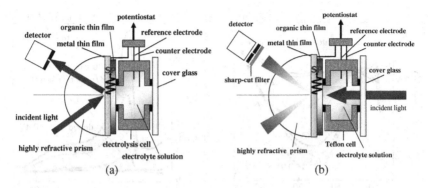

Fig. 26. Experimental setup for the electrochemical measurements of ATR (a) and SP emission lights (b).

A constant potential was applied to the P3HT thin film and the change of SP emission light properties was observed upon doping and dedoping of the P3HT thin film. The SP emission light is observed due to the molecular luminescence of P3HT.

The electrochemical ATR properties of the P3HT thin film applied at constant potentials are shown in Fig. 27(a) for the doped-state P3HT thin film.[37] It is found that the dip of the ATR curve becomes deeper with increase of the applied potential and the resonant angle changes with the doping of the P3HT thin film. From the ATR properties, the dielectric constants of the P3HT films are found to change remarkably with the doping. On the contrary, the ATR curve hardly changes with the dedoping. These ATR properties correspond to the optical absorption properties well.

Figure 27(b) shows the electrochemical SP emission light properties of the P3HT thin film applied at constant potentials.[37] The SP emission light in the undoped state is observed at the emission angle of around 52°. The intensity of the SP emission light decreases in the doped state at 0.2 V. At 0.4 V, the SP emission light is hardly observed. The emission light properties correspond to the resonant conditions of SPs in the Kretschmann configuration, and it is considered that multiple SPs are induced due to the excitation of P3HT molecules by the incident light and that the SPs are converted into emission light through near-field coupling. It is considered that the SP emission light excited by the

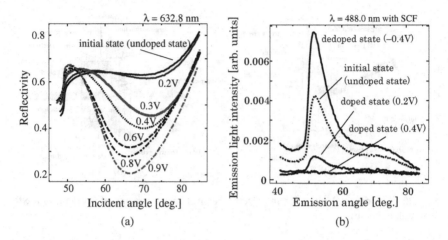

Fig. 27. Electrochemical ATR and SP emission light properties of the doped-state P3HT thin film applied at constant potentials. (a) ATR properties, (b) SP emission light properties.

molecular luminescence decreased since the molecular luminescence of P3HT became weaker at the doped state. On the other hand, since a strong molecular luminescence of P3HT is obtained in the dedoped state, it is considered that the SP emission light in the dedoped state increased as can be seen in Fig. 27(b). The reversible change of the SP emission light was observed by the doping and the dedoping. The SP emission light excited by molecular luminescence can be controlled by the control of doping–dedoping state.

4.5. *Grating Coupling SP Excitations and SP Emission Lights*

Figure 28 shows the configuration for the diffraction grating SP excitation. The SP dispersion branches at the interface between the metal grating and air. These branches are obtained by the following equation:

$$k_{sp} = k_{px} + G = \frac{2\pi}{\lambda}\sqrt{\varepsilon_m(\omega)}\sin\theta + \frac{2\pi}{\Lambda}m \qquad (6)$$

where Λ is the diffraction grating pitch, λ is the wavelength, m is the diffraction orders, and $\varepsilon_m(\omega)$ is the wavelength-dependent dielectric

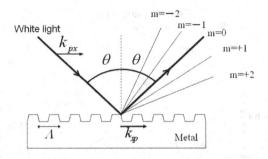

Fig. 28. Configuration for the diffraction grating SP excitation.

constant of the metal grating. Grating coupling SP excitation was studied upon an irradiation of white light on CD-R metallic diffraction gratings/plastic substrates.[38] As the grating pitch was large enough, multimode SP excitations could be obtained. The experimentally obtained SP dispersion was corresponded well with theoretical SPR dispersion curves. Grating coupling SP excitations are useful for the application to organic devices such as sensors and new SP enhanced devices.

SP emission light due to molecular luminescence, that is, SP-coupled photoluminescence was studied on MEH-PPV thin film/silver grating samples. Figure 29 shows the photoluminescence properties of MEH-PPV thin films on the silver grating and flat silver substrates (without grating).[38] As schematically shown in the inset, a laser at 543 nm was used as the excitation source at an incident angle of 0°, and the photoluminescence was detected as a function of emission angle. As shown in the figure, an increased photoluminescence peak was clearly observed on the silver grating at approximately 49°, while no peak was observed on the flat silver surface. It should be noted that the light intensity on the silver grating increased up to fivefold more than that on the flat surface. In order to investigate the origin of the increased photoluminescence intensity, the emission angle was compared with angular SPR reflectivity curves on the MEH-PPV/silver grating. The angular SPR reflectivity was measured at 632.8 nm, which roughly corresponds to the photoluminescence wavelength of MEH-PPV. The angular SPR reflectivity at the excitation wavelength of 543 nm was also

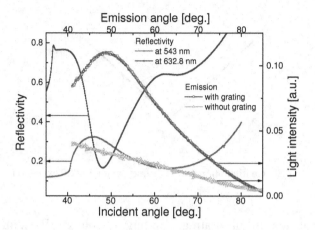

Fig. 29. Photoluminescence properties of MEH-PPV thin films on silver grating and flat silver substrates (without grating) and angular SPR reflectivity properties of MEH-PPV on silver grating measured at 543 nm and 632.8 nm.

measured. As can be observed from these figures, the SPR excitation angle at 632.8 nm almost corresponds to the emission angle. This indicates that the increased photoluminescence intensity is responsible for the coupling and emission of the SP excitation with photoluminescence from MEH-PPV.

5. Conclusions

ATR method utilizing SP excitations and the evaluation of nanostructured organic films were summarized in detail. The SP emission light properties strongly depend on the incident angles of laser lights, the emission angles of emission lights, and the thicknesses of dielectric and metal thin films. The SP emission lights are also related to the photoluminescence of the luminescent dyes. The SP emission lights are useful for application to new optical devices. The SP emission lights can be widely used for the sensing or plasmonic devices. The grating coupling SP excitations are also useful for the application to organic flexible devices such as sensors and novel SP enhanced devices.

Acknowledgments

This research was supported partly by Grant-in-Aid for Scientific Research from the Japan Society of the Promotion of Science. The author also expresses his thanks to Prof. F. Kaneko, Prof. K. Shinbo, Prof. A. Baba and Prof. Y. Ohdaira of Niigata University for this research.

References

1. V. M. Agranovich and D. L. Mills (eds.), "Surface Polaritons", North-Holland, Amsterdam (1982).
2. W. Knoll, Ann. Rev. Phys. Chem. 49, 569 (1998).
3. S. Kawata (ed.), "Near-Field Optics and Surface Plasmon Polaritons", Springer, Berlin (2001).
4. I. Pockrand, A. Brillante and D. Möbius, Chem. Phys. Lett. 69, 499 (1980).
5. S. Hayashi, T. Kume, T. Amano and K. Yamamoto, Jpn. J. Appl. Phys. 35, L331 (1996).
6. K. Kato, M. Terakado, K. Shinbo, F. Kaneko and T. Wakamatsu, Thin Solid Films, 393, 97 (2001).
7. T. Nakano, M. Terakado, K. Shinbo, K. Kato, F. Kaneko, T. Kawakami and T. Wakamatsu, Jpn. J. Appl. Phys. 41, 2774 (2002).
8. F. Kaneko, T. Sato, M. Terakado, T. Nakano, K. Shinbo, K. Kato, N. Tsuboi, T. Wakamatsu and R. C. Advincula, Jpn. J. Appl. Phys. 42, 2551 (2003).
9. F. Kaneko, K. Wakui, H. Hatakeyama, S. Toyoshima, K. Shinbo, K. Kato, T. Kawakami, Y. Ohdaira and T. Wakamatsu, Jpn. J. Appl. Phys. 43, 2335 (2004).
10. K. Shinbo, S. Toyoshima, Y. Ohdaira, K. Kato and F. Kaneko, Jpn. J. Appl. Phys. 44, 599 (2005).
11. W. L. Barnes, A. Dereux and T. W. Ebbesen, Nature 424, 824 (2003).
12. F. C. Chien, C. Y. Lin, J. N. Yih, K. L. Lee, C. W. Change, P. K. Wei, C. C. Suna and S. J. Chen, Biosens. Bioelectron. 22, 2737 (2007).
13. N. F. Chiu, C. W. Lin, J. H. Lee, C. H. Kuan, K. C. Wu and C. K. Lee, Appl. Phys. Lett. 91, 083114 (2007).
14. S. Massenot, R. Chevallier, J.-L. de Bougrenet de la Tocnaye and O. Parriaux, Opt. Commun. 275, 318 (2007).
15. B. K. Singh and A. C. Hillier, Anal. Chem. 78, 2009 (2006).
16. B. K. Singh and A. C. Hillier, Anal. Chem. 80, 3803 (2008).
17. Y. Aoki, K. Kato, K. Shinbo, F. Kaneko and T. Wakamatsu, IEICE Trans. Electron. E81-C, 1098 (1998).

18. Y. Aoki, K. Kato, K. Shinbo, F. Kaneko and T. Wakamatsu, Thin Solid Films, 327-329, 360 (1998).
19. Y. Aoki, K. Kato, K. Shinbo, F. Kaneko and T. Wakamatsu, Mol. Cryst. Liq. Cryst. 327, 127 (1999).
20. E. Kröger and E. Kretschmann, Z. Phys. 237, 1 (1970).
21. Y. Naoi and M. Fukui, J. Phys. Soc. Jpn. 58, 4511 (1989).
22. A. Baba, F. Kaneko, K. Kato, S. Kobayashi and T. Wakamatu, Jpn. J. Appl. Phys. 37, 2581 (1998).
23. H. Knobloch, H. Orendi, M. Buchel, T. Seki, S. Ito and W. Knoll, J. Appl. Phys. 77, 481 (1995).
24. A. Baba, F. Kaneko, K. Shinbo, K. Kato, S. Kobayashi and T. Wakamatsu, Mater. Sci. Eng. C8-9, 145 (1999).
25. R. C. Advincula, D. Roitman, C. Frank, W. Knoll, A. Baba and F. Kanako, Polymer Preprints 40, 467 (1999).
26. J. Ishikawa, A. Baba, F. Kaneko, K. Shinbo, K. Kato and R . C. Advincula, Colloids and Surfaces A, 198-200, 917 (2002).
27. T. Wakamatsu, K. Saito, Y. Sakakibara and H. Yokoyama, Jpn. J. Appl. Phys. 34, L1467 (1995).
28. T. Wakamatsu, K. Saito, Y. Sakakibara and H. Yokoyama, Jpn. J. Appl. Phys. 36, 155 (1997).
29. K. Shinbo, T. Ebe, F. Kaneko, K. Kato and T. Wakamatsu, IEICE Trans. Electron. E83-C, 1081 (2000).
30. K. Kato, M. Hirano, F. Takahashi, K, Shinbo, F. Kaneko and T. Wakamatsu, Trans. Mat. Res. Soc. Jpn., 29, p.775-778 (2004).
31. A. P. Piechowski, G. R. Bird, D. L. Morel and E. L. Stogryn, J. Phys. Chem. 88, 934 (1984).
32. B. Valeur, "Molecular Fluorescence", Wiley-VCH, Weinheim (2002).
33. K. Kato, Y. Saito, Y. Ohdaira, K. Shinbo and F. Kaneko, Thin Solid Films, 499, 174 (2006).
34. Y. Kobayashi, M. Fukushima, A. Baba, Y. Ohdaira, K. Shinbo, K. Kato and F. Kaneko, Jpn. J. Appl. Phys. 47, 1110 (2008).
35. K. Kato, M. Imai, Y. Ohdaira, K. Shinbo and F. Kaneko, Mol. Cryst. Liq. Cryst. 472, 441 (2007).
36. M. Hafuka, M. Minagawa, Y. Ohdaira, A. Baba, K. Shinbo, K. Kato and F. Kaneko, IEICE Trans. Electron. E91-C, 1883 (2008).
37. K. Kato, K. Yamashita, Y. Ohdaira, A. Baba, K. Shinbo and F. Kaneko, Thin Solid Films 518, 758 (2009).
38. A. Baba, K. Kanda, T. Ohno, Y. Ohdaira, K. Shinbo, K. Kato and F. Kaneko, Jpn. J. Appl. Phys. 49, 01AE02 (2010).

CHAPTER 13

MORPHOLOGY CONTROL OF NANOSTRUCTURED CONJUGATED POLYMER FILMS

Mitsuyoshi Onoda* and Kazuya Tada

*Department of Electrical Engineering and Computer Sciences,
Graduate School of Engineering, University of Hyogo, Himeji Shosha Campus,
2167 Shosha, Himeji, Hyogo 671-2280, Japan
E-mail: onoda@eng.u-hyogo.ac.jp or ss-syu02@eng.u-hyogo.ac.jp

Some experimental results on the electrophoretic deposition of conjugated polymer are reported. The scanning electron micrographs clearly indicate that the morphology of films of a fluorine-based polymer PDOF-MEHPV by the electrophoretic deposition strongly depends on the toluene/acetonitrile ratio of the parent suspensions. It is also shown that the suspension prepared from a 0.1 g/l toluene solution of the polymer, can give flat and dense polymer films with a thickness in the order of 100 nm by using the electrophoretic deposition. The electrical current originating from the movement of colloidal particles has been measured for the suspensions containing polymer more than 0.2 g/l. However, the current due to impurities in the solvent makes it difficult to detect the electrical current corresponding to the particle movement in the suspensions containing smaller amount of the polymer. Polymer light-emitting devices having PDOF-MEHPV films with various thickness prepared by electrophoretic deposition have also been demonstrated.

1. Introduction

The compatibility with wet-processes which can be performed under the atmospheric pressure is one of the most fascinating features of polymeric materials as semiconductors.[1-3] The spin-coating technique is very familiar to the researchers working in the laboratory, since the apparatus required is simple and cheap. However, the poorness in the materials efficiency of this method may be a problem for the mass-production. Thus, it is desirable to develop alternative methods to coat the semiconducting polymers. The electrophoretic deposition is a classical coating method widely used in the industrial coating process.[4,5] It has been demonstrated that the electrophoretic deposition yields nano-porous, films of conjugated polymer by us.[6] One of the most important features of the electrophoretic deposition is high material efficiency. This method also enables the deposition of films with several 10 to 100 nm thickness from dilute polymer solutions.

Although the ability to install the nanoporosity in the polymer film is a unique feature of this method, dense and flat films are suitable for most electronic devices such as light-emitting device, photocell, and field-effect transistors. Recently, we have shown that the electrophoretic deposition from suspension containing comparable amount of good and poor solvent for the polymer can generate such films.[7]

In this chapter, we show some experimental results on the electrophoretic deposition of conjugated polymer such as scanning electron microscopy and the current measurement during the deposition. It is also shown that the electrophoretic deposition from dilute suspensions, which are too dilute to be employed in spin-coating method, can yield the polymer films applicable to light-emitting device.

2. Conjugated Polymer and Experimental Procedure

A fluorine-based conjugated polymer poly(9,9-dioctyl-2,7-divinylene-fluorenylene)-alt-{2-methoxy-5-(2-ethylhexyloxy)-1,4-phenylene}], (PDOF-MEHPV), the molecular structure of which is shown in Fig. 1, purchased from American Dye Source Inc., Canada, and toluene and acetonitrile purchased from Wako Pure Chemical Industries Ltd., Japan, were used as received.

The polymer suspensions were prepared by mixing the toluene solution of the polymer and the acetonitrile at various volume ratios. The electrophore-tic deposition is carried out by applying DC voltage between a couple of indium-tin-oxide (ITO) electrodes coated on glass plates which are soaked in the suspension charged in a glass cuvette as

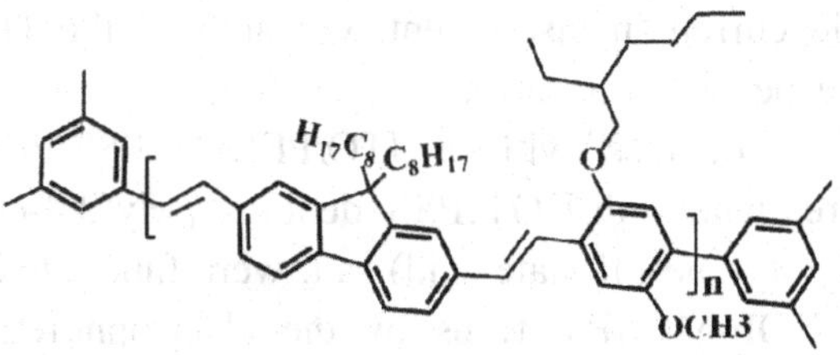

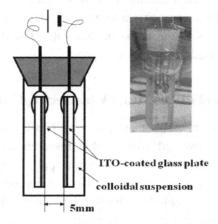

Fig. 1. Molecular structure of poly(9,9-dioctyl-2,7-divinylenefluore-nylene)-alt-{2-methoxy-5-(2-ethylhexyloxy)-1,4-phenylene}], (PDOF-MEHPV) used in this study.

Fig. 2. Setup for the electrophoretic deposition. The inset shows the snapshot photograph of the glass cuvette.

shown in Fig. 2. Typically, DC200-300V was applied between the electrodes 5 mm apart. The deposition of PDOF-MEHPV occurs on the positively biased electrode or the anode, indicating that the colloidal particles in the suspension are negatively charged.

The optical absorption spectra were collected by using a Hitachi U-3410 spectrophotometer. The thickness of the film was estimated from the fact that the PDOF-MEHPV film with unit peak absorbance observed at around 455 nm is approximately 120 nm in thickness. Scanning electron micrographs were taken with a Keyence VE-9800 system without metal coatings. The electric current during the electrophoretic deposition was measured by using a Keithley model 6517A electrometer.

For the electric current measurement, the area of the ITO electrodes soaked in the suspension was approximately 9 mm × 15 mm rectangle.

Light-emitting devices with a ITO/PEDOT:PSS/PDOF-MEHPV/ MgAg structure, where PEDOT:PSS denotes poly(3,4-ethylenedioxythiophene):poly(styrenesulfonate acid) salt, were fabricated with various thickness of PDOF-MEHPV films by the electrophoretic deposition. PDOF-MEHPV films were deposited on the PEDOT:PSS-coated ITO films from suspensions containing 0.05 g/l of the polymer for 4-10 s, resulting in the PDOF-MEHPV films with thicknesses ranging from 80 nm to 250 nm. The emission area of the devices is approximately 3 mm × 3 mm square, and a Si-photodiode attached on the emission face of the devices serves to the emission intensity. The device structure is similar to that reported in our previous paper.[8]

3. Uniform Film of PDOF-MEHPV by Electrophoretic Deposition

Conjugated polymer is a soft material as well as a semiconductor. This unique feature has induced considerable research work for electronic engineering in the next generation. Numerous publication has emerged in the field of polymer-based electronic devices including light-emitting devices, photocell and field-effect transistors.[9] One of the most important features of conjugated polymer motivating the researchers must be solubility, because it is a great advantage for the realization of so-called printed electronics using the wet-process under atmospheric pressure.

The spin-coating technique is frequently used to make polymer films for laboratory-scale fabrication, because of the low initial cost as well as the high quality of the resultant films. However, this method requires polymer solutions with relatively high concentration, restricting the applicability of the technique. Moreover, the material efficiency of the spin-coating technique is usually poor, because most of the polymer solutions placed on the substrate are blown out during spinning.

With the aim of establishing the wet-process for polymer film deposition which can utilize a relatively dilute solution and realize high material efficiency, several methods including ink-jet printing,[2,10] spraying[11] and electrophoretic deposition from the suspension[12] have been explored. Among them, electrophoretic deposition has a unique feature because it can yield nanostructured films of conjugated polymers, thanks to the separation of the stage of film formation from the stage of solidification of material. The suspension of conjugated polymers can be prepared simply by mixing the polymer dissolved in a good solvent, i.e. the polymer solution, and a poor solvent for the polymer. We have reported that the light-emitting device using a nanostructured poly(3-octadecylthiophene) film shows a unique emission pattern owing to the unique surface morphology of the film, which can be used as artificial fingerprint devices for anticounterfeit technology.[8,12] Although the ability to generate nanostructured films is an important feature of the suspension-based deposition methods including electrophoretic deposition, uniform films are generally preferable for electronic devices. For example, the heat treatment is known to make the conjugated polymer films consisting of colloidal particles uniform.[13]

This paragraph reports preliminary results on the electrophoretic deposition from suspensions of a fluorine-based conjugated polymer, PDOF-MEHPV, with various good/poor solvent ratios. It is found that the suspension containing more good solvents than poor solvents can be used for electrophoretic deposition and yield a uniform film without heat treatment.

3.1. *Optical Absorption Spectra of PDOF-MEHPV Films*

Figure 3(a) shows the optical absorption spectra of PDOF-MEHPV films prepared by electrophoretic deposition for 2 s from suspensions with various acetonitrile contents. Film deposition was observed when the content of acetonitrile is more than 30%. The optical absorption spectrum of the ITO-coated glass substrate after the application of 100 V for 10 s in a liquid containing 30% of acetonitrile was the same as that of another ITO-coated glass substrate which was just soaked in the liquid and dried, indicating that no electrophoretic deposition occurred. Therefore, from the viewpoint of electrophoretic deposition, the liquids containing more than 40% of acetonitrile are regarded as suspensions of the polymer, and the others are solutions. It is also confirmed that the UV-VIS spectra of the latter liquids are almost the same as the toluene solution of PDOF-MEHPV, while the others resemble the solid film of the polymer as the acetonitrile content increases. As shown in Fig. 3(b), the morphology of the resultant film strongly depends on the composition of parent suspension. In the case of the acetonitrile content being more than 80%, the films are apparently turbid and show strong scattering which results in the modification of the corresponding spectra shown in Fig. 3(a), indicating that the films has roughness of the micrometer scale. When the acetonitrile content is between 40% and 80%, the deposited films are apparently transparent.

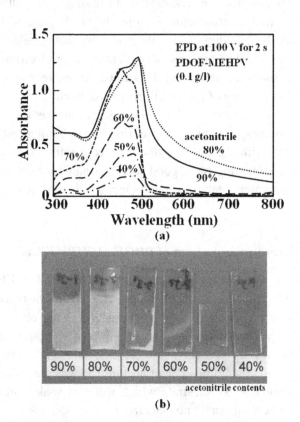

Fig. 3. (a) Optical absorption spectra of PDOF-MEHPV films prepared by electrophoretic deposition. (b) Photographs of PDOF-MEHPV films prepared by electrophoretic deposition from suspensions with various acetonitrile contents. The films were deposited 2 s by applying dc 100 V between a pair of ITO electrodes separated by 5 mm. The size of substrates used is approximately 25 mm × 9 mm.

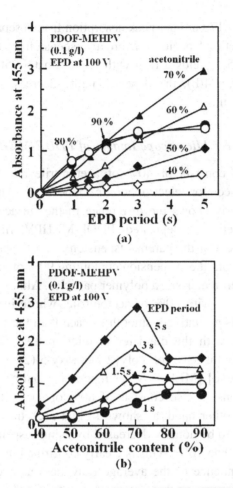

Fig. 4. (a) Dependence of absorbance at 455 nm of PDOF-MEHPV films prepared by electrophoretic deposition on the on the deposition period at various acetonitrile contents. The same data are plotted in (b) to show the dependence on the acetonitrile content.

Figure 4(a) shows the dependence of the absorbance at 455 nm, at which the absorption peak of the PDOF-MEHPV film appears, on the deposition period. It is found that the film thickness by electrophoretic deposition in the suspension containing more than 80% of acetonitrile tends to saturate, while the film thickness is almost the proportional to the deposition period in the case of the suspension with acetonitrile less than 70%. Since the film thickness of PDOF-MEHPV was estimated to be approximately 170 nm per peak absorption from a separate experiment, the electrophoretic deposition from a suspension containing 70% of acetonitrile for 5 s yields a film with approximately 500 nm thickness. As shown in Fig. 4(b), the deposition rate decreases with increasing acetonitrile content.

3.2. *Atomic Force Microscope Images of PDOF-MEHPV Films*

The suspension consists of 40% of acetonitrile and 60% of toluene can still be used for electrophoretic deposition. The atomic force microscopy clearly shows the difference in the surface morphology of the electrophoretically deposited PDOF-MEHPV films due to the acetonitrile content of the parent suspensions. As shown in Fig. 5(a), the film deposited from the suspension containing 90% of acetonitrile seems to consist of micrometer-sized polymer particles. Although the film made from a suspension with 80% of acetonitrile, whose image is shown in Fig. 5(b), reveals a relatively smooth surface than the former case, a lot of holes appear with the diameter in micrometres. This phenomenon is also found in the case of poly{2-methoxy-5-(2-ethylhexyloxy)-1,4-phenylene vinylene} (MEH-PPV) as reported in our previous paper,[14] in which the suspension also consists of 80% of acetonitrile and 20% of toluene. On the other hand, as shown in Fig. 5(c), the roughness of the films was found to be notably decreased by using suspensions containing acetonitrile less than 70%. This is clearly confirmed by Fig. 5(d), which shows the dependence of the average roughness and root mean square roughness of the films on the acetonitrile content in the suspensions. For example, electrophoretic deposition for 2 s in the suspension containing 60% of acetonitrile yields a 110 nm thick film with an average roughness of 6 nm, which may be uniform enough to be used for polymer devices such as light-emitting diodes and photocell.

The difference in the surface morphology can be roughly explained by taking the difference in the evaporation rate between acetonitrile and toluene. Just after the deposition, the polymer film contains and is covered with a mixture of acetonitrile and toluene. Since the acetonitrile evaporates considerably faster than toluene, the suspension containing 90% of acetonitrile quickly dries and leaves condensed toluene in the polymer film. In this case, the amount of residual toluene is so small and may partially dissolve the polymer particles forming the film to bond them. However, with decreasing acetonitrile content, the uniformity of the film improves. The holes appearing in Fig. 5(b) may be imprints of the micrometer-sized droplets of toluene formed on the polymer film at this intermediate acetonitrile concentration. Further reduction of acetonitrile resulted in uniform toluene film covering the polymer film, which fully dissolves the polymer film to make it uniform. The dissolution of polymer particles into a residual solvent in the course of drying is clearly observed by the naked eye as the changing in the transparency of the film. That is, a turbid film initially deposited on the substrate turns into a clear film during drying.

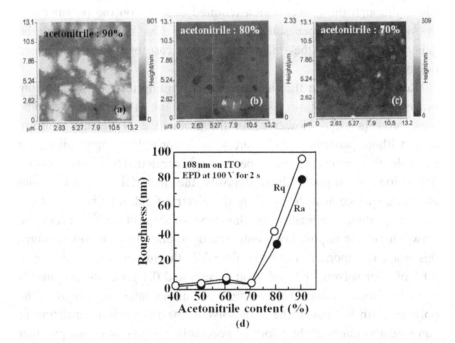

Fig. 5. Atomic force microscope images of PDOF-MEHPV films by electrophoretic deposition from suspensions containing (a) 90%, (b) 80% and (c) 70% of acetonitrile. (d) The dependence of average roughness (Ra) as well as root mean square roughness (Rq) on the acetonitrile content calculated from atomic force microscope images. The data were collected from films deposited for 2 s by applying dc 100 V.

4. Preparation of Flat and Dense Conjugated Polymer Films from Dilute Solution

Since the spin-coating technique only requires simple and cheap apparatus, this technique widely used for the laboratory-scale preparation of the polymer electronic devices. However, other than the incompatibleness with the patterning, this technique seems to possess some disadvantages for the commercial production. For example, the material efficiency is poor because most of materials placed on the substrate is blown out during the spinning. Moreover, the polymer solution for this technique must be relatively thick if one targets this type of polymer films suitable for sandwich-type electronic devices. To our typical experience, a chloroform solution containing several g/l of a conjugated polymer is appropriate to obtain a 100 nm-thick polymer film. Although the concentration required depends on the polymer and the solvent, a polymer solution containing less than 1 g/l may be generally useless for the spin-coating in the case of typical conjugated polymers.

Recent progress in the field of ink-jet printing technology allowing precise positioning of very small amount of ink is going to overcome these advantages.[2,10] However, for the production of flat and uniform films without patterning, which are suitable for lighting applications for example, this method seems to be too sophisticated. The electrophoretic deposition technique,[5] which deposits the material over the entire electrode surface at once by using the electric field, may be one of the most important candidates for this type of deposition.[6,8,14] From the viewpoint of the required concentration of solution to make suspensions, this result is important. For example, while the suspension consisting of 90% of poor solvent, 10% of good solvent and 0.1 g/l of the polymer is made by mixing a unit of the polymer solution containing 1.0 g/l of the polymer with 9 units of the poor solvent, the suspension consisting of equivalent volumes of the poor and good solvents with the same polymer content requires a solution containing just 0.2 g/l of the polymer.

In this paragraph, it is shown that this technique enables to prepare a few 100 nm-thick polymer films from very dilute polymer solutions, say 0.1 g/l. The effect of the coating of the electrodes with poly(3,4-ethylenedioxythiophene):poly(styrenesulfonate) salt (PEDOT:PSS) on the deposition behavior is also mentioned.

4.1. Scanning Electron Micrographs of PDOF-MEHPV

As mentioned in paragraph 3, the ratio of toluene and acetonitrile in the suspensions strongly influences the morphology of the polymer films generated by means of electrophoretic deposition. Briefly, the suspension containing more than 80% of acetonitrile yields rough and frosted films, and flat and transparent films can be obtained by using those with less than 70% of acetonitrile. Figure 6 shows the scanning electron micrograph of the PDOF-MEHPV films deposited from the suspensions with various acetonitrile/toluene ratios. It is noticed that the surface morphology of the film from the suspension containing 90% of acetonitrile is quite different from the film obtained from the suspension containing 80% of acetonitrile. The former seems like stacked nanopartiles, while the latter is a flat film with many circular holes. In the case of the suspensions containing less than 70% of acetonitrile, the surface of the film becomes flat, except for the accidental defects.

From these results, the relationship between the content of the acetonitrile in the suspensions and the morphology of the films obtained by the electrophoretic deposition from the suspensions can be explained as follows. A film as deposited or an accumulation of colloidal particles on the electrode is covered with the suspension used. Since the evaporation of acetonitrile is faster than toluene, the toluene condenses in the films during the drying. When the amount of the condensed toluene covering the polymer films is large enough to form a continuous liquid film covering the polymer film, the accumulated colloidal particles are once fully dissolved in the condensed toluene to form a unifotm film. On the other hand, small amount of the condensed toluene dissolves solely interfacial parts of the colloidal particles, or forms droplets to generate holes on the polymer film. The generation of holes similar to that found in Fig. 6(c) has been observed in the case of poly[2-methoxy-5-(2'ethyl hexyloxy)-1,4-phenylene vinylene.[14]

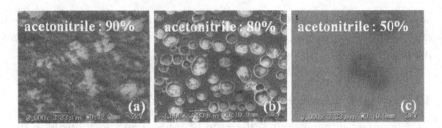

Fig. 6. Scannning electron micrographs of PDOF-MEHPV films deposited from the suspensions containing (a) 90%, (b) 80% and (c) 50% of acetonitrile.

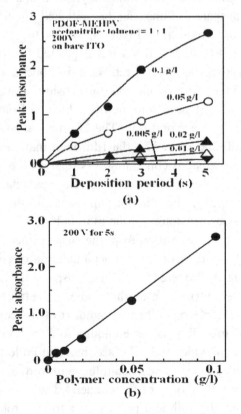

Fig. 7. (a) Dependence of the peak absorbance of PDOF-MEHPV films deposited on bare ITO electrodes on the deposition period by using suspensions with various polymer concentrations. (b) Dependence of the absorbance on the polymer concentration of the suspension in the case of 200 V was applied for 5 s. The line shows the linear fitting.

4.2. Deposition of PDOF-MEHPV from Dilute Solution

Figure 7(a) shows the dependence of the peak absorbance of the PDOF-MEHPV films on the deposition period. The peak absorbance is almost proportional to the deposition period at the early stage, and gradually saturates by the prolonged deposition. The saturation corresponds to the exhaustion of the colloidal particles between the electrodes. It has been found that the electrophoretic deposition in the suspension containing 5.0×10^{-2} g/l of the polymer, which is derived from a 0.1 g/l toluene solution of the polymer, could give the polymer film with a thickness of a few 100 nm. It should be worth to note that the spin-coating of the toluene solution containing 1.0 g/l of PDOF-MEHPV did not give any film detectable by the spectrophotometer on a glass plate. As shown in Fig. 7(b), the linear relationship has been found in the polymer concentration dependence of the peak absorbance at a constant voltage as well as a constant deposition period. The linearity is held down to the 5.0×10 g/l suspension, which was derived from a 1.0×10^{-2} g/l solution.

4.3. PEDOT Coating Effect on Deposition

Since it is well known that the insertion of a thin layer of PEDOT:PSS between the ITO and the active semiconducting polymer layer much improves the performance of the polymer-based devices such as light-emitting devices and photocells, the deposition on the ITO electrode coated with PEDOT:PSS was surveyed. Figure 8 shows the deposition period dependence of the peak absorbance of PDOF-MEHPV films on PEDOT:PSS coated ITO electrodes. Approximately 50 nm-thick PEDOT:PSS layer was deposited on the ITO by spin-coating the aqueous suspension purchased from Aldrich, in comparison with the data shown in Fig. 7, the peak absorbance and thus the film thickness increases nonlinearly with the deposition period, and no notable deposition was found within 5 s from the suspension containing the polymer less than or equals to 2.0×10^{-2} g/l. Moreover, the film is apparently inhomogeneous when the deposition period is less than 2 s.

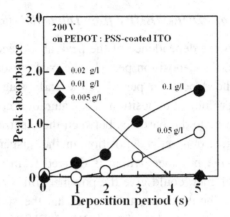

Fig. 8. Dependence of the peak absorbance of PDOF-MEHPV films deposited on PEDOT:PSS- coated ITO electrodes on the deposition period by using suspensions with various polymer concentrations.

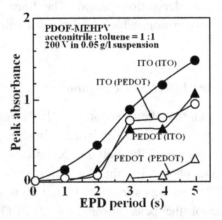

Fig. 9. Dependence of the peak absorbance of PDOF-MEHPV films on the deposition period with various electrode conditions. The counter electrodes are indicated in the parentheses.

The difference in the above-mentioned deposition behavior suggests that the profile of the voltage and thus the electric field between the electrodes is modulated by the PEDOT:PSS film. As shown in Fig. 9, when the PEDOT:PSS is coated on the counter ITO electrode, the behavior of the deposition on bare ITO electrodes mimics that on the

PEDOT:PSS coated ITO electrode with a bare ITO counter electrode. The nonlinearity is pronounced when both electrodes are coated with PEDOT:PSS. The data shown in Fig. 9 suggest that there is a threshold time for the deposition when one of the ITO electrodes is coated with PEDOT:PSS, and the threshold time is almost doubled when both electrodes are coated. The delayed deposition by the PEDOT:PSS coating may come from the reduced electric field to accelerate the colloidal particles in the suspension. In the case of electrochemical systems filled with electrolyte, it is well known that the electric double layer is formed nearby the electrode to flatten the potential profile at the offshore area between the working and the counter electrodes. In our case, ionic portions detached from the PEDOT:PSS layer or the semiconducting nature of PEDOT:PSS can induce the reduction of the electric field at the offshore area. However, the explanation for the nonlinearity in the time dependence as well as the threshold periods may require another scenario, which is now under study.

5. Fabrication of PDOF-MEHPV Light-Emitting Devices by Electrophoretic Deposition

Major advantages of conjugated polymers as semiconductors against conventional inorganic semiconductors may be in their unique properties such as mechanical flexibility and solubility. Especially, the latter feature enables the preparation of semiconductor films under the atmospheric pressure by wet-processes, motivating a number of researches on the application of conjugated polymers for light-emitting devices,[1,9,15] photocells,[16,17] field-effect transistors[18,19] and other electronic devices, or "printed electronics".

For the application of the conjugated polymers to large-area electronic devices, it is important to develop high-throughput and efficient technologies of coating. Although the electrophoretic deposition technology is widely used in the industrial coating process, it has not caught the adequate attentions of the researchers of organic electronics for many years. The principle of electrophoretic deposition is quite simple; the electric field accelerates the colloidal particles in the suspension of material, and the particles reached to the electrode form deposit. It is obvious that the resultant films by electrophoretic deposition are particulate, since they are just accumulated colloidal particles. We have reported the electrophoretic deposition of conjugated polymers, from suspensions which are prepared by simple re-precipitation technique, as a method to obtain nanostructured conjugated polymer films.[6,14]

Recently, we have found that the morphology of the films of a polyfluorene-type conjugated polymer PDOF-MEHPV by electrophoretic deposition strongly depends on the content of good solvent of suspension used.[7] The polymer films from suspensions containing less than 20% of toluene has apparently rough surface and are frosted. On the other hand, the films from the suspensions with more than 30% of toluene are treatment. The atomic force microscopy study has indicated that the latter films with approximately 100 nm in thickness have the rms-roughness below 10 nm. It has been also confirmed that the light-emitting devices with ITO/PEDOT:PSS/PDOF-MEHPV/MgAg structure, where PEDOT:PSS denotes poly(3,4-ethylenedioxythiophene):poly(styrenesulfonate) salt, using the latter type of films show uniform emission.

From the viewpoint of the required concentration of the polymer solution in good solvent to make suspensions, this result is important. For example, while the suspension consisting of 90% of poor solvent, 10% of good solvent and 0.1 g/l of the polymer is made by mixing a unit of the polymer solution containing 1.0 g/l of the polymer with 9 units of the poor solvent, the suspension consisting of equivalent volume of the poor and good solvents with the same polymer content requires the polymer solution containing just 0.2 g/l, 5 times thinner than the aforementioned one, of the polymer.

In this study, the preparation of the light-emitting devices with various thickness of the emission layer from dilute polymer soluteons by using the electrophoretic deposition technique has been performed.

5.1. *Structure of PDOF-MEHPV Light-Emitting Devices*

Since the suspension used contains equivalent volumes of good and poor solvents, dense and homogeneous thin films are obtained by the natural drying in air, as mentioned in the previous paper. To improve the device performance, the ITO electrodes served as anode are coated with approximately 50 nm-thick PEDOT:PSS layer by spin-coating the aqueous suspension.

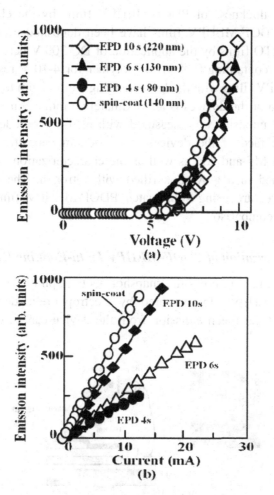

Fig. 10. (a) Emission intensity-voltage characteristics of the light-emitting devices using PDOF-MEHPV films prepared by the electrophoretic deposition with various deposition periods. The case of spin-coated PDOF-MEHPV film is also indicated. The thickness of PDOF-MEHPV are indicated in the parentheses. Accompanying emission intensity-current characteristics are shown in (b). The emission area of the devices is approximately 3 mm × 3 mm.

The thickness of the PDOF-MEHPV films is estimated by using the relationship that unit peak absorbance approximately corresponds to 120 nm in thickness. The device structure is similar to that reported in our previous paper as shown in Fig. 10.[8] The light-emitting devices with a ITO/PEDOT:PSS/ PDOF-MEHPV/ MgAg structure were fabricated with various thickness of PDOF-MEHPV films by the electrophoretic deposition. PDOF-MEHPV films have been deposited on the PEDOT: PSS-coated ITO films by the application of DC 300 V between 5 mm in a suspension containing 0.05 g/l of polymer for 4-10 s, resulting in the PDOF-MEHPV films with thickness ranging from 80 nm to 250 nm. The emission area of the devices is approximately 3 mm × 3 mm square, and the emission intensity was measured with a Si-photodiode attached on the emission face of the devices. The vacuum deposition of cathode composed of Mg and Ag, as well as the characterization of the devices was performed in a glove-box filled with nitrogen. The devices with the same structure using spin-coated PDOF-MEHPV films were also prepared for comparison.

5.2. *Chracterization of PDOF-MEHPV Light-Emitting Devices*

The inset of Fig. 11 shows tha snapshot photograph of the device using the PDOF-MEHPV film prepared by electrophoretic deposition. The uniformity of the green emission from the device can be confirmed by this snapshot.

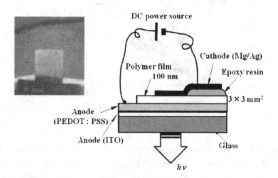

Fig. 11. Structure of PDOF-MEHPV Light-Emitting Diodes Prepared by electrophoretic Deposition. The inset shows tha snapshot of the device using the PDOF-MEHPV film prepared by electrophoretic deposition.

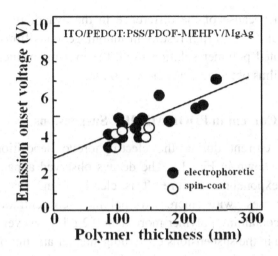

Fig. 12. Dependence of the emission onset voltage on the PDOF-MEHPV thickness of the light-emitting devices using PDOF-MEHPV films prepared by the electrophoretic deposition. The cases of spin-coated PDOF-MEHPV films are also indicated. The line shows the linear fitting.

Typical emission intensity-voltage and the emission intensity-current characteristics of the devices are shown in Figs. 10(a) and 10(b), respectively. Since the emission onset voltage as well as the quantum efficiency of the devices does not seem to seriously depend on the preparation method. It can be concluded that the characteristics of the devices fabricated by the electrophoretic deposition are comparable to those by the spin-coating.

The dependence of the emission onset voltage on the polymer thickness is plotted in Fig. 12. The onset voltage seems to be determined by the film thickness, and there is no significant difference due to the difference in the preparation method. The onset voltage is monotonically increased with the increasing thickness of the PDOF-MEHPV layer, as commonly observed in the study of polymer light-emitting devices.

For the preparation of the devices by the spin-coating technique, a chloroform solution containing 10 g/l of the polymer and the spin-speeds ranging from 1000-5000 rpm have been employed. Generally, the spin-coating from the chloroform solution of a polymer gives thicker film easily than that from the toluene solution at a constant polymer concentration because of the difference in the evaporation rate, and a spin-speed below 1000 rpm results in inhomogeneous films. Despite of the concentrated polymer solution used, the maximum thickness of the spin-coated films obtained was below 150 nm.

6. Electric Current in PDOF-MEHPV Suspensions

The electric current during the electrophoretic deposition has been measured as shown in Fig. 13. The decays observed apparently do not have single-exponential feature. It is clearly found that the electric current increases with increased polymer concentration for the suspensions containing polymer more than 0.2 g/l. However, the electric current found in the suspensions containing smaller amount of polymer is almost identical to that in pure toluene/acetonitrile mixture. The electric current in the pure mixture of solvents, which tends to change from trial to trial, may mainly come from dissolved impurities such as oxygen and water. The suppression of this kind of electric current must be a key to reduce the power consumption during the deposition. Detailed measurement and analysis of the electric current during the electrophoretic deposition are under way.

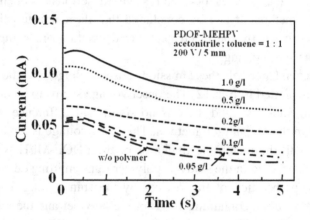

Fig. 13. Electrical current during the electrophoretic deposition in the suspensions with various PDOF-MEHPV concentrations.

7. Conclusions

In this chapter, we have reported some experimental results on the lectrophoretic deposition of the conjugated polymer. The electrophoretic deposition of a fluorene-based conjugated polymer, poly(9,9-dioctyl-2,7-divinylene-fluorenylene)-alt-{2-methoxy-5-(2-ethylhexyloxy)-1,4-phenylene}], PDOF-MEHPV has been carried out from suspensions with various good/poor solvent ratios.

The optical absorption spectra and atomic force microscopy show that the roughness of the resultant film strongly depends on the good/poor solvent ratio of the parent suspension. That is, rough and turbid films are obtained when the content of the poor solvent is high, while uniform and transparent films, which may be suitable for polymer devices such as light-emitting devices and photocells, can be obtained in the case of low poor solvent content.

The scanning electron micrographs clearly indicate the dependence of the morphology of the PDOF-MEHPV films by the electrophoretic deposition on the toluene/acetonitrile ratio of the parent suspensions. It is also shown that the suspension prepared from a 0.1 g/l toluene solution of the polymer, can give the polymer film with a thickness of a few 100 nm.

That is, the PDOF-MEHPV films prepared by the electrophoretic deposition in a suspension prepared by mixing equivalent volumes of 0.1 g/l toluene solution of the polymer with pure acetonitrile can be more than 200 nm in thickness, which could not be achieved by the conventional spin-coating technique using 100-times thick polymer solution. It has been shown that the coating of ITO electrodes with PEDOT:PSS thin layer on the results in low and nonlinear deposition rates.

It has been mentioned that the electrophoretic deposition using a suspension derived from a toluene solution containing only 0.1 g/l of the polymer can yield films with thickness >200 nm, which were not obtained by a single spin-coating shot with a 10 g/l chloroform solution.

The electrical current originating from the movement of colloidal particles is clearly detectable for the suspensions containing polymer more than 0.2 g/l. However, the electrical current due to the impurities dissolved in the solvent makes it difficult to detect the electrical current corresponding to the particle movement in the suspensions containing smaller amount of the polymer.

The demonstration of the polymer light-emitting device confirmed that the electrophoretic deposition technique can yield uniform films instantly from dilute polymer solutions.

Thus, it has been shown that the electrophoretic deposition is a useful technique to obtain conjugated polymer films with thickness of few 100 nm from dilute polymer solution. This method can provide an alternative way for producing large-area polymer electronic devices. These features indicate that the electrophoretic deposition can be an important candidate for the film deposition technique for polymer-based electronic devices.

Acknowledgments

We acknowledge the supports by Grant-in-Aid for Young Scientists from JSPS, the research-grant from the IKETANI Science and Technology Foundation and the research-grant from the Hyogo Science and Technology Association.

References

1. J. H. Burroughes, D. D. C. Bradley, A. R. Brown, R. N. Marks, K. Mackay, R. H. Friend, P. L. Burns and A. B. Hilmes, *Nature*, **347**, 539 (1990).
2. H. Sirringhaus, T. Kawase, R. H. Friend, T. Shimoda, M. Inbasekaran, W. Wu and E. P. Woo, *Science*, **290**, 2123 (2000).
3. G. E. Jabbour, R. Radspinner and N. Peyghambarian, *IEEE J. Sel. Top. Quant. Electron.*, **7**, 769 (2001).
4. P. Sarkar and P. S. Nicholson, *J. Amer. Ceram. Soc.*, **79**, 1987 (1996).
5. O. O. Van der Biest and L. J. Vandeperre, *Annu. Rev. Mater. Sci.*, **29**, 327 (1999).
6. K. Tada and M. Onoda, *Adv. Funct. Mater.*, **12**, 420 (2002).
7. K. Tada and M. Onoda, *J. Phy. D: Appl.Phys.*, **41**, 032001 (2008).
8. K. Tada and M. Onoda, *Japan. J. Appl. Phys.*, **42**, L1093 (2003).
9. H. S. Nalwa and L. S. Lohwer, Ed., *Handbook of Luminescence, Display Materials and Divices* (Valencia, CA: American Scientific Publishers, 2003).
10. J. Bharathan and Y. Yang, *Appl. Phys. Lett.*, **72**, 2660 (1998).
11. K. Fujita, T. Ishikawa and T. Tsutsui, *Japan. J. Appl. Phys.*, **41**, L70 (2002).

12. M. Onoda and K. Tada, *J. Sci. Con. Proceedings.*, **1**, 27 (2008).
13. K. Landfester, R. Montenegro, U. Scherf, R. Guentner, U. Asawapirom, S. Patil, D. Neher and T. Kietzke T, *Adv. Mater.*, **14**, 651 (2002).
14. K. Tada and M. Onoda, *Adv. Funct. Mater.*, **14**, 139 (2004).
15. K. Tada and M. Onoda, *Appl. Phys. Lett.*, **89**, 043508 (2006).
16. G. Yu, K. Pakbaz and A. J. Heeger, *Appl. Phys. Lett.*, **64**, 3422 (1994).
17. K. Tada, M. Onoda, A. A. Zakhidov and K. Yoshino, *Japan. J. Appl. Phys.*, **36**, L306 (1997).
18. A. Tsumura, H. Koezuka and T. Ando, *Appl. Phys. Lett.*, **49**, 1210 (1986).
19. K. Tada, H. Harada and K. Yoshino, *Japan. J. Appl. Phys.*, **36**, L718 (1997).
20. I. D. Parker, *J. Appl. Phys.*, **75**, 1656 (1994).

CHAPTER 14

WAY OF ROLL-TO-ROLL PRINTED 13.56 MHz OPERATED RFID TAGS

Minhun Jung[1,2], Jinsoo Noh[3], Hwangyou Oh[4], Hwiwon Kang[2], Dongsun Yeom[2], Donghwan Kim[2,3] and Gyoujin Cho[1,3,4,*]

[1]*Department of Chemical Engineering, Sunchon National University,*
315 Maegok-dong, Sunchon 540-742, Korea
[2]*Printed Electronics Research Institute,*
Paru Co. 42-4 Sunchon Industrial Complex, Sunchon 540-813, Korea
[3]*Department of Printed Electronics Engineering in World Class University*
(WCU) Program, Sunchon National University,
315 Maegok-dong, Sunchon 540-742, Korea
[4]*Green Technology Fused Advanced Materials (GTFAM)*
Regional Innovation Center (RIC), Sunchon National University,
315 Maegok-dong, Sunchon 540-742, Korea
*E-mail: gcho@sunchon.ac.kr

Roll-to-roll printing process has been recently considered as a key technology for the realization of a penny RFID tag. To employ the current commercially avaiable R2R printing process to print RFID tags including antenna, rectifier, and digital processors on plastic or paper foils, materials and device circuits should be first well optimized to the R2R printers and substrates. In this chapter, we would like to briefly discuss about our own device circuits with our designed materials for R2R gravure printing RFID tags on plastic foils.

1. Introduction

Radio frequency idenification (RFID) tags will take a key role for bringing the ubquitous society at 21st century.[1] For the realization of the true ubiqutous society, all goods should be able to sense their surroundings and exchange information *via* wireless communication instaneously. In other words, all goods should have their own RFID tags in which sensors, displays, CPU etc. may be integrated depending on their usages. In fact, current Si and MEMs technologies are fully able to produce the RFID tags with the integration of sensors, displays or CPU if the cost of the tags is not a matter. However, in markets, the cost is everthing to be considered so that 20 cent of an eraser can not have 20 cent of RFID tag.[2,3] Therefore, the manufacturing cost of RFID tags should be first resolved for the realiztion of ubiqutous society.

Based on the current Si and MEMs technologies, ultra-low cost of the tags can not be attained beacuse about 80% of the manufacturing cost is originated from Si chip bonding and labelling processes, not from the cost of Si chips. Futhermore, if we want to integrate sensors, displays or CPU on the tags, the portion of bonding and labelling cost will be drastically increased. Therefore, in recent, a roll to roll (R2R) printing process has been drawn a lot of attention to manufacture the tags through an inline printing process so that chips, sensors, CPU and antenna can be directly R2R printed on palstic or paper foils without any bonding and labelling processes.

To employ R2R printing process to manufacture only RFID tags without sensors, displays and CPU, four different major parts should be simultaneously developed. First, materials for R2R printing antenna, wires, thin film transistors, diodes, capacitors and resistors should be first available. Second, R2R printers with overlay printing registration accuray (OPA) of less than 10 µm to integrate a number of transistors for multi-bit digital circuits should be developed. Third, inexpensive rollable substrates with a relatively good surface roughness and less coefficient of thermal expansion (CTE) should be available. Finally, a new circuit design to efficiently utilize a less number of integrated transistors to overcome the limit of the OPA of R2R printers should be developed.

In this chapter, among four major parts for R2R printed RFID tags, the new circuit design and performance of the designed circuit for R2R printed RFID tags will be discussed based on our own R2R printed thin films transistors (TFTs) on poly(ethylene terephtalate) (PET) foils with a brief introduction on our materials, R2R printer and substrates to print the digital circuit.

2. Matrerials, Substrates and R2R Printer

2.1. Conducting Silver Ink

For R2R printing gate and drain-source electrodes of TFTs, nanoparticle based silver inks was formulated by using the following procedures. 40 g of $AgNO_3$ and 15 g of poly(vinylpyrrolidone) was stirred into 1200 mL of double distilled water under ambient conditions. After completely dissolving all additives, 14 mL of hydrazine hydrate (reagent grade, 55%, Aldrich) was added to the solution for 1 min. After 5 min, 2800 mL of acetone was added to the mixture and it was then further stirred for 30 mins. The resulting mixture was centrifuged at 5000 rpm for 5 min at 10°C. After the solution was slowly decanted, 20 g of silver gel was attained. The resulting silver gel was used for the manufacture of gravure inks by adding appropriate amounts of ethylene glycol and hexanol to control surface tension, as shown Table 1.

The resistivity of printed silver lines after R2R gravure printing on PET foils using the formulated silver inks, depending the drying temperatures and times, are shown in Fig 1. Furthermore, the R2R gravure-printed silver patterns achieved good adhesion (5B) on PET foils even when dried for ≤ 10 s using ASTM D3359 test.

2.2. Dielectric Ink

The gate dielectric layer is critical material in R2R printed TFTs since the gate insulator capacitance will controll density of carriers in the conducting channel and the electrical chracteristics of TFTs Eq. (1). Therefore, R2R printed pinhole-free dielectric layers with the high capacitance are highly desireable to attain high drain current values of printed TFTs at low gate voltages. Since R2R printing process would often generate many pin-holes at thinner layers, relatively thick dielectric layers are mandatory to prevent the possible short. Therefore, high dielectric materials are needed for providing the comparable capacitance to the ultra-thin dielectric layers to compensate the R2R printed thick dielectric layers.

Table 1. The surface tension and viscosity depending on conducting gravure ink formulation.

	Formulation 1	Formulation 2	Formulation 3
Silver nanoparticles	60 wt%	60 wt%	50 wt%
Ethylene glycol	27 wt%	32 wt%	50 wt%
Hexanol	13 wt%	8 wt%	
Surface tension	23 mN/m	36 mN/m	44 mN/m
Viscosity	200 cp	200 cp	200 cp

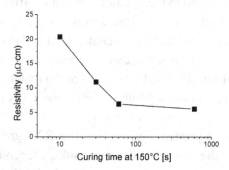

Fig. 1. Temperature and drying time dependence of resistivity for gravure printed silver patterns on PET foils.

In this work, a high κ insulating ink was manufactured with a viscosity of 200 cp and surface tension of 30 mN/m using various amounts of barium titanate (BT) nanopowders (50 nm) and 5 g of poly(methylmethacrylate) (PMMA) with 30 mL of MOE (methoxy ethanol) through vigorous mechanical stirring for 2 h and then further dispersed using ultrasonic for 4 h at 30°C. To determine the dielectric constant of the printed insulating films with a loading ratio of BT, parallel plate capacitors were fabricated using PET substrates and printed silver electrodes. Capacitance measurements were carried out using a LCR meter (Agilent LA5673) from 100 Hz to 100 KHz. Depending on the amount of BT loaded, a dielectric constant from 13 to 40 could be attained as shown in Fig. 2(A). Dielectric constant values were calculated from each of the capacitances C using Eq. (1). The fabricated capacitors had a 2.45 μm dielectric thickness as measured with a surface profilometer as shown in Fig. 2(B).

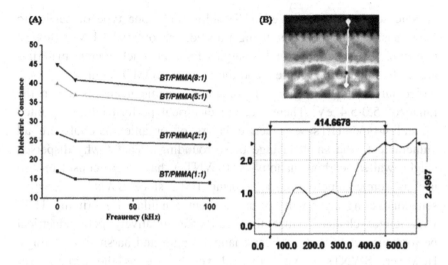

Fig. 2. (A), Dielectric constant measured between 10 Hz to 100 KHz with various amounts of barium titanate (BT) nanopowders loaded into PMMA solutions and (B), 3-D image of surface profilometer of R2R gravure-printed dielectric layer on gate electrodes.

$$C = (\varepsilon_r \times \varepsilon_o)/d \times A \qquad (1)$$

C: Capacitance of gate dielectrics, ε_r: Relative permittivity of dielectric layer, ε_o: permittivity of free space, d: thickness of dielectric layer, A: Area.

Considering the brittleness of printed dielectric layers, insulating ink with 29 wt% BT loading to total loading [A 1:1 ratio of BT to PMMA is 50% BT and 50% PMMA] was chosen in this R2R printing work. Furthermore, the insulating ink was stable under ambient conditions as no precipitation was observed for 20 days.

2.3. Semiconductive Ink

A large number of inorganic and organic materials are available to print as active layers for TFTs. Among them, if we consider the flexibility and low temperature curing (<150°C) of R2R printing process, organic semiconductors are the best option to employ. However, since the organic semiconductor based TFTs allow only one type of carrier to move easily with the other being trapped, the optimal LUMO (lowest unoccupied molecular orbital) energies for n-channel organic materials are in the range of 3.8-4.4 eV and the optimal HOMO (highest occupied molecular orbital) energies for p-channel organic materials are in the range of 5.0-5.4 eV. Therefore, the electrical performance of printed p- and n-type TFTs with active layers of organic materials are all vulnerable to the small change of surroundings. That's why dispersed single walled carbon nanotubes (SWNTs) has been considered to use as semiconductive inks to print TFTs since SWNTs as active semiconducting layer will be an excellent candidate for printing TFTs with longer channels since they can offer relatively good electrical performance even under longer channel length and harsh surroundings. However, SWNTs is not practical yet to use as the active layer because the SWNTs has a major intrinsic problem of heterogeneous bandstructures that span widely from metal to semiconductor when they synthesized. Therefore, to attain reasonable on-current with a good on-off ratio of SWNT-TFTs,[4-9] the metallic SWNTs need to be modulated before employing as a semiconducting layer in printed TFTs.

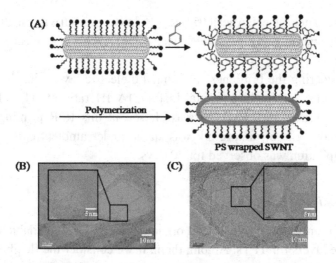

Fig. 3. (A) Schematic illustration of the process of admicellar polymerization of styrene on SDS/SWNTs. TEM images for bare SWNT (B) and PS wrapped SWNT (C).

In order to escape the trade off between high on-state conductance and switching ability, it is therefore necessary to alleviate the metallic pathways in printed SWNT-TFTs. As shown in Fig. 3(A), we develop the wrapping method to disconnect the metallic pathways in SWNT-TFTs. To wrap SWNTs with ultra-thin polystyrene films, first, various amounts of purified styrene monomers were added to four different aqueous solutions (1 mL) of SDS-wrapped SWNTs, and the mixtures were allowed to equilibrate for 6 h at 25°C while being stirred. After this time, 0.08 mg (0.5 μmole) of 2,2'-azobis(isobutyronitrile) was added to the four different mixtures to initiate the polymerization and the mixtures were heated to 60°C in an oil bath for 12 h with continued stirring. The resulting PS wrapped SWNTs (PS-SWNTs) were characterized using TEM (JEM 2100F) (Figs. 3(B) and 3(C)). A sample for TEM analysis was prepared by placing a single drop of the PS-SWNTs solution onto a holey carbon coated-copper grid and then, blotting with filter paper to form a thin deposition. The PS-SWNTs deposited grids were rinsed 5 times to remove surfactants using a consecutive process of dropping double distilled water onto them and blotting with filter paper. The rinsed grids were dried under ambient conditions.

2.4. Substrate

For R2R printing process, the substrate must be rollable foils with the less thermal deformation. Furthermore, since electronic devices with various inks are printed and cured at maximum 150°C, a stress field arises due to the difference in thermal contraction between the printed devices and the substrate upon winding and cooling. Therefore, the thickness of substrate should be considered with the wettability and Young's modulus. If the stress of the printed devices become too large or the adhesion between the printed devices and the substrate is weak, the printed devices may crack or peel off the substrate. In our works, R2R printing is performed on corona treated poly(ethylene terephtalate) (PET) foils (SKC, Korea) having a thickness of 75 μm and width of 240 mm with CTE of 30-65 × 10^{-6} K^{-1}. The surface roughness of PET was 10 nm RMS by measuring AFM (Fig. 4).

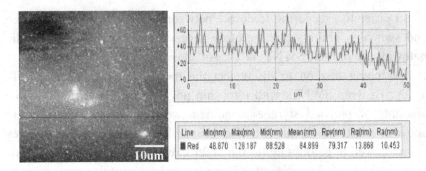

Fig. 4. AFM image of PET surface.

2.5. *R2R Gravure*

R2R printing techniques have received significant attention for the manufacturing of various macro-electronic devices such as RFID tags,[2] sensors,[10] flexible signage[11,12] and e-paper[12,13] with an ultra-low cost. Among commercially available R2R printing techniques, R2R gravure has drawn a lot of attention as a practical production process for printing macro-electronic devices since R2R gravure has a high throughput with a mechanically simple process and fewer controlling variables than other printing processes such as flexography, offset, gravure-offset, screen, etc.[14-19] To enable the use of gravure printing to realize properly scaled and integrated electronic systems, however, the microscopic feature production of R2R gravure printing needs to be studied in detail. As a typical example, a commercially available R2R gravure should not generate any problems for printing graphic arts and packaging where microscopic defects and surface roughness (<100 μm) are acceptable. However, to print macro-electronic devices, those microscopic defects will completely damage the devices and will drastically limit circuit yield. Therefore, the R2R gravure printing process has been first evaluated to determine whether the R2R gravure is suitable for rendering defect-free, reliable and constant microscopic uniformity (<100 μm) under the relationship between the various print, ink, and substrate-related variables.[20] In this work, we are employing the evaluated 4 color units of R2R gravure (Fig. 5) to print TFTs and digital circuits on plastic foils with the printing speed of 12 m/min, roll pressure of 1.5 MPa on the PET

Fig. 5. Actual image of R2R gravure with 4 color units manufactured by AVACO, Korea.

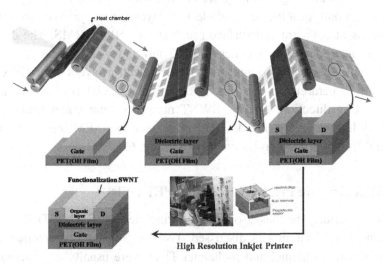

Scheme 1. Schematic illustration of R2R gravure printing process to print SWNT-TFTs on PET foils.

web, and a sufficiently high pressure of the doctor blade with an angle of 60°.

By using R2R gravure in Fig. 5, SWNT-TFTs, inverters, NAND gate, NOR gate and 4-bit digital circuits were respectively printed on PET foils by using above-disscussed inks as shown in Scheme 1. At the first

Fig. 6. The cross-sectional SEM image of all printed gate, gate dielectric layer and drain-source electrodes.

printing unit, silver gate lines were printed with 200 μm width and then, at the second printing unit, dielectric layers were printed with the thickness of 3.48 μm and surface roughness of 50 nm RMS. The drain-source electrodes with thickness of 357 nm and weaveniness of 1 μm were then selectively printed on printed dielectric layers with OPA of ±50 μm in pararelle to the printing direction and ±10 μm in vertical to the printing direction. Finally, SWNT networks were inkjet printed on printed drain-source electrodes. The cross-sectional SEM images for the three different printed layers are shown in Fig. 6.

3. R2R Gravure Printed Circuits on PET Foils

In this section, the basic operating principles of SWNT-TFT based *p*-channel TFT circuits is discussed. Usually Si based complementary circuits with *n*-channel and *p*-channel TFTs were usually considered to have many advantages over circuits constructed by either only *p*-channel TFTs or *n*-channel TFTs. Among those advantages, better noise margins, lower power dissipation and stable performances with fluctuated TFTs are most prominent. However, under current technologies in R2R printed TFTs on plastic foils, the R2R printed RFID tags can not employ the all printed complementary circuits because R2R printed *n*-channel TFTs on plastic foils are not practically ready yet at this moment. Therefore, in the following sub-sections, only *p*-channel SWNT-TFTs based ciruits will be discussed with operating principles, design simulations, and performance characteristics of SWNT network based R2R gravure printed inverters, ring oscillators, NAND gate, OR gate and D flip-flop circuits on PET foils. In fact, *p*-channel based circuits has small benefits that only one type of semiconductor is needed so that printing and materials stability issues are minimized. However, it has to sacrifice static power dissipation and rely on ratioed logic.

3.1. R2R Gravure Printed p-Channel SWNT-TFTs for Circuit Design Simulations

Typical characteristics of printed SWNT-TFTs used in the circuits presented here are shown in Fig. 7. Devices have 2.5 µm of gate dielectrics, channel width of $W = 3900$ µm, and channel length $L = 200$ µm. All electrical measurements of the printed SWNT-TFTs were carried out under ambient conditions using a semiconductor parameter analyzer (Keithly 4200). Characteristic *p*-channel transistor behavior was seen in the drain-source current (I_{DS}) versus drain-source voltage (V_{DS}) curve at various gate voltages (V_{GS}) (Fig. 7(A)). I_{DS} increased linearly with saturation at higher V_{DS}. The observed field effect originated from the field dependence of carrier concentration in the PS wrapped semiconducting SWNTs not from charge accumulation due to water molecules (Fig. 7(B)). In fact, as shown in Fig. 7(A), almost no hysteresis effect of SWNT-TFTs was observed under ambient conditions due to the PS wrapping of the SWNTs.

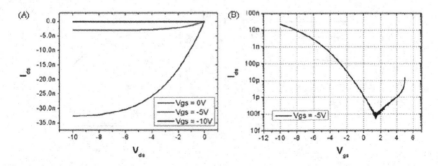

Fig. 7. (A), *I-V* characteristics of a printed SWCNT-TFT and (B), the corresponding transfer characteristics at V_{DS} of -5 V.

For practical applications, the transistor parameters that govern the devices' performance should be evaluated. Among them, transconductance, mobility, the on/off ratio, subthreshold swing, and threshold voltage need to be screened for device applications. The mobility was 0.04 cm^2/Vs at an on/off ratio of 10^6. Extrapolation of the linear region of transconductance plot results in a V_{th} of -6.5 V. The slope in the linear region of I_{DS} versus V_{GS} gives transconductance g_m = d I_{DS}/d V_{GS} = 5.83 nS (Fig. 7(B)), and the subthreshold swing for the printed SWNT-TFT was 2.6 V/decade.

Since the mobility of the printed SWNT-TFTs show too low to operate the printed devices at a reasonable switching speed, we reprinted active layers on printed SWNT-TFTs to increase the current levels of I_{ds} but decrease the on-off ratio (Fig. 8). The reprinted SWNT-TFTs show a mobility of 9.5 cm^2/Vs at an on-off ratio of 1000, but they are good enough to operate circuits at a switching speed of 60 Hz. The slope in the linear region of I_{DS} versus V_{GS} gives transconductance g_m = d I_{DS}/d V_{GS} = 1.399 μS (Fig. 8(B)). The subthreshold swing for the printed SWNT-TFT was 4.5 V/decade and a threshold voltage was 0.5 V.

Based on those data from Fig. 8, the simulations of SWNT-TFTs were carried out using Spice software. The software includes various TFT models for amorphous as well as for poly-silicon. For our purpose the Spice Level 2 was used, which was originally developed for amorphous silicon TFTs. The parameters for the amorphous silicon TFT model were determined in two steps. Firstly, the mobility was calculated

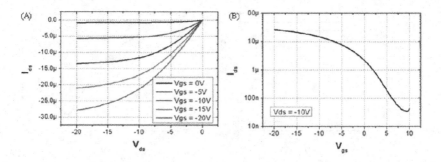

Fig. 8. (A), *I-V* characteristics of a reprinted active layers in printed SWCNT-TFTs and (B), the corresponding transfer characteristics at V_{DS} of -10 V.

from the measured output and transfer characteristics. In order to do so, the mobility is derived from Eqs. (2) and (3):

$$\mu = (\partial I_{DS} \cdot L)/(\partial V_G \cdot W \cdot C_i \cdot V_{DS}) \text{ for } |V_{GS} - V_{TH}| > |V_{DS}| \quad (2)$$

and

$$\mu = (\partial \sqrt{(I_{DS})} \cdot 2L)/(\partial V_G \cdot W \cdot C_i) \text{ for } |V_{GS} - V_{TH}| < |V_{DS}| \quad (3)$$

In a second step the parameter extractor of Spice was used to fit the whole parameter set for the Level 2. While the Level 2 is valid for n-semiconductors, it had to be taken into account that printed SWNT-TFTs are a *p*-type semiconductor, i.e. all currents and voltages had to change sign. Based on the extracted value from SWNT-TFTs (Table 2), a good agreement between simulation and experimental results could be achieved. Figure 9 shows the *IV* characteristic curves of the SWNT-TFT simulation model and the measured values of the driver transistor.

Table 2. Amorphous-Silicon TFT model parameter fit for a printed SWNT-TFT.

Parameter	Value
W	3900 um
L	200 um
F	50 um
R_I	195 KΩm^2
C_I	7 nF/cm^2
Sigma$_I$	50 pA

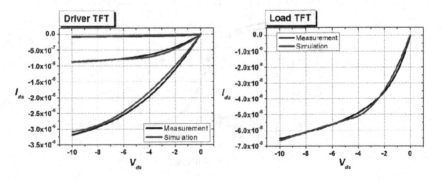

Fig. 9. Simulated and measured IV characteristic curves of a printed SWNT-TFT respectively for the driver and load TFTs.

3.2. 2R Gravure Printed Inverters

The inverter is the most basic building block for the construction of circuits, and its *p*-channel based configuration is shown in Fig. 10. Figure 10 shows the inverter design with an enhancement *p*-type load TFT since the threshold voltage of SWNT-TFT is negative.

Figure 11 shows the measured (black line) and calculated (red line) characteristics for *p*-channel SWNT-TFT based inverter. The inverter is composed of driver *p*-channel SWNT-TFT and load *p*-channel SWNT-TFT with same dimension. In this work, instead of having different dimensions for each driver and load TFTs, we simply change the on current by enhancing the network density of SWNT at the channel.

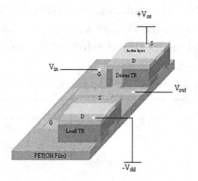

Fig. 10. Schematics of inverter design with an enhancement mode SWNT-TFTs.

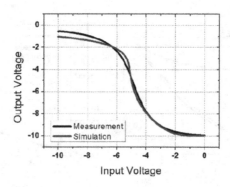

Fig. 11. Measured and calculated VTC (voltage transfer characteristic) of printed *p*-SWNT-TFTs based inverter.

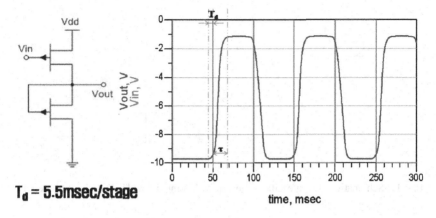

Fig. 12. Transient chracteristics of printed *p*-channel SWNT-TFT based inverter.

Figure 12 shows the schematic and measured response characteristics to a pulse wave input. There are two important features of response. First, there is a delay t_{d1} (t_{d2}) between the step in the input and the response of the out put. Second, there is a slow rise (fall) in the output characterized by the time τ_r (τ_f). These times are just given by simple RC expressions.

The delay t_{d1} (t_{d2}) were observed by the channel resistance of the switching transistor in the second transistor when it charges (discharges) the gate and overlap capacitances of the transistors in the third inverter. A simple expression for this is obtained from the drain current (I_D) equation for the rising and falling pulse delays t_{d1} and t_{d2} respectively. The delay time will limit the circuit speed. This means to increase speed one should keep capaciances (overlap, channel, interconnect and parasitic) and fanout to a minimum.

3.3. *R2R Gravure Printed Ring Oscillators*

Printed ring oscillators are indispensible for on-circuit clock generation in printed RFID tags. Furthermore, the ring oscillators are useful in determining the speed of a given design tools with accounting for the effects of semiconductor, dielectric, and geometrical design rules. The schematic of the ring oscillator is shown in Fig. 13. This circuit consists of five inverters, each composed of two *p*-channel SWNT-TFTs. The

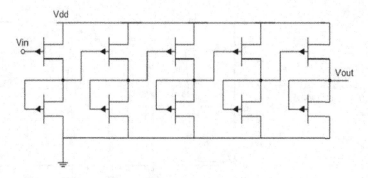

Fig. 13. Schematics of ring oscillator design with printed SWNT-TFTs based inverter.

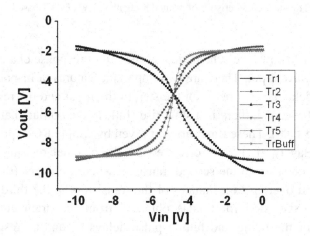

Fig. 14. Simulated transfer characteristics of 5 interconnected inverters at Vds = -10 V (sample with 3 times printed active layer).

simulation result of 5 chain interconnected inverters is shown in Fig. 14 for supply voltages of -10 V. If the inverter gain (the slope of the output characteristics, dV_{out}/dV_{in}) is higher than 1, the inverter chain work as an amplifier and oscillation can be sustained.

Using the printed SWNT-TFT model, we are able to simulate the printed ring oscillator circuit and predict the oscillation frequency changed by channel mobility (depending on the jetting times of active layer). The calculated output signals are shown in Fig. 15.

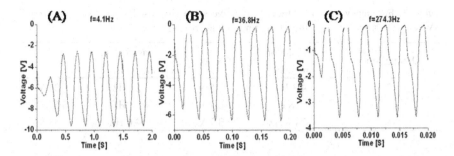

Fig. 15. Spice calculated signals of the five-stage ring oscillators with different jetting times of active layers at Vds = -10 V; (A) 2 times, (B) 3 times and (C) 4 times.

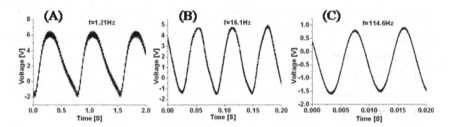

Fig. 16. Measured output signal of the five-stage ring oscillators with different jetting times of active layer at Vds = -10 V; (A) 2 times, (B) 3 times and (C) 4 times.

The amplitude of the oscillation is in a good agreement with the measured output signal (Fig. 16). However the amplitude of the measured output signal is lower than calculated results. The reason for this deviation could be originated from the input resistance of our measurement equipment. This load on the output of the oscillator can be responsible for the lower oscillation amplitude and frequency.

Table 3 shows the simulated and measured frequencies and amplitudes of the ring oscillators based on printed SWNT-TFTs for different jetting times of active layer. Unfortunately there is a large variation between calculated and measured oscillation frequency, which is due to the mobility and the overlap capacitance of the printed SWNT-TFTs. To analyze the differentiation between simulated and measured oscillation frequency, drift velocity is adopted to calculate a delay time of an inverter and an oscillation frequency.

Table 3. Frequency and amplitude of the five-stage ring oscillators based on printed SWNTn-TFTs according to the jetting times of active layer.

Jetting time	Simulation Frequency	Vout	Measurement Frequency	Vout
2	4.1 Hz	7.5 V	1.21 Hz	8.24 V
3	36.8 Hz	6.1 V	16.1 Hz	6.57 V
4	274.3 Hz	3.7 V	114.6 Hz	2.58 V

Delay time of MOSFET inverters is given by drift velocity Eq. (4) and oscillation frequency of five-stage ring oscillator, Eq. (5), is derived from Eq. (4)

$$v_d = L/t = E \cdot u \qquad (4)$$

where
v_d: drift velocity
L: Channel Length
t: carrier drift time
E: Electric Field
u: mobility

$$f_0 = 1/(10 \cdot t) = (u \cdot V_{DS})/(10 \cdot L^2) \qquad (5)$$

The calculated oscillation frequency using Eq. (5) is five times bigger than measured frequency. This discrepancy would originate from the uneven overlap capacitances between gate and drain/source. To invest the effect of the overlap capacitance of printed SWNT-TFT, oscillation frequency was calculated by using the RC (resistor-capacitor) charging/discharging circuit which can be written as Eq. (6)

$$f_0 = 1/(10 \cdot RC) \qquad (6)$$

Table 4. Measured and calculated oscillation frequency based on Eq. (5).

Jetting times	I_{DS} [A]	$R_{channel}$ [ohm]	$C_{gs} + C_{gd}$ [F]	$f_{measure}$ [Hz]	$f_{calculation}$ [Hz]
2	5.69E-9	8.79E+8	2.80E-11	1.21E+0	4.06E+0
3	8.35E-8	5.99E+7	2.80E-11	1.61E+1	5.96E+1
4	7.61E-7	6.57E+6	2.80E-11	1.15E+2	5.44E+2

In the Table 4, calculated oscillation frequency using Eq. (6) is 4 times bigger than measured frequency. As the result of five-stage ring oscillator simulation using printed SWNT-TFT model, if the overlap capacitance between gate and drain/source is reduced to 25%, simulation results are well matched with experimental results. We consider the reason of higher overlap capacitance in the real five-stage ring oscillator is the channel capacitance generated from printed SWNT.

3.4. R2R Gravure Printed D Flip-Flop

Figure 17(A) shows schematic NAND and OR gates based on *p*-type TFTs, and Fig. 17(B) shows measured results for R2R printed NAND and OR gates. The materials parameters for simulations are those shown in Fig. 9.

The positive edge-triggered D flip-flop, the data is loaded on the raising edge of the clock, including 8 NAND and 4 NOT gates has been

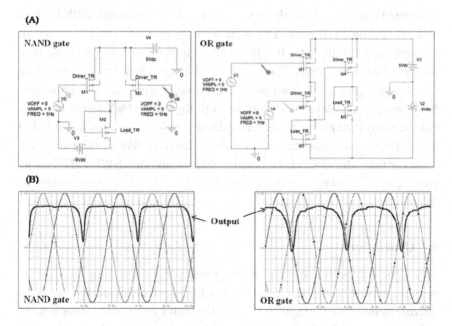

Fig. 17. Schematics of NAND/OR gates (A) and characteristics (B) of R2R printed NAND/OR gates.

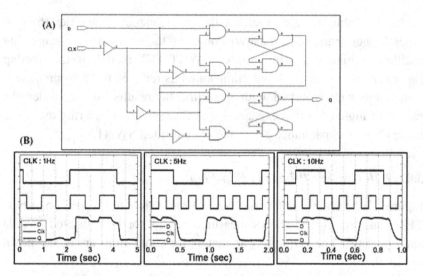

Fig. 18. Schematics of D flip-flop (A) and characteristics (B) of R2R printed D flip-flop with a various of input clock signals.

designed (Fig. 18(A)) which cirucit consist of 32 p-channel SWNT-TFTs and printed on PET foils. The R2R gravure printed D flip-flop circuit was measured with 1 to 10 Hz clock frequency to analyze frequency performance and power consumption. Figure 18(B) shows the data (black line), clock (rede line) and the Q outputs (blue line) of each stages. No propagation delays are observed untill the input clock signal of 3 Hz, but the delays are about 75 ms when the input clock signal is increased to 5 Hz. The propagation delays are originated from SWNT-TFTs at each stages where the SWNT-TFT must switch another SWNT-TFT. The static current drawn by the flip-flop was 10 µA, for a static power of 0.1 mW.

4. Conclusions

Since commercially available R2R gravures have the limit of overlay printing registration accuracy of ±50 µm, all of cicuit design for R2R gravure-printed RFID tags on PET foils should employ the device units to immune those fluctuations generated from the uneven channel lengths. Therefore, in this chapter, we introduced 200 µm of channel length for

R2R gravure-printede TFTs using SWNT as semiconducting material to overcome a lower operation frequency due to the lengthy channel. Based on R2R gravure-printed SWNT-TFTs, inverters, NAND/NOR gates and D flip-flop on PET foils were respectively designed and printed on PET foils. To attain full of R2R gravure-printed RFID tags, the R2R gravure-printed D flip-flop will be further utilized for 96 bit of decoder.

Acknowledgments

We are grateful to the World Class University Program at Sunchon National University and MKE for their support in the research consortium for developing roll-to-roll printed RFID tags. JMT acknowledges NASA for support of the work at Rice University. GC specially thanks to Dr. Dustin James for his valuable suggestions and comments in preparing this manuscript. Further, this work is supported by the Ministry of Knowledge Economy (MKE) through the project of GTFAM Regional Innovation Center (RIC).

References

1. S. R. Forrest. *Science* 428, 911 (2004).
2. V. Subramanian, P. C. Chang, J. B. Lee, S. E. Molesa, S. K. Volkman. *IEEE Trans. on Comp. & Pack. Tech.* 28, 742 (2005).
3. V. Subramanian, J. M. J. Frechet, P. C. Chang, D. C. Huang, J. B. Lee, S. E. Molesa, A. R. Murphy, D. R. Redinger, S. K. Volkman. *Proceedings of the IEEE* 93, 1330 (2005).
4. N. Saran, K. Parikh, D. S. Suh, E. Muñoz, H. Kolla, S. K. Manohar. *J. Am. Chem. Soc.* 126, 4462 (2004).
5. S. Trans, S. Verschueren, C. Dekker. *Nature* 393, 49 (1998).
6. P. Avouris, J. Appenzeller, R. Martel, S. J. Wind. *Proc. IEEE* 91, 1772 (2003); Y. Zhou, A. Gaur, S.-H. Hur, C. Kocabas, M. A. Meitl, M. Shim, J. A. Rogers. *Nano Lett.* 4, 2031 (2004); E. Artukovic, M. Kaempgen, H. S. Roth, G. Grűner. *Nano Lett.* 5, 757 (2005).
7. T. Dűrkop, S. A. Getty, E. Cobas, M. S. Fuhrer. *Nano Lett.* 4, 35 (2004).
8. K. Bradley, J. P. Gabriel, G. Grűner. *Nano Lett.* 3, 1353 (2003).
9. E. S. Snow, J. P. Novak, P. M. Campbell, D. Park. *Appl. Phys. Lett.* 82, 2145 (2003).
10. U. Tomas, E. N. Hans. *IEEE Sensor* 9, 922 (2009).
11. H. Huitema, G. Gelinck, J. Putten, K. Kuijk, C. Hart, E. Cantatore, P. Herwig, A. Breemen, D. Leeuw. *Nature* 414, 599 (2001).

12. G. Gelinck. *Nat. Mat.* 3, 106 (2004).
13. P. Andersson, D. Nilsson, P. Svensson, M. Chen, A. Malmstrom, T. Remonen, T. Kugler, M. Berggren. *Adv. Mat.* 14, 1460 (2002).
14. A. Huebler, F. Doetz, H. Kempa, H. Katz, M. Bartzsch, N. Brandt, I. Hennig, U. Fuegmann, S. Vaidyanathan, J. Granstrom, S. Liu, A. Sydorenko, T. Zillger, G. Schmidt, K. Preissler, E. Reichmanis, P. Eckerle, F. Richter, T. Fischer, U. Hahn. *Organic Electronics* 8, 480 (2007).
15. T. Kelley, P. Baude, C. Gerlach, D. Ender, D. Muyres, M. Haase, D. Vogel, S. Theiss. *Chem. Mater.* 16, 4413 (2004).
16. N. Lim, J. Kim, S. Lee, N. Kim, G. Cho. *IEEE Trans. Adv. Pack.* 32, 72 (2009).
17. K. Reuter, H. Kempa, N. Brandt, M. Bartzsch, A. C. Huebler. *Prog. Org. Coat.* 58, 312 (2007).
18. T. Fischer, U. Hahn, M. Dinter, M. Bartzsch, G. Schmidt, H. Kempa, A. C. Huebler. *Org. Elect.* 10, 547 (2009).
19. D. Zielke, A. C. Huebler, U. Hahn, N. Brandt, M. Bartzsch, U. Fuegmann, T. Fischer. *Appl. Phys. Lett.* 87, 123508 (2005).
20. Y. Xiuyan, K. Satish. *Chem. Eng. Sci.* 61, 1146 (2006).

CHAPTER 15

PHYSICAL VAPOR DEPOSITION OF POLYMER THIN FILMS AND ITS APPLICATION TO ORGANIC DEVICES

Hiroaki Usui

*Department of Organic and Polymer Materials Chemistry,
Tokyo University of Agriculture and Technology, 2-24-16 Naka-cho,
Koganei, Tokyo 183-8538, Japan
E-mail: h_usui@cc.tuat.ac.jp*

This chapter describes novel method for preparing polymeric thin films by way of physical vapor deposition (PVD) instead of the wet-coating techniques that have been used conventionally. The PVD has advantages in preparing highly pure nanometric thin films and their multilayered structures, and is suitable for constructing electronic and optical devices. Four methods of polymer PVD, including direct evaporation of polymers, coevaporation of bifunctional monomers, radical polymerization from single monomer, and surface-initiated deposition polymerization are described in this chapter. The type of polymers includes polyethylene, polytetrafluoroethylene, polyimide, polyurea, vinyl or acryl polymers, and polypeptide. It was found that the deposition polymerization provides organic thin films that are superior in stability and uniformity compared to the conventional vapor-deposited thin films of small molecules. In addition, the surface-initiated deposition polymerization have possibility of controlling the interface between inorganic substrate and polymer thin films. These characteristics can be utilized for improving the device characteristics such as the lifetime of organic light emitting diodes (OLEDs).

1. Introduction

Film formation techniques have been playing an important role in the development of optical and electronic devices. For example, organic light-emitting diodes (OLEDs) have not been realized until physical vapor deposition (PVD) was utilized to prepare uniform organic thin films, thereby achieving sufficient current flow through highly resistive organic semiconductors, although electroluminescence of certain organic materials has been known for a long time. The PVD was also convenient for preparing multilayered device structures, which enabled efficient carrier recombination at the heterojunction interface.[1] Many of small molecules can be vapor-deposited without difficulty, and PVD has been used as a standard technique for fabricating organic devices such as OLEDs and organic thin film transistors (OTFTs). In general, PVD has advantages in preparing high-purity thin films and multi-layered structures with high controllability. PVD is also compatible with the standard semiconductor processes such as patterning through shadow mask and formation of electrical contacts by depositing metal electrodes.

On the other hand, polymer thin films have been prepared mainly by wet-coating processes. The wet-coating has advantages in productivity and cost-effectiveness. However, there is still a way to use the wet-coating methods for constructing well-controlled microstructures that are required for devices. Moreover, organic solvents used for the wet-coating can cause environmental problems, i.e. emission of volatile organic compounds (VOCs). In this respect, it is significant to develop a technique for preparing polymeric thin films by vacuum-based dry coating methods. Since polymers do not evaporate like metals or small molecules, PVD of polymers is not straightforward. However, there are several strategies to make PVD of polymers possible. This chapter introduces some examples of PVD of polymer thin films and their applications to organic devices.

2. PVD of Polymer Thin Films

PVD is a technique that accumulates thin films by condensing the vapor of film materials on the substrate surface. The film materials are vaporized generally by heating to an appropriate temperature that gives a saturated vapor pressure of the order of 0.1 Pa or higher. However, most of the polymers cannot be vaporized, and are unable to be handled by PVD. Nevertheless, there are several strategies to achieve polymer deposition by means of PVD as illustrated in Fig. 1.

The simplest method of polymer PVD is to apply the conventional vacuum deposition technique as it is by using the polymer material for the evaporation source as illustrated in Fig. 1(a). This method can be applied for few polymers whose intermolecular interaction is sufficiently weak and molecular weight is not very high. Since the direct evaporation can be utilized only in limited cases, a practical method for PVD of polymer is to evaporate its monomers and let them react on the substrate surface to yield polymeric thin films. Such a method is called deposition polymerization, and classified as a solventless polymerization method.

The vapor deposition polymerization was initially reported for preparation of polyimide and polyurea thin films by codeposition of two bifunctional monomers as schematically illustrated in Fig. 1(b).

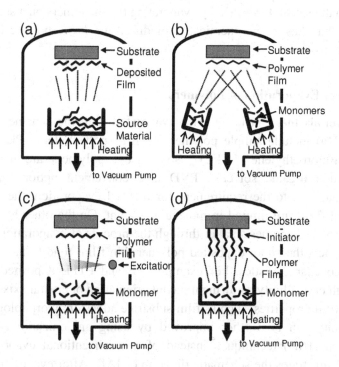

Fig. 1. Typical methods for physical vapor deposition of polymer thin films. (a) direct evaporation, (b) stepwise reaction by coevaporation of two monomers, (c) chain reaction of single monomer assisted by electron or UV excitation, and (d) surface-initiated deposition polymerization.

This method can be applied for various materials that polymerize by condensation or polyaddition reaction. Another strategy for vapor deposition polymerization is to make use of chain reaction by evaporating such monomers as vinyl or acryl compounds. The polymerization can be initiated by generating radicals using electron or UV irradiation in the course of deposition as illustrated in Fig. 1(c).

The films deposited by PVD, including the three methods mentioned above, are in general attached to the substrate surface by physical adsorption, which does not involve specific chemical bonding to the substrate. Therefore, PVD can form thin films on any kind of substrate. However, the adhesion at the film/substrate interface is not complete in many cases. The surface-initiated deposition polymerization, represented by Fig. 1(d), attempts to grow polymeric thin films that are chemically bound to the substrate surface by evaporating the monomers on a substrate surface that has been chemically modified to have polymerization initiating groups.

3. Direct Evaporation of Polymers

The materials that can be directly evaporated with the scheme shown in Fig. 1(a) include simple polymers such as polyethylene (PE)[2,3] and polytetrafluoroethylene (PTFE).[4] Since these polymers are insoluble to common organic solvents, PVD is a convenient option for film deposition. Due to the requirement for thermal evaporation, the vapor-deposited films are limited in molecular weight. On the other hand, the polydispersity becomes smaller through the process of evaporation. As a consequence, the vapor-deposited polymer thin films tend to have well-controlled characteristics. For example, PE and PTFE deposited films have high crystallinity, showing uniaxial orientation of crystal axis.

The film properties, such as film-substrate adhesion, morphology and crystallinity can be further improved by using an ionization-assisted deposition (IAD) technique,[5] instead of the conventional evaporation. Figure 2 illustrates the schematic diagram of IAD. After evaporating the source material, a part of the vapor is ionized by electron irradiation. The ions can be accelerated to an arbitral kinetic energy by applying an ion acceleration voltage to the substrate. This gives a distinct advantage

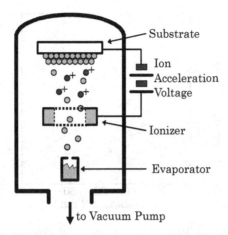

Fig. 2. Schematic diagram of the ionization-assisted vapor deposition system.

compared to the conventional vapor deposition, in which the kinetic energy of the evaporated molecules is limited to the order of kT, where k is Boltzmann constant and T is the evaporation temperature.[6] The fraction of ionization is generally of the order of 1% or smaller. However, the ion concentration and the ion acceleration voltage can be used as useful parameters for controlling the film characteristics.

In the case of PTFE, a source material having number-averaged molecular weight M_n of 8500 can be evaporated at a temperature of 470°C. The deposited film showed preferential crystal orientation with the (100) plane parallel to the substrate surface. Figure 3 shows the pinhole density of the PTFE films deposited by IAD at the ion acceleration voltage of 0 and 500 V.[7] The pinhole density was measured by using a copper decoration method. Generally, the pinhole density decreases with increasing film thickness. In addition, the ion acceleration during the film deposition was effective in reducing the pinhole density. It was also found that the surface roughness of the film decreases by applying the ion acceleration voltage, which led to the formation of uniform and pinhole-free thin films.

PTFE is unique in low refractive index, low dielectric constant, high electrical insulation, as well as high thermal and chemical stability. PTFE is also known to have extremely low surface energy, giving poor

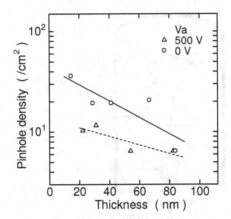

Fig. 3. Pinhole density of PTFE films of various thicknesses deposited with ion acceleration voltage V_a of 0 and 500 V.

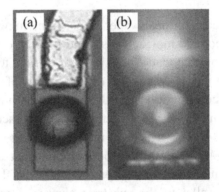

Fig. 4. Plan view (a) and stereoscopic view (b) of spherical plastic microlense prepared on semiconductor device covered with PTFE thin film. Lens diameter is 10 μm.

wettablity to most of the liquids. This characteristics was utilized for the preparation of a spherical plastic microlense as shown in Fig. 4.[8] The microlense was constructed by melting phenol resin on a PTFE thin film that has small opening at the center of lens location. The molten phenol resin formed spheric shape and aligned its position to the center of opening spontaneously. The patterning of vapor-deposited PTFE films can be easily achieved by the lift-off technique of photolithography. An array of spheric microlenses can be fabricated without difficulty.

4. Deposition Polymerization by Coevaporation Method

4.1. *Stepwise Polymerization by Coevaporation of Two Monomers*

Although the direct evaporation scheme described in previous section is simple, only few polymers can be deposited by that method. Moreover, the limitation on molecular weight diminishes the advantage of using polymers. Polymers of higher molecular weight can be obtained only by using the deposition polymerization method, which polymerizes the evaporated monomers on the substrate surface. Polymer deposition by coevaporation of two monomers, as illustrated in Fig. 1(b), was first developed for preparing polyimide[9,10] or polyurea thin films.[11] The films can be obtained by coevaporating two bifunctional monomers; diamine with acid dianhydride for polyimide, and diamine with diisocyanate for polyurea. Since the polymerization proceeds by step reaction, it is essential to balance the rate of monomer supply to obtain a high degree of polymerization. The normal PVD is operated in high-vacuum condition, where the mean free path of the evaporated monomers is much longer than the dimension of vacuum chamber. As a consequence, the polymerization proceeds on the substrate surface by the collision of monomers that are migrating on the surface. This method resembles to the chemical vapor deposition (CVD) if operated under higher pressure.

4.2. *Deposition Polymerization of Polyimide*

As an example, polyimide that has perylene unit in the backbone can be prepared by codepositing perylenetetracarboxylic dianhydride (PTCDA) and diaminonaphthalene (DAN) according to the reaction (1).

$$\text{PTCDA} + \text{DAN} \xrightarrow{\text{codeposition}} \text{polyamic acid} \xrightarrow{\text{annealing}} \text{polyimide} \quad (1)$$

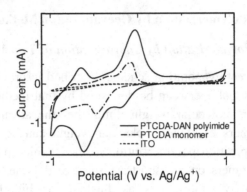

Fig. 5. Cyclic voltammogram of PTCDA-DAN polyimide and PTCDA monomer films deposited on ITO surface.

This polymerization proceeds by condensation reaction. Codeposition of PTCDA and DAN produces a thin film of polyamic acid, which can be annealed in air at 100°C for 1 h to yield the polyimide. PTCDA has poor solubility in organic solvents. Moreover, the product polyimide has rigid backbone and is also insoluble. Under these circumstances, PVD can make the most of its solvent-free advantage.

Figure 5 shows a cyclic voltammogram of the perylene polyimide film deposited on an indium-tin-oxide (ITO) substrate measured in 0.5 M Na_2SO_4 solution. This film showed reversible two-step reduction at -0.6 and -1.0 V, changing its color from orange to purple and further to yellowish green, respectively. Similar electrochromism was also observed for a vapor-deposited film of PTCDA monomer, but the monomer film started to disintegrate after repeating the redox cycle, whereas the polyimide had higher durability against the electrochemical process. A displacement current measurement showed that the PTCDA polyimide has electron transport characteristics.[12]

4.3. *Deposition Polymerization of Polyurea*

Polyurea (PU) can be synthesized by polyaddition reaction, and the films can be prepared by coevaporating diamine and diisocyanate monomers in as-deposited state. No post-deposition annealing is required to finish the polymerization. Reaction (2) gives an example of polyurea formation from 1, 3-di(4-piperidyl)propane (PIP) and 4, 4'-diphenylmethane diisocyanate (MDI).[13,14]

$$O{=}C{=}N{-}{-}CH_2{-}{-}N{=}C{=}O + HN{-}CH_2CH_2CH_2{-}NH \atop \text{MDI} \text{PIP}$$

$$\xrightarrow{\text{codeposition}} {\Large\{} \underset{H}{\overset{O}{C{-}N}}{-}{-}\underset{H_2}{C}{-}{-}\underset{H}{\overset{O}{N{-}C{-}N}}{-}CH_2CH_2CH_2{-}N {\Large\}}_n \quad (2)$$
$$\text{PU}$$

Polyurea is attractive not only as a transparent and thermally stable polymer, but also as an optically nonlinear or piezoelectric material when the dipole moment of urea bond is aligned to non-centrosymetric orientation. Conventionally, the dipole alignment has been achieved by the poling technique, where a preformed film is heated to its glass transition temperature under a high electric field. However, this method has difficulty in compromising the poling capability and thermal stability of dipole orientation. With the vapor deposition polymerization, dipole orientation can be controlled efficiently by using the IAD method shown in Fig. 2, since the electric field generated by the substrate bias voltage exerts its force directly to the mobile monomers and growing end of polymers during the film growth process, instead of post-deposition poling of densely-packed solid polymer.

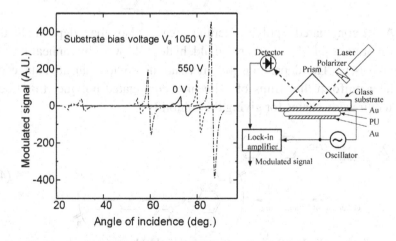

Fig. 6. Schematic diagram of ATR measurement system (right) and electrically modulated ATR spectra for PU films deposited with different V_a (left).

The dipole orientation of the PU of reaction product in (2) was measured by using an attenuated total reflection (ATR) method as illustrated by the diagram in Fig. 6.[15] The PU film deposited by IAD was stacked between two semitransparent gold electrodes. The voltage applied to the electrodes modulates the optical constants of the PU film through electro-optic and piezoelectric effects, which can be monitored by measuring the ATR-coupled optical reflection signal. The chart in Fig. 6 shows the amplitude of electrically modulated signal for the PU films deposited under different substrate bias voltage V_a. Larger modulation was observed for the films prepared with higher bias voltage, indicating enhanced dipole orientation during the deposition process.

4.4. *Other Polymers Prepared by Codeposition Method*

Polyurethane thin films can be prepared by codeposition of diol and diisocyanate as shown in reaction (3). This polyurethane of zinc-complex has an electron-transporting and light-emitting characteristics.[16]

$$\text{(3)}$$

A π-conjugated polyazomethine can also be deposited by coevaporation of diamine and dialdehyde followed by annealing as reaction (4).[17] The deposition polymerization provides an advantage of forming uniform thin films of a linear π-conjugated polymer that does not require side chains for giving the solubility.

$$\text{(4)}$$

Another example of practical interest is the deposition of epoxy resin, which can be applied for electrical insulating and packaging coatings.

Codeposition of diepoxide and diamine, which works as a crosslinking reagent, forms a network polymer according to reaction (5). When the monomer structure has sufficient flexibility, the reaction proceeds at room temperature without post-deposition annealing.

$$2 \underset{R}{\triangle\!\!\!\!\triangle} + H_2N-R'-NH_2 \xrightarrow{\text{codeposition}} \left(\begin{array}{c} R \underset{OH}{\diagup} \underset{N}{\diagdown} \\ OH \quad R' \\ R \underset{OH}{\diagup} N \\ OH \end{array} \right)_n \quad (5)$$

5. Deposition Polymerization by Radical Reaction

5.1. *Deposition Polymerization of Vinyl and Acryl Monomers*

In general, radical reaction is more suitable for obtaining higher polymers compared to the stepwise reaction described in the previous section. Moreover, polymer can be prepared from single species of monomer, which is convenient for practical application. The radical reaction can be applied for vapor deposition polymerization by evaporating vinyl or acryl monomers as shown in reaction (6).

$$H_2C=CH \atop R' \longrightarrow \left(C-C \atop H_2 \; R' \right)_n \quad \text{or} \quad H_2C=CH \atop \substack{C=O \\ O \\ R''} \longrightarrow \left(C-C \atop H_2 \; C=O \atop O \atop R'' \right)_n \quad (6)$$

vinyl monomer $\qquad\qquad\qquad$ acryl monomer

This system also has advantage in versatility of designing molecular structure simply by attaching the functional units as R' or R" in (6). On the other hand, the simple vapor deposition is not capable of polymerizing monomers unless active radical species are provided to initiate the polymerization. This can be achieved, for example, by electron irradiation,[18] UV exposure,[19,20] radiation from a hot filament,[21] and in some cases by heating the substrate,[22] as illustrated in Fig. 1(c). The vinyl and acryl polymers produced in this way have flexible backbone. Structurally stable network polymers can be obtained by depositing multifunctional monomers, or coevaporating with a multifunctional monomer as a crosslinking reagent.

5.2. *Preparation of Fluoropolymer by Radical Polymerization*

The electron-assisted deposition polymerization was applied for preparing fluoropolymer thin films from a fluorinated alkylacrylate monomer, 1H, 1H, 2H, 2H-heneicosafluorododecylacrylate (Rf-10). Its product polymer has comb-like structure, where Teflon-like side chains are densely attached to the main chain.[23] Its film growth process was observed by *in-situ* IR absorption spectra using the reflection-absorption method on gold surfaces. Figure 7 shows the IR spectra during film growth of Rf-10 by conventional vapor deposition (a) and by electron-assisted deposition (b). Since Rf-10 is volatile, the deposition was performed at a substrate temperature of 0°C.

In Fig. 7, the characteristic absorption bands of C-F stretching vibrations appeared at 1230 and around 1160 cm^{-1}. The bands associated

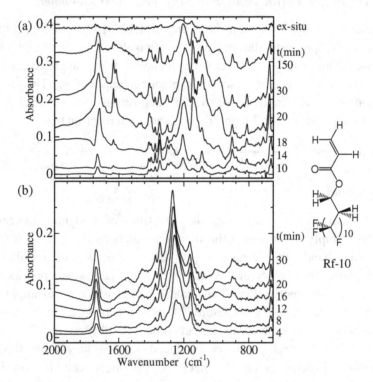

Fig. 7. IR spectra during film growth of Rf-10 by conventional vapor deposition (a) and by electron-assisted deposition (b). Each spectrum has been shifted for clarity.

with the vinyl group, such as C=C stretching at 1630 cm^{-1}, CH$_2$ in-plane deformation at 1410 cm^{-1}, and CH in- and out-of-plane deformations at 980 and 810 cm^{-1}, were observed for the vapor deposited film. However these bands of vinyl group were not observed when the film was deposited by the electron-assisted method. These results suggest that the film deposited by the conventional vapor deposition consists of Rf-10 monomers, while the electron-assisted deposition accumulates a polymer thin film by opening the vinyl bond of Rf-10. It is noteworthy that the film by vapor deposition started to reevaporate after stopping the deposition, while the electron-assisted deposition formed a stable film. In addition, the loss of vinyl peak was observed from the initial stage of electron-assisted deposition, suggesting that the polymerization occurs at the moment of film accumulation.

It is known that the alkyl chain length largely influences the molecular packing and surface characteristics.[24] The fluoropolymer films were deposited using monomers of different alkyl chain length, and their characteristics were investigated by a dynamic contact angle measurement. In general, the film growth rate decreased with decreasing chain length since the smaller molecules are more volatile. Figure 8 shows advancing and receding contact angles as a function of alkyl chain length of the monomer. The contact angle increased, and the hysteresis

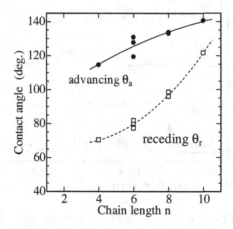

Fig. 8. Advancing and receding contact angles of fluoropolymer films deposited from monomers of different fluorinated alkyl chains.

of contact angle decreased with increasing chain length, suggesting that a longer alkyl side chain leads to higher stability of molecular packing. The surface energy estimated by Owens-Wendt formula was about 6 mN/m for n = 10, which is comparable to the standard PTFE. The surface energy increased with reducing the chain length of side group.

5.3. Deposition Polymerization of Carbazole Polymers

Polymer thin films for electronic devices can be prepared by attaching semiconducting functional units as the pendant groups of the vinyl or acrylate monomers. The functional units, as small molecules, might be vapor deposited for film formation by themselves. However, vapor deposited films of small molecules frequently suffer lack of thermal stability. On the contrary, vapor deposition polymerization is effective in obtaining thermally and mechanically stable thin films.

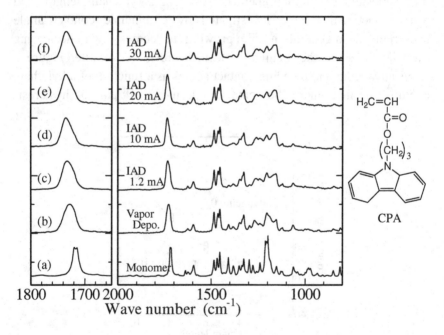

Fig. 9. IR spectra of CPA monomer (a), a film by conventional PVD (b), and films by electron-assisted deposition with different electron assist current I_e (c-f). The left chart shows detail of C=O stretching band. Molecular structure of CPA is also shown on right.

One of the simplest examples is the formation of carbazole polymer by electron-assisted deposition of 3-(*N*-carbazolyl)propyl acrylate (CPA). Figure 9 shows the IR spectra of CPA monomer (a), a film by conventional PVD (b), and the films by electron-assisted deposition with different electron assist current Ie (c-f). The left chart shows detail of the C=O stretching band.[25]

The absorption by vinyl group, such as C=C stretching (1636 cm^{-1}), CH$_2$ in-plane deformation (1404 cm^{-1}), and C-H out-of-plane deformation (982 cm^{-1}), were observed in the conventional PVD film, as well as in the CPA monomer. However, these bands of vinyl group became considerably weaker by the electron-assisted deposition at $I_e = 1.2$ mA, and almost disappeared at I_e larger than 10 mA. In addition, the C=O stretching band of ester group, which appeared as a sharp peak at 1717 cm^{-1}, broadened and shifted toward larger wavenumber by electron-assisted deposition as a result of loosing conjugation with the vinyl group. These results indicate that the vinyl polymerization was enhanced with increasing electron irradiation during the vapor deposition.

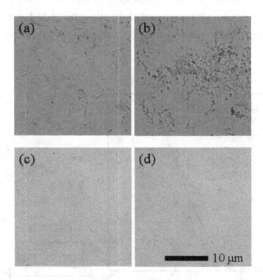

Fig. 10. Differential interference optical micrographs of CPA film deposited by conventional vapor deposition in as deposited state (a) and after annealing at 125°C for 30 min (b). A CPA film by electron-assisted deposition polymerization is also shown before (c) and after (d) annealing, respectively.

Figure 10 compares optical micrographs of CPA films deposited by the conventional vapor deposition (a) and by the electron-assisted deposition polymerization (c). The former film had considerable amount of needle-like hillocks due to crystallization of CPA monomers. By annealing at 125°C for 30 min, this film underwent severe coagulation due to extensive crystallization (b). On the other hand, the film prepared by deposition polymerization had few such structures, and was not damaged by the annealing (d).[26]

5.4. *Application to Light-Emitting Diodes*

Organic light-emitting diodes (OLEDs) can be prepared by the vapor deposition of monomers having charge transport and luminescent functional units. As a simple example, a hole transport layer (HTL) was prepared by deposition polymerization of *N*-(4-acryloyloxymethylphenyl)-*N'*-phenyl-*N*,*N'*-bis(4-methylphenyl)-[1,1'-biphenyl]-4,4'-diamine (TPD-Ac). TPD-Ac was deposited by the electron-assisted method on indium-tin-oxide (ITO) substrate at different substrate temperatures, on which tris(8-hydroxyquinolinate)aluminumn (Alq$_3$) emissive layer (EML) and aluminum cathode were vapor-deposited to construct an OLED.[27]

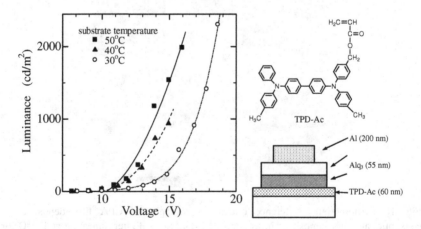

Fig. 11. Characteristics of OLEDs using TPD-Ac HTL. TPD-Ac was deposition-polymerized at different substrate temperature. Structures of TPD-Ac and OLED are also shown.

Figure 11 shows the luminescence characteristics of the devices whose HTL were deposition-polymerized at different substrate temperatures. With increasing substrate temperature from 30 to 50°C, the TPD-Ac film increased its molecular weight from 4.4×10^3 to 1.6×10^4, and polymer yield from 27 to 71%. Correspondingly, the luminance of the device also increased with increasing substrate temperature. This result indicates that the deposition-polymerization is an effective method for constructing organic devices. Polymerization reaction can be further enhanced by increasing the substrate temperature, but excessive increase of substrate temperature resulted in rough film surface.

The vapor deposition-polymerization can make the advantage of its solventless feature in constructing multilayered structures of polymer thin films, which is effective in developing OLED due to the possibility of optimizing carrier balance and recombination at the heterojunctions. Figure 12 shows an example of OLED structure prepared by deposition polymerization of HTL and EML susscessively.[22] A divinyl derivative of tetraphenyldiaminobiphenyl DvTPD was deposited to form a HTL on an ITO substrate coated with poly(3,4-ethylene dioxythiophene)-poly (styrene sulfonate) (PEDOT:PSS), on which an EML was deposited by coevaporation of DvTPD and a vinyl derivative of phosphorescent dopant Ir(piq)$_2$acac-vb. After depositing the HTL and the EML, the substrate was annealed in the vacuum chamber at 100°C for 1 h to thermally polymerize these layers. The device was finished by depositing an electron transport layer (ETL) of bathocuproin (BCP), an electron injection layer of lithium fluoride, and then an aluminum cathode. The

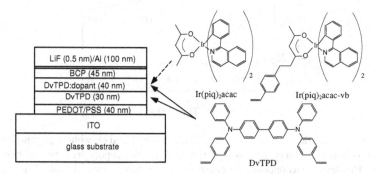

Fig. 12. OLED structure and monomer materials for vapor deposition polymerization.

polymerization was effective especially in improving the thermal stability, thereby extending the device lifetime.

Devices of identical structures were prepared by combining both the conventional dopant Ir(piq)$_2$acac or its vinyl derivative Ir(piq)$_2$acac-vb with the DvTPD host material for the EML, with and without thermal polymerization. Figure 13 shows the luminance decay under a constant current operation, starting from the initial luminance of 500 cd/m^2. The control device (a) prepared without polymerizing DvTPD:Ir(piq)$_2$acac had the shortest lifetime, while the lifetime increased by polymerizing only the host material by annealing DvTPD:Ir(piq)$_2$acac EML (b). Further improvement in lifetime was achieved by copolymerizing host and dopant in EML (c) that was prepared by annealing EML of DvTPD:Ir(piq)$_2$acac-vb. This result indicates that the successive vapor deposition polymerization can prepare stable polymer multilayers, leading to the improvement in device characteristics. Using the same procedure, an OLED having wide emission spectra can be prepared, which has multiple EMLs stacked by the vapor deposition polymerization.[28] It was found that the polymerization was effective in stabilizing the device characteristics such as emission spectra, as well as in improving the quantum efficiency.

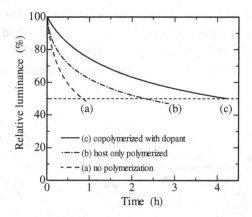

Fig. 13. Luminance decay of OLEDs whose EML was prepared (a) without polymerization, (b) with polymerizing only the host material, and (c) with copolymerization of host and dopant. The device was operated by constant current starting from the initial luminance of 500 cd/m^2.

5.5. Application to Photopatterning

Vapor deposited films have been conveniently patterned by evaporating through a shadow mask. On the other hand, the shadow mask bears such problems as contamination or alignment, especially when the mask size is enlarged. With a purpose to develop a new method for patterning polymeric thin films, photosensitive thin films were prepared by vapor deposition of vinyl monomer mixed with photoinitiator, which can be selectively polymerized by UV exposure and patterned according to the photolithographic technique after the film deposition.[29]

Figure 14 schematically explain the concept. A carbazole monomer 9H-carbazole-9-ethylmethacrylate (CEMA) was coevaporated with a photoinitiator of 4-dimethylamino benzophenone (DABP) to deposit a photosensitive film. The film was exposed to a UV light of 355 nm for 30 s through a photomask in the air, and then rinsed in tetrahydrofuran (THF) for 1 min to remove the non-irradiated, i.e. non-polymerized part. Pattern formation was observed for the DABP concentration higher than 1% and the UV power higher than 12 mW/cm^2. Figure 15 shows the examples of 10 μm-diameter dots and 10 μm-wide line-and-space patterns of CEMA polymer obtained by this process. This method has an advantage of eliminating the shadow mask during the vapor deposition process, and forming the patterns in the photolithographic technique,

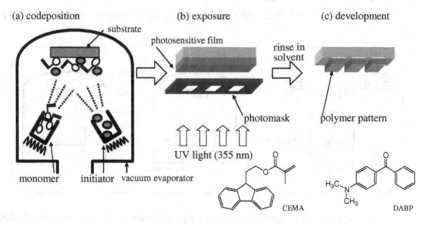

Fig. 14. The concept of photopatterning by vapor deposition of photosensitive film. Molecular structures of monomer and photoinitiator are also shown.

Fig. 15. 10 μm-diameter dots and 10 μm-wide line-space patterns obtained by the photoresponsive CEMA film.

which is already well established. It was also confirmed that the photopatterning process does not damage the electrical characteristics of the thin films.[30]

6. Surface-Initiated Deposition Polymerization

6.1. *The Concept of Surface-Initiated Deposition Polymerization*

One of the largest difficulties in device application of organic thin films is the lack of thermal stability. It comes partly from the small intermolecular interaction, which can be alleviated by polymerizing the material. The instability also comes from the incommensurateness at the interfaces, especially between organic films and inorganic substrates. Unlike the inorganic semiconductor thin films that can be epitaxially grown or alloyed with substrates, most of the organic films are weakly physisorbed on inorganic substrates without forming stable chemical

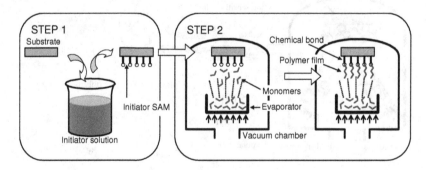

Fig. 16. Concept of surface-initiated deposition polymerization, which consists of SAM formation on substrate surface in solution (STEP 1), and physical vapor deposition of monomer on the surface of SAM (STEP2).

bonds at the interface. The surface-initiated deposition polymerization shown in Fig. 1(d) was devised to solve this problem by combining the deposition polymerization with a self-assembled monolayer (SAM) that bears a polymerization initiating unit as the end group.

Figure 16 illustrates the concept of surface-initiated deposition polymerization, which consists of two steps, modification of substrate surface with a polymerization initiator, followed by vapor deposition of monomers. The initiator is attached to the substrate surface as a SAM by using a combination that can form specific chemical bonds between the molecules and the substrate. The SAM can be prepared simply by dipping the substrate into an appropriate solution, but can also be prepared by vapor deposition of the initiator.

6.2. Interface Control by Surface-Initiated Deposition Polymerization

Formation of stable covalent bonds at the film-substrate interface is expected to solve various problems at the electrode contacts to organic devices. An example is the growth of carbazole thin films on ITO substrates, which can be achieved as shown in route (7).[31]

An ITO substrate was immersed in solutions containing trimethoxy(3-aminopropyl)silane (APS), succinic anhydride, pentafluorophenol, and finally Vazo 56 (DuPont) successively to form a SAM layer that has azo terminal as polymerization initiator.[32] The SAM-modified substrate was introduced into a vacuum chamber, where CPA was deposited by the electron-assisted method under irradiation of UV light from a high-pressure mercury lamp. The electron-assisted deposition leads to polymerization of CPA by itself as described in Sec. 5.3. However, the film deposited on a bare substrate was soluble to THF due to the flexible nature of acryl polymer backbone. On the other hand, the same film deposited on the azo-terminated SAM dissolved only partly to the solvent, suggesting the formation of polymers that are chemically bound to the ITO surface through the SAM molecules.

OLEDs were prepared by stacking an Alq_3 EML, a lithium fluoride electron injection layer, and an aluminum cathode on the CPA HTL. Figure 17 compares the characteristics of the devices prepared with and without the SAM layer. The SAM modification was effective in increasing the device current and reducing the onset voltage for luminescence.[33] It is reported that the SAM layer at the interface lowers the carrier injection barrier by forming a dipole layer.[34] Moreover, the surface-initiated deposition polymerization is expected to form a tight contact at the interface, thereby alleviating the problems inherent to the interface, as well as improving thermal and mechanical stability of the organic layer.

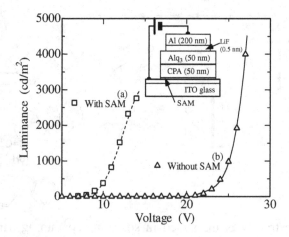

Fig. 17. Voltage-luminance characteristics of OLEDs prepared with (a) and without (b) SAM layer on ITO surface.

6.3. Formation of Polypeptide Thin Films

Polypeptide tethered on a substrate has potential applications in biosensors and optical devices, and may also address important biocompatibility issues. Polypeptide can be obtained by a ring opening polymerization of aminoacid *N*-carboxy anhydrides (NCA) using amine initiator.[35] However, the reaction in solution often suffer from the lack of reaction rate and insufficient degree of polymerization. It is reported that the reaction proceeds much efficiently in vapor phase due to effective removal of impurities and moisture.[36] In this respect, the surface-initiated deposition polymerization can be applied conveniently for growing polypeptide thin films tethered on substrate surface.[37,38] The film was obtained according to scheme (8), first preparing a SAM of APS on an oxide surface and then vapor depositing NCA monomers.

(8)

The reaction proceeds by decarboxylation at room temperature concurrently with the NCA deposition. Figure 18 shows IR absorption spectra of the benzylserine NCA monomer (a), and the films deposited on bare Al (b) and on SAM-modified Al (c). The Al surface has native oxide, which enables the SAM formation of APS. Carbonyl peaks in the spectrum of the NCA monomer appeared at both 1860 and 1785 cm^{-1}. The lack of a change in location of the carbonyl peaks for the film deposited on the bare aluminum shows that no polymerization has occurred and only accumulation of the monomer has been carried out. On the other hand, the film deposited on the APS-modified surface showed the loss of the anhydride carbonyl peaks and development of new peaks at 1635 and 1530 cm^{-1} corresponding to amide I (backbone carbonyl stretch) and amide II (C-N stretch and N-H deformation) bands,

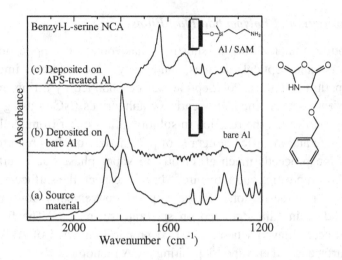

Fig. 18. IR absorption spectra of (a) NCA monomer, and the films deposited (b) on bare Al surface and (c) on APS-modified Al surface. Structure of monomer is shown on right.

respectively. Optical micrographs of the films deposited on the bare substrates showed that a crystalline film was produced, whereas the films deposited on the SAM showed homogeneous morphologies.

Polypeptide is known to have conformational variation according to the balance of inter- and intramolecular hydrogen bonding. This secondary structure is primarily determined by the molecular structure, but can also be influenced by the film growth process. The possibility of controlling the secondary structure by way of surface-initiated deposition polymerization was investigated by controlling the number density of amino groups on the substrate, which was achieved by preparing mixed SAMs of amino ethane thiol (AET) and ethyl mercaptan (EM) in various ratios on gold substrates. Figure 19 shows IR spectra of the amide I band for the benzylserine NCA deposited on these SAMs. The absorption of amide I is known to reflect the secondary structure of polypeptide.[39] The result in Fig. 19 indicates that no film was accumulated on the surface that does not have amino groups. With increasing density of amino group, the conformation changed from random coil (AET:EM = 1:2), α-helix (1:1 to 2:1), and then to β-sheet (EM only). These results show that the surface-initiated deposition polymerization is effective not only to grow surface-tethered polymer thin films but also in controlling the higher

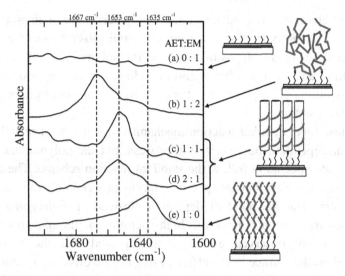

Fig. 19. IR absorption spectra of benzylserine NCA deposited on mixed SAM layers having different ratio of AET:ET.

structure of the deposited films. *In-situ* surface plasmon resonance observation of the polypeptide film growth process has shown that the film growth morphology can be largely influenced by the SAM modification.[40,41]

7. Conclusion

This chapter intended to bring new light to bear on the PVD process from the standpoint of polymer materials. The chapter has elucidated a wide possibility of polymer film deposition by the dry vacuum processes. Four methods of polymer PVD, including direct evaporation of polymer, coevaporation of reactive monomers, radical polymerization assisted by some activation method, and surface-initiated deposition polymerization have been introduced. The direct evaporation has rather limited applicability, but the coevaporation method can be applicable to wide range of materials that are formed by condensation polymerization or by polyaddition reaction. The radical polymerization forms vinyl or acryl polymers, and has technical advantage in polymer formation from single monomer species. The surface-initiated polymerization can initiate variety of chain reaction, such as radical polymerization or ring-opening, depending on the type of initiator bound to the substrate surface. There are other types of dry deposition method for polymer, including sputter deposition[42] and plasma polymerization.[43] However, the vapor-deposition polymerization is superior to these methods in controllability of molecular structure.

At this stage, detailed reaction mechanism has not been revealed for the vapor-deposition polymerization process, but the polymerization on the substrate appears to follow the standard reaction scheme. The results of analyses indicate that the polymerization proceeds on the substrate surface almost at the point of deposition, instead of undergoing solid-phase polymerization. Under the high-vacuum condition, vapor phase reaction is not observed either. It is expected that the molecules deposited on the surface have sufficient mobility to drive the reaction.

The major issue in PVD of polymer, from the standpoint of industrial fabrication, is the insufficient productivity. The vapor deposition apparatus is considered to be more costly compared to the wet-coating machines. This problem can be overcome by using a roll-to-roll deposition system and by using an efficient evaporation source such as slit nozzle evaporator, or by reducing the distance between the substrate and the evaporator. On the other hand, the PVD process can spare the drying process that is required after the wet coating.

Nevertheless, the PVD has unique advantages, such as high controllability, formation of ultra-thin layers and multilayers, access to interface phenomena, and even to higher order of polymer structure such as molecular conformation. For some of the insoluble materials, PVD could be the only option for film formation. These characteristics make the PVD attractive especially for device applications. Also, the PVD would have potential applicability to those fields that require solvent free process. Finally, from the standpoint of environmental issue, the PVD process can reduce the emission of volatile organic compounds.

Acknowledgments

This research was supported partly by Grant-in-Aid for Scientific Research from the Japan Society of the Promotion of Science and by

Innovative Seeds Development Project from the Japan Science and Technology Agency. The surface-initiated deposition polymerization technique was developed in collaboration with Prof. R. Advincula of University of Houston. The author also expresses his thanks to Prof. H. Sato and Prof. K. Ogino for synthesizing the monomers specially designed for this research.

References

1. C. W. Tang and S. A. VanSlyke, *Appl. Phys. Lett.* **51**, 913 (1987).
2. H. Usui, I. Yamada and T. Takagi, *J. Vac. Sci. Technol.* **A4**, 52 (1986).
3. H. Usui, K. Numata, H. Dohmoto, I. Yamada and T. Takagi, *Mat. Res. Soc. Symp. Proc.* **108**, 201 (1988).
4. H. Usui, H. Koshikawa and K. Tanaka, *J. Vac. Sci. Technol.* **A13**, 2318 (1995).
5. T. Takagi, in *Physics of Thin Films*, Vol. 13, Eds. M. H. Francombe and J. L. Vossen (Academic Press, New York, 1987), p. 1.
6. T. Takagi, *Thin Solid Films* **92**, 1 (1982).
7. H. Usui, H. Koshikawa and K. Tanaka, *IEICE Trans. Electron.* **E-81-C**, 1083 (1998).
8. H. Tamura, R. Kojima and H. Usui, *Appl. Opt.* **42**, 4008 (2003).
9. J. R. Salem, F. O. Sequeda, J. Duran, W. Y. Lee and R. M. Yang, *J. Vac. Sci. Technol.* **A4**, 369 (1986).
10. Y. Takahashi, M. Iijima, K. Inagawa and A. Itoh, *J. Vac. Sci. Technol.* **A5**, 2253 (1987).
11. Y. Takahashi, M. Iijima and E. Fukada, *Jpn. J. Appl. Phys.* **28**, L2245 (1989).
12. H. Usui, M. Watanabe, C. Arai, K. Hibi and K. Tanaka, *Jpn. J. Appl. Phys.* **44**, 2810 (2005).
13. H. Usui, H. Kikuchi, K. Tanaka, S. Miyata and T. Watanabe, *J. Vac. Sci. Technol.* **A16**, 108 (1998).
14. H. Usui, F. Kikuchi, K. Tanaka, T. Watanabe and S. Miyata, *IEICE Trans. Electron.* **E85-C**, 1270 (2002).
15. H. Usui, F. Kikuchi, K. Tanaka, T. Watanabe, S. Miyata, H. Bock and W. Knoll, *Nonlinear Optics* **22**, 135 (1999).
16. X. Wang, K. Ogino, K. Tanaka and H. Usui, *IEICE Trans. Electron.* **E87-C**, 2122 (2004).
17. X. Wang, K. Ogino, K. Tanaka and H. Usui, *Thin Solid Films* **438/439**, 75 (2003).
18. H. Usui, *Thin Solid Films* **365**, 22 (2000).
19. M. Tamada, H. Koshikawa, F. Hosoi, T. Suwa, H. Usui, A. Kosaka and H. Sato, *Polymer* **40**, 3061 (1999).
20. M. Tamada, H. Koshikawa, T. Suwa, T. Yoshioka, H. Usui and H. Sato, *Polymer* **41**, 5661 (2000).

21. M. Tamada, H. Omichi and N. Okui, *Thin Solid Films* **251**, 36 (1994).
22. A. Kawakami, E. Otsuki, M. Fujieda, H. Kita, H. Taka, H. Sato and H. Usui, *Jpn. J. Appl. Phys.* **47**, 1279 (2008).
23. H. Usui, A. Katayama, T. Honda and K. Tanaka, *Mat. Res. Soc. Symp. Proc.* **734**, 321 (2003).
24. K. Honda, M. Morita, H. Otsuka and A. Takahara, *Macromol.* **38**, 5699 (2005).
25. K. Katsuki, A. Kawakami, K. Ogino, K. Tanaka and H. Usui, *Jpn. J. Appl. Phys.* **44**, 4182 (2005).
26. K. Katsuki, H. Bekku, A. Kawakami, J. Locklin, D. Patton, K. Tanaka, R. Advincula and H. Usui, *Jpn. J. Appl. Phys.* **44**, 504 (2005).
27. H. Usui, T. Yoshioka, T. Katayama, K. Tanaka and H. Sato, *Mat. Res. Soc. Symp. Proc.* **710**, DD12.16.1 (2002).
28. E. Otsuki, H. Sato, A. Kawakami, H. Taka, H. Kita and H. Usui, *Thin Solid Films* **518**, 703 (2009).
29. M. Muroyama, I. Saito, S. Yokokura, K. Tanaka and H. Usui, *Jpn. J. Appl. Phys.* **48**, 04C163 (2009).
30. M. Muroyama, S. Yokokura, K. Tanaka and H. Usui, *Jpn. J. Appl. Phys.* **49**, 01AE03 (2010).
31. K. Katsuki, H. Bekku, A. Kawakami, J. Locklin, D. Patton, K. Tanaka, R. Advincula and H. Usui, *Jpn. J. Appl. Phys.* **44**, 504 (2005).
32. J. Hyun and A. Chilkoti, *Macromolecules* **34**, 5644 (2001).
33. A. Kawakami, K. Katsuki, R. C. Advincula, K. Tanaka, K. Ogino and H. Usui, *Jpn. J. Appl. Phys.* **47**, 3156 (2008).
34. S. F. J. Appleyard, S. R. Day, R. D. Pickford and M. R. Willis, *J. Mater. Chem.* **10**, 169 (2000).
35. R. H. Wieringa and A. J. Schouten, *Macromolecules* **29**, 3032 (1996).
36. Y. C. Chang and C. W. Frank, *Langmuir* **14**, 326 (1998).
37. T. M. Fulghum, H. Yamagami, K. Tanaka, H. Usui, K. Shigehara and R. C. Advincula, *Mat. Res. Soc. Symp. Proc.* **711**, 251 (2002).
38. T. M. Fulghum, H. Yamagami, K. Tanaka, H. Usui, K. Shigehara and R. C. Advincula, *Polymeric Materials Science and Engineering* **86**, 196 (2002).
39. T. Buffeteau, E. Le Calvez, S. Castano, B. Desbat, D. Blaudez and J. Dufourcq, *J. Phys. Chem.* **B104**, 4537 (2000).
40. K. Ogura, K. Tanaka, R. C. Advincula and H. Usui, *Polymer Preprints* **47**, 72 (2006).
41. H. Duran, K. Ogura, K. Nakao, S. D. B. Vianna, H. Usui, R. C. Advincula and W. Knoll, *Langmuir* **25**, 10711 (2009).
42. M. Rost, H. J. Erler, H. Giegengack, O. Fiedler and Chr. Weissmantel, *Thin Solid Films* **20**, S15 (1974).
43. J. M. Tibbitt, M. Shen and A. T. Bell, *Thin Solid Films* **29**, L43 (1975).

CHAPTER 16

NANOSCALE BIOELECTRONIC DEVICE CONSISTING OF BIOMOLECULES

Jeong-Woo Choi[1,2,*], Taek Lee[1] and Junhong Min[3]

[1]*Department of Chemical and Biomolecular Engineering and*
[2]*Interdisciplinary Program of Integrated Biotechnology,*
Sogang University, #1 Shinsu-Dong, Mapo-Gu, Seoul 121-742, Korea
[3]*College of Bionanotechnology, Kyungwon University, Bokjung-Dong,*
Sujung-Gu, Seongnam 461-701, Korea
*E-mail: jwchoi@ccs.sogang.ac.kr

Currently, a nanoscle bioelectronic devices consisting of biomolecules and the interaction mechanisms of biomolecules have been developed in various industrial fields such as clinical diagnosis, electronic device, pharmaceutical screening, bioprocess, photonic device, environmental pollution detection, and etc. Biomaterials such as protein, DNA/RNA, pigment, and cells have been introduced into inorganic electronic structure in order to construct the biochip devices such as bio-photodiode, bio-information storage device and bioelectroluminescence device, protein chip, DNA chip, and cell chip. Several nanobio technologies (nano-scaled immobilization technology of recombinant biomolecules, nanopatterning technology, detection technology and micro- or nano-electromechanical systems technology) have been applied in order to provide new functions to a nanobio device with advantages such as high density immobilization, orientation and control of immobilized biomolecules. In this review, we describe the nano scale fabrication of biomolecules based on electronic device using immobilizing technology (langmuir-blodgett and self-assembly) and also briefly prospect a bioelectronic device as one of the alternatives of current inorganic electronic devices.

1. Introduction

In 1965, Gordon Moore anticipated that the number of transistors and memory devices per unit area on integrated circuit had been doubled per year since functionality chip was proposed. With his observation, advances in memory devices, transistors and information processors have gave the micro-sized computing machines with useful processing capabilities with traditional 'top-down' lithography method. However, nowadays, semiconductor industry faced some serious problem that brings increasing technological and fundamental limitations as well as economical barriers. For example, when device features are pushed towards the deep sub-100 nm regime, then, a fabrication cost was drastically increased with lithography equipment and related facilities.[1] In this time, some genius scientists and engineers suggested the new approaching method which calls 'bottom-up' method. This method give arise molecular electronics with bio or organic molecules. This new method and new type approach have been considered as a potential alternative technology. And it may overcome the problems conventional technologies currently consider in commercial fields and Fig. 1 is shown to this trend.[2,3]

Over the past decade several concepts for molecular electronics, defined as molecular level electronic systems that only use a few electrons for operation, have been introduced. Several concepts have been suggested for developing the molecular electronic devices including spin based and ferromagnetism based electronic devices.[4-6] This novel electronic device design has the potential to solve the technological challenges which are currently facing the industry and provide new strategies for enhancing the device-integration density.[7]

For this purpose, one strategy to increase electronic device density such as transistor and memory would be to impose bioelectronics approach wherein the charge storage elements contain biomolecules such as DNA, proteins and the other biomolecules. These various concepts were proposed to alternate current electronic technique and it is called a bioelectronics. The bioelectronics comes from molecular electronics, and which elements are contained and consisted of biomolecules. Keren *et al.* was suggested DNA-templated field-effect transistor based on carbon

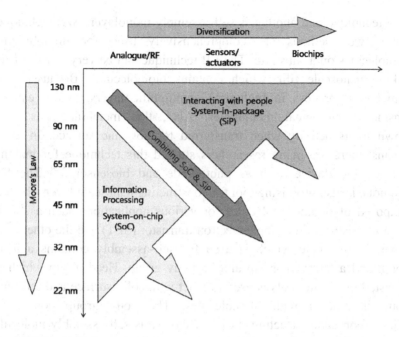

Fig. 1. Schematic diagram of Moore's Law and diversification concerning semiconductors taken from the international technology roadmap.

nanotube and Seeman *et al.* invented DNA mechanical device. Also, Tseng *et al.* developed a memory device which was composed of tobacco mosaic virus. Fugibayashi *et al.* makes DNA-based algorithm systems and computing device. Kershner *et al.* also developed placement and orientation of individual DNA shapes on lithographically patterned surfaces for making integrated chip. The other genius researchers developed valuable concepts and ideas regarding bioelectronics.[8-17] Thus, electronic switching for electronic device and for increasing electronic device density is known to exist in a wide variety of biomaterials.

To make the bioelectronic device which was composed of biomaterials, it needs an important consideration of thin film preparation methods. Recently, thin layer preparation technology is regarded as important role in the fabrication of bioelectronic device. Conventionally, when immobilizing the biomaterial on an inorganic surface, immobilizing methods are generally classified two types. One is langmuir-blodgett

(LB) technique and another is self-assembly monolayer (SA) technique. These two methods are comprehensively used for immobilizing biomolecules on a substrate. The LB technique can be very powerful for making nanoscale film which contains biomolecules. Because, it is available to process in biological amphiphilic layers, and it can be considered biological membranes. Also, this membrane has been shown to its activity when transferred to inorganic surface. In these reasons, there are many researchers studied this technique for making bioelectronic device such as photodiode and biosensor.[18-24] The SA technique is also promising tool for constructing nanoscale layer which is composed of organic molecules for various applications such as field-effect transistor (FET), single-electron transistor (SET) and the others.[25-29] Generally molecule which is used for self-assembly contains a head group, and a terminal group and its body group. Head group which is consisted of alkanethiols is used to react with solid surface and terminal group is reacted with biomolecules. The body group constitutes hydrocarbon chain structure. In many case, this self-assembly molecule was regarded as a chemical linker for immobilizing the biomaterial. In many case, it has been shown that sulfur compounds coordinate strongly many metal substrates, for example, Au, Ag, Ti, Cu and Pt. In nowadays, Au substrate has been widely used to make SAM. It is hard to oxidize in the ambient conditions.

In previous time, our group proposed some bioelectronic device composed of biomolecular hetero langmuir-blodgett (LB) film. The purpose of this electronic device was to accomplish electronic functions of the molecular photodiode, switching device for photocurrent generation and rectifier function.[30-33] Recently, our group introduced the redox protein for making biomemory device by self-assembly technique and the principles of this protein based biomemory device were demonstrated in a previous study by our research group.[34-39]

Here, we review the working principle, fabrication technologies, and electronic function proofs of biomolecules based on bioelectronic device and also briefly survey nanobio elelctronic device which could be one of alternative standard device format in human-mechanics interfacing system.

2. Nanoscale Biomolecule Layer Fabrications

2.1. *Langmuir-Blodgett (LB) Technique*

The Langmuir-Blodgett (LB) technique is named after Irving Langmuir and Katharine B. Blodgett. The Langmuir-blodgett film is composed of layers of organic molecules, deposited from the substrate surface of a liquid onto a solid by dipping the solid surface into the liquid. And then, the organic molecular layer is adsorbed with each dipping step. The thin films are assembled vertically and are usually contained amphiphilic molecules with a hydrophilic head group and a hydrophobic tail group such as fatty acids.[40-44]

2.1.1. *Optimal Deposition Condition of Chlorophyll a Langmuir-Blodgett Film*

To determine the optimal deposition conditions, the p–a isotherm of chlorophyll a was analyzed at various temperatures and pH conditions. The pressure/area isotherm (p–a isotherm) is called the dependence of the surface pressure on the unit area per molecule. This isotherm usually depicts the formation of the thin layer and phase transitions.[45] The p–a isotherm of chlorophyll a is described in Fig. 2. When it reached about

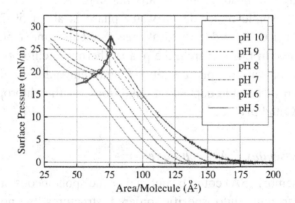

Fig. 2. p–a isotherms of chlorophyll a langmuir-blodgget layers at various pH conditions. Reprinted with permission from Ref. 48 Y. S. Nam *et al.*, Optimal deposition condition of chlorophyll a Langmuir–Blodgett film, *Mater. Sci. Eng. C.*, 24, 35 (2004), Copyright Elsevier (2009).

25-35 mN/m as the monolayer was compressed, the isotherm depicts that the surface pressure constantly increased. This first condensed monolayer region was originated by a transition region, which showed a rapid increase in compressibility, when it formed a second condensed monolayer region. At first time, compression didn't converse the surface pressure. In this case, the molecules are nearly removed from each other and do not interact. This area is referred as a two dimensional (2D) gaseous state (GS). As compression is continued, surface pressure increases. This area is called as the 2D liquid state (LS). In this state, each molecule begins to interact with each other. A 2D solid-like (SL) state will be reached when the compression of the thin layer is continued.

Molecules are closely condensed and the thin layer can be regarded as a 2D crystal. The p–a isotherm of Fig. 2 at various phase pH conditions are presented to this phenomena in Fig. 2. Here, GS appeared in the area above 110-170 $Å^2$/molecule, LS from 110 to 170 $Å^2$/molecule to around 50 $Å^2$/molecule and SL in the area below 50 $Å^2$/molecule. The area/molecule and the surface pressure of the transition regions of the chlorophyll a 2D monolayer of the GS, LS and SL states were all increased when increasing the pH condition from 5 to 10. It is provided that chlorophyll a molecules offer a promising basis for the applications of LB layers due to characteristics with respect to pH. Since the diameter of chlorophyll a head group is around 8-10 $Å^2$.[46,47] The region of chlorophyll a single molecule on water phase is about 50-75 $Å^2$. The chlorophyll a LB layer deposited at area/molecules under 50 $Å^2$ forms the aggregates. In this research, the p–a isotherms of chlorophyll a were investigated at various pH conditions and temperatures. Thus, it is demonstrated that stable deposition conditions for the chlorophyll a LB layer is 25°C and pH 7.0.[48]

2.2. *Self-Assembly (SA) Technique*

The self-assembly (SA) could be defined as the spontaneous organization of molecular units into specific ordered structures by non-covalent interactions. The interactions were responsible for the formation of the self-assembled system act on a strictly local level in the nanostructure builds itself.[49-54]

2.2.1. ph Effect of Recombinant Protein Monolayer on Au Surface by Self-Assembly Technique

The redox property of cysteine-modified protein layer was investigated by cyclic voltammetry method. The redox reaction was reversible, suggesting that the self-assembled protein maintain the reduction-oxidation properties.[55] We take an optimal pH conditions by pH-control experiments from pH 5.0 to the pH 9.0. The result was described in Fig. 3(A). As seen in Fig. 3(B), the reduction potential was 146 mV and

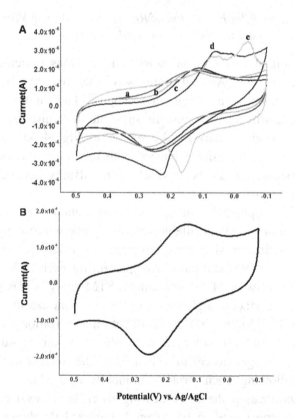

Fig. 3. Cyclic voltammogram of cysteine-modified azurin. (A) a: pH 5.0, b: pH 6.0, c: pH 7.0, d: pH 8.0, e: pH 9.0 (10 mM HEPES, pH 7.0, azurin sample concentration = 0.1 mg/ml, electrode area = 0.25 cm^2). (B) Cyclic voltammetry at pH 7.0. Reprinted with permission from Ref. 56 S.-U. Kim *et al.*, Direct immobilization of cupredoxin azurin modified by site-directed mutagenesis on gold surface, Ultramicroscopy, 108, 1390 (2008), Copyright Elsevier (2009).

the oxidation potential was 265 mV. That is matched with our previous results and reference.[56] The obtained results were remarkable. These stable electrical properties of azurin layer by the direct immobilization of cysteine-modified azurin imply that the direct immobilization technology can offer very stable and well oriented protein layer. In case of the use of the linker in the process of protein immobilization, not-completed immobilized protein existed in the protein layer and this protein seems to be detached under a harsh condition.[57]

2.2.2. *Nanoscale Film Formation of Recombinant Azurin Variants with Various Cysteine Residues on Gold Substrate*

The orientation of recombinant azurin with various cysteine residues immobilized on the Au substrate was analyzed by fluorescence microscope, scanning tunneling microscope.[58] The binding ability of recombinant azurin variants on Au substrate was also examined by fluorescence imaging analysis. As shown in Fig. 4(A), the relative fluorescence signal intensity enhanced with increasing number of thiol group on cysteine residue because of strong affinity between thiol of cysteine and Au surface.

The surface topography of the prepared metalloprotein layer was obtained by commercially available scanning probe microscopy. In order to measure the increased relative coverage, we calculated the average surface roughness (Ra) and number of protein for each STM 2D image with provided software of the instrument. STM analysis was performed to examine the surface morphologies of recombinant azurin layers on a gold surface. The data might provide direct immobilization on how each recombinant azurin layer with cysteine residue forms on the Au substrate. The surface coverage was calculated based on the changes of the surface roughness following each protein immobilization. For recombinant protein immobilization, the areas showing lower Ra values than the Ra of bare gold are considered to be covered. Figure 4(B) shows the STM images of the recombinant layers and the changed surfaces by covalent bond with corresponding Ra (nm) and coverage (number of protein) values. We calculated the coverage (number of protein) for each STM image with provided software of the instrument.

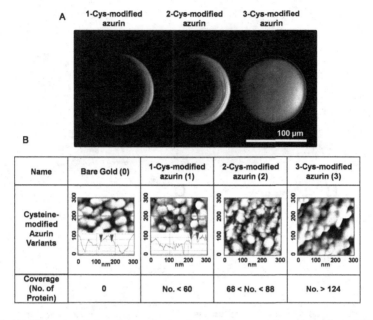

Fig. 4. (A) Fluorescence imaging analyses of recombinant azurin variants immobilization on gold surface. (B) STM topography images of the recombinant azurin variants on gold surface.

2.2.3. *Temperature Dependent Redox Reaction of Recombinant Protein Thin Film*

Thermally induced unfolding and denaturation of ferredoxin was monitored by differential scanning calorimetry because biomolecules has affected by heat or thermal turbulence.[36] The conductivity (σ) was determined, at a given temperature from the plot of current (I) versus voltage (V) or from log I versus log V when the current and voltage range was large in Fig. 5(A).[59,60] As observed in general electrochemical phenomena, the increase in conductance of the protein with increase in temperature is because it acquires extra thermal energy but after 320 K the conductance decreases and this due to the thermal unfolding of the Fe-S cluster in the protein.

The complete denaturation of this protein was estimated by differential scanning calorimetry at 342 K (Fig. 5(B)). The thermally induced unfolding reaction of the recombinant protein is irreversible.[61]

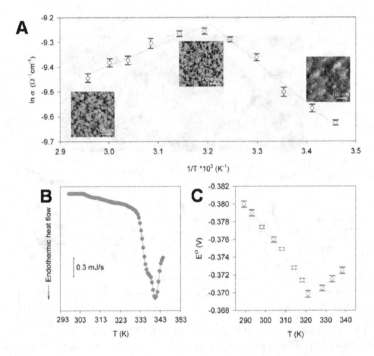

Fig. 5. Temperature dependent working performance of protein-based device: (A) Arrhenius plot for the increase in temperature with two distinctly different regions. (B) DSC thermogram of recombinant protein. The sample pans contained 0.1 mg/ml solution, DSC heating rate was 5°C/min. (C) E° versus T plot of recombinant protein on Au electrode in 10 mM Tris-HCl buffer solution (pH 7.4) respectively. Dotted lines are least square fit to the data points. Reprinted with permission from Ref. 36. A. K. Yagati et al., Write-Once–Read-Many-Times (WORM) biomemory device consisting of cysteine modified ferredoxin, Electrochem. Commun., 11, 854 (2009), Copyright Elsevier (2009).

Since the irreversibility is most likely due to degradation of the cluster in the unfolding state.[62]

The temperature dependence of standard potential (E°) was recognized were shown in Fig. 5(C). Initially, the E° potentials were observed to decrease with increasing temperature because protein itself has some resistance but as the temperature increases its resistance decreases and consequently the E° decreases up to 320 K, this change in E° potential could be the structural fluctuations that occur within the protein.[63] Nevertheless, at 320 K, slope change in entropy ($\Delta S° = dE°/dT$)

is due to the non-spontaneous disruption of the Iron-sulfur cluster is followed by the spontaneous complete unfolding of the protein. Finally, this behavior can be understood because of the loss of Iron-sulfur cluster and denaturation of protein.

2.2.4. Surfactant CHAPS Effect of Bioelectronic Device

We showed previously that this protein layer was formed as cohesive clusters on Au surface through physical adsorption.[64-67] To reduce the formation of these cohesion clusters, a zwitterionic surfactant, (3-[(3-cholamidopropyl) dimethylammonio]-1-propanesulfonate) (CHAPS) was introduced to modify the surface properties.[68-70] The surface morphology of the fabricated electrode surface was investigated by AFM to determine the effect of CHAPS treatment.

Figure 6(a) depicts the recombinant azurin layer immobilized on a Au surface without any treatment. As shown in this image, in the absence of CHAPS the surface contains several azurin aggregates. Treatment with 10.0 mM CHAPS resulted in recombinant azurin clusters that were highly segregated and existed as near single molecules (Fig. 6(d)). However, any further increase in CHAPS concentration over 10.0 mM

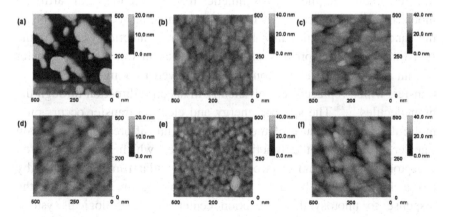

Fig. 6. Surface morphology of the recombinant azurin layer after CHAPS treatment by atomic force microscopy. (b): 2.5 mM CHAPS treatment, (c): 5.0 mM CHAPS treatment, (d): 10.0 mM CHAPS treatment, (e): 15.0 mM CHAPS treatment, (f): 20.0 mM CHAPS treatment.

did not result in an increased reduction in protein aggregation, as shown in Figs. 6(e) and 6(f).

3. Nanoscale Bioelectronic Devices based on Biomolecules

3.1. *Nanoscale Biophoto Diode*

In a biological system, an electron transfer from one side of a molecule to the other parts or between molecules is one of the most fundamental and general processes.[71] The control and study of this electron transfer phenomena in organized molecular systems are major destination for bioelectronics.[72] Progress in bioelectronic devices engineering is still rather moderate, owing to some serious problems associated with the effective control of such as its structures and interactions at the nanoscale level.

The electron transport from photoinduced processes in nature is known to occur very efficiently as well as undirectionally such as photoelectric conversion and long-range electron transfer in photosynthetic organisms guided by molecular functional groups.[71,73] The notions for the advance of bioelectronic devices can be originated from biological systems, for example, a TCA cycle or the electron transfer chain or the photosynthetic reaction center. An artificial bioelectronic device can be achieved by mimicking the organization of the functional molecules in a biological photosynthetic system. At the first time, an initial process of photosynthesis in the electron transfer system occur photoelectric conversion followed by long-range electron transfer, which replaces very efficiently in specific direction through the biomolecules.[74,75] The specific energy and electron transfer occurs on a molecular scale, because of the redox potential difference and electron transfer property of the functional molecules which are the electron acceptor, sensitizer and electron donor. Molecular films, fabricated by LB techniques, can be used as in vitro biological systems for the response to photosynthetic reaction center in the biological system. Recently, a considerable interest has focused on nanoscale layer fabrication or the formation of biomaterials thin layers on solid surfaces.[76] Based on these techniques, various bioelectronic devices have been developed to emulate the electron transport function of biological photosynthesis. Isoda et al. proposed a biophotodiode consisted of flavin-porphyrin hetero LB films and investigated its electrical properties.[77]

They use respectively flavin and porphyrin as a sensitizer (S) and an electron acceptor (A). Fujihira et al. also developed an electrochemical photodiode which was composed of the LB films of three functional biomolecules.[78] Monitoring the electron transfer between the electrode and the excited dye molecules were also performed, in which ferrocene; pyrene and viologen were used respectively as the electron donor (D), S and A units. The metal/insulator/metal (MIM) structured device was fabricated composed of hetero-type LB films of D, S and A. Also, the photoinduced electron transfer observed. Currently, a biomolecular photodiode consisted of electron D/S/Relay (R)/A type 4 component MIM devices has been developed by Choi et al.[79]

Figure 7 depicts scheme of MIM device, and photoswitching characteristics of biomolecular photodiode. A biophotodiode which is composed of LB films of ferrocene, flavin, viologen and TCNQ as the D, S, R and A units, respectively, was proposed based on the photoinduced electron transport in a biological system.[80] By introducing two acceptor molecules (R and A), the time for the separated charge state (A-/R/S/D+) can be retained longer than those of the A-/S+ hetero system. Charge recombination from R and A, to the ground state S, can be decreased

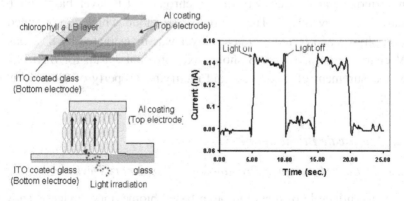

Fig. 7. Schematic diagram of MIM biomolecular photodiode and photoswitching property of biomolecular device.

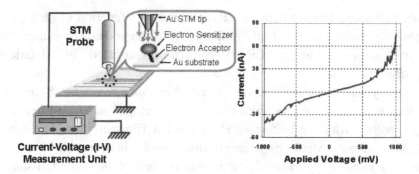

Fig. 8. Scheme of Experimental set-up for I-V measurement of bioelectronic device and rectifying property of biomolecular diode.

because of the rapid electron transport from S* to A, via R, and the extended distance between S and A in the existence of R. The biophotodiode consisted of D/S/R/A hetero LB films is predicted to display better diode and switching properties than those of S/A and D/S/A hetero LB films based on these effects. The backward electron transport of excited S can be reduced by adding the D molecules and the charge recombination from A to ground state S can be reduced by the addition of the R molecules. Choi *et al.* developed some biophotodiodes consisting of green fluorescent protein (GFP) and cytochrome c.[81,82] Recently, researchers have studied the nanoscale molecular diode using scanning probe microscopy (SPM).[83] Khomutov *et al.* measured the single molecular conductivity of cytochrome c LB layer based on I-V measurement by STS.[84] The biomolecular diode which is contained a chlorophyll a and ferredoxin heterolayer was investigated by STS based I-V characteristics. Figure 8 shows experimental set-up for STS based I-V measurement of biodevice, and rectifying property of biomolecular diode.

3.2. Nanoscale Biomemory Devices

3.2.1. A Basic Concept of Protein-Based Biomemory Device

The basic principle of metalloprotein based biomemory device is shown in Fig. 9. The redox property of metalloprotein is controlled upon

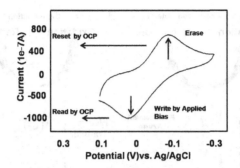

Fig. 9. Schematic representation of cyclic voltammogram of azurin assembled layer on gold working electrode, which depicts the memory function upon application of proper bias potentials. Reprinted with permission from Ref. 35 S.-U. Kim *et al.*, Charge storage investigation in self-assembled monolayer of redox-active recombinant azurin, Curr. Appl. Phys. 9, e71 (2009), Copyright Elsevier (2009).

application of an external potential to possess two electrically distinct positions.[34] When an oxidation potential is applied to metalloprotein, the metalloprotein loses an electron to the inorganic base, and thereby entrapment of positive charge occurs in the protein biofilm, as shown in Fig. 10(A). This trap process of positive charge inside the metalloprotein layer corresponds to the function of storage (writing) of information in a conventional silicon-based memory device. The trapped charge (written information) in the protein biofilm is measured (read) when an open-circuit potential (OCP) is applied to the fabricated electrode.

When a reduction potential is applied to the fabricated electrode after the initial step of oxidation the inorganic base gives back the electron to the metalloprotein monolayer, as shown in Fig. 10(B). Thus, the initial charge trapped in the protein biofilm during the time of oxidation, is neutralized (erasing). The sequential application of oxidation potential, OCP and reduction potential makes the developed protein-based biomemory device write, read and erase the information as a conventional inorganic memory device.

However, efficient combination techniques of recombinant protein in conventional electronic devices have not yet been reported; nevertheless, it is a key process to sustain and control the electrochemical properties of the recombinant protein to mimic the biological memory system.

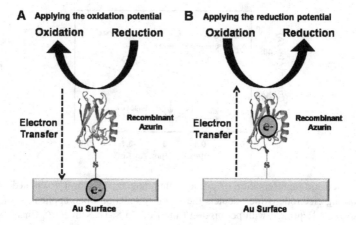

Fig. 10. Schematic diagram showing the electron transfer mechanism of recombinant azurin on Au electrode. (A) Application of oxidation voltage causes the electron to move from protein to Au electrode. (B) Application of reduction voltage sends the electron back to the protein. Reprinted with permission from Ref. 35 S.-U. Kim *et al.*, Charge storage investigation in self-assembled monolayer of redox-active recombinant azurin, Curr. Appl. Phys. 9, e71 (2009), Copyright Elsevier (2009).

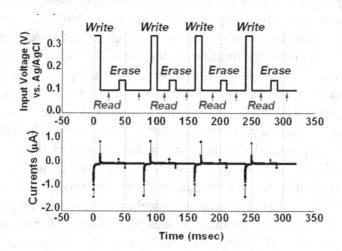

Fig. 11. Write-read-erase cycles of a protein-based biomemory device. The top curve shows the applied sequence of pulses for "write" "read" and "erase" functions and the bottom curve is corresponding current response. Reprinted with permission from Ref. 34 J.-W. Choi *et al.*, Protein-based biomemory device consisting of the cysteine-modified azurin, Appl. Phys. Lett. 91, 263902 (2007), Copyright American Institute of Physics (2009).

Information about the reduction and oxidation of protein molecules can be obtained from the measurement of OCPA (open circuit potential amperometry) with time. The potential decay in solution in open circuit conditions with time has been recorded. This potential was used to read the stored charge. OCPA is qualitatively similar to conventional chronoamperometry. However, the method differs by the measurement of the charge associated with the oxidized protein. The OCPA method is an electrochemical technique that is used in pulse mode to write-read-erase cycles, as shown in Fig. 11. In each memory cycle, an oxidation voltage of -500 mV was applied to store the charge in the protein layer; reduction voltage of -260 mV was applied to erase all the stored charge in the device as described in our previous report.[35,38]

3.2.2. Write-Once-Read-Many-Times (WORM) Biomemory Device

To read the stored charge multiple times we applied an OCP of -200 mV at very small disconnecting time and the observed current reveals that the

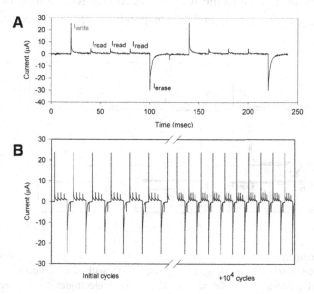

Fig. 12. (A) Multiple times reading test. (B) Current response validating the memory switching cycles. Reprinted with permission from Ref. 36. A. K. Yagati *et al.*, Write-Once–Read-Many-Times (WORM) biomemory device consisting of cysteine modified ferredoxin, Electrochem. Commun., 11, 854 (2009), Copyright Elsevier (2009).

stored charge can be read multiple times. Oxidation voltage of -260 mV (20 ms) can reproducibly charges on the device. A set of three voltage pulses of -200 mV (OCP) of 20 ms duration with small disconnecting times showed the current switchings thereby maintaining the oxidized state of the memory device. Finally, a voltage pulse of -500 mV erases all the stored charge. The detailed change of current I as a function of time for two memory cycles were shown in Fig. 12. Further studies will be required to determine lower limit for the switching time.[36]

3.2.3. *Multi-Bit Biomemory Device*

Protein based multi-bit biomemory device consisting of recombinant azurin with a cysteine residue modified by site directed mutagenesis has been developed.[37] The recombinant azurin was directly immobilized on four different gold (Au) electrodes patterned on a single silicon substrate. Figure 13 represents an experimental system of a protein based device to store a 4-bit data. Input bias potentials of oxidation and open circuit potential were applied to the four Au electrodes for duration of 25 msec consisting azurin SAM and corresponding charging currents were

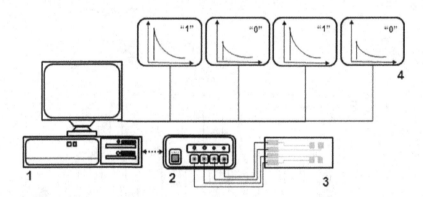

Fig. 13. Schematics of the experimental system for measuring charging currents from azurin adsorbed on gold working electrodes; 1, PC with data acquisition system; 2, multichannel potentiostat; 3, fabricated gold working electrodes (active area 1 mm × 1 mm) on a Si surface (24 mm × 13 mm); 4, signals obtained from the measuring device. Reprinted with permission from Ref. 37. A. K. Yagati *et al.*, Multi-bit biomemory consisting of recombinant protein variants, azurin, Biosens. Bioelectron. 24, 1503, (2009), Copyright Elsevier (2009).

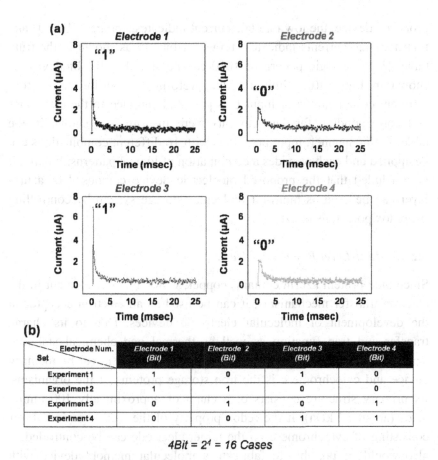

Fig. 14. Faradaic currents observed form a four bit memory device (a) application of oxidation and open circuit potentials in tandem to the four electrodes (high current represents storage of "1" and low current represents "0") represents a 1010 logic system (b) table represents the results obtained from 4 experiments for storing bits by these combinations similarly this can be extended for all 16 combinations. Reprinted with permission from Ref. 37. A. K. Yagati *et al.*, Multi-bit biomemory consisting of recombinant protein variants, azurin, Biosens. Bioelectron. 24, 1503, (2009), Copyright Elsevier (2009).

measured as shown in Fig. 14(a). In experiment 1, when the electrodes no. 1 and 3 were applied with oxidizing potential, which produces a high faradaic current nearly 7 µA whereas the electrodes no. 2 and 4 were connected with OCP produces practically a low current of 2 µA. In the proposed device, the low faradaic current indicates storage of bit "0" and high faradaic currents indicates storage of bit "1" as shown in the truth table. Hence a logic pattern of 1010 can be stored in the biodevice as shown in Fig. 14(b). Similarly, we preformed experiments for four different combinations of input bias potentials applied to the electrodes and corresponding faradaic currents were measured as shown in the table. In this manner, input potentials with 16 different combinations can be applied and each provides a combination of logical patterns. Finally, it is concluded that the proposed bioelectric device composed of azurin layer can be used as multi-bit molecular storage system by controlling the redox potentials of azurin.

3.2.4. *Multi-Level Biomemory Device*

Since blue protein azurin contains copper, which is a key element in the electron transfer mechanisms, it can be used as an electron acceptor in the development of molecular electronic devices. Due to its charge transfer and trap function as well as thermal and chemical stability, azurin was directly applied in the development of a novel biomemory device and cytochrome c is the iron storage proteins of the organisms. Its primary structure consists of a chain of approximately 100 amino acids (about 12 kDa). If the redox property of the well-organized layer consisting of cytochrome c on the target electrode can be controlled, it also would be possible to fabricate a molecular memory device with these biomolecules.[39]

Recombinant azurin containing cysteine residues was immobilized directly on Au surface and cytochrome c was adsorbed onto the immobilized azurin layer by electrostatic bonding. This heterolayer composed of recombinant azurin and cytochrome c was capable of storing the two pairs of information, which is referred to as the multi-state memory. Figure 15(A) depicts schematic diagram for the electron transfer mechanism of a cytochrome c/recombinant azurin layer on Au surface, and This multi-state memory displayed two pairs non-volatile write, read and erase function and it is described in Fig. 15(B).

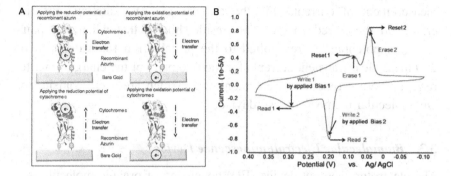

Fig. 15. (A) The electron transfer mechanism of a cytochrome c/recombinant azurin layer on Au surface. (B) Schematic representation for cyclic voltammogram of hetero layers consisting of recombinant azurin/cytochrome c on gold surface which depicts the memory function upon each applications of proper bias potentials.

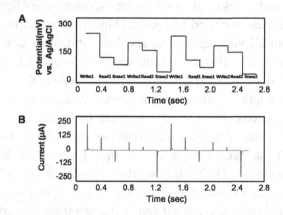

Fig. 16. The memory function characteristics of multi-state protein based biomemory device. (A) Two pairs of input pulse potentials of oxidation (charge write 1, 2), open circuit (charge read 1, 2) and reduction (charge erase 1, 2) potentials were applied to gold electrodes simultaneously, having a pulse width of 20 msec. (B) the corresponding charging currents were measured for a total duration of 260 msec.

Figures 16(A) and 16(B) show the faradaic currents obtained upon applying proper bias potentials to the protein-based molecular memory device. Here, applying the oxidation voltage is the writing step, whereas applying the open circuit voltage is the reading step with respect to the measurement of currents. Finally, the use of the reduction potential erased all the stored charge. This result showed that when two pairs of redox potentials were applied to the protein film for a duration of 280 msec clear transient currents for the two pairs of the charge to write, read, and erase functions were monitored, which are a prerequisite for any molecular memory storage device.

3.3. *Biomolecular Electroluminescence Device*

The electroluminescent device (ED) consisting of organic molecule is a light emitting diode (LED) which is based on carbon-based molecules instead of current inorganic semiconductors. It is totally different from the conventional inorganic semiconductors. According to recently research, the organic EL device is better than the current normal liquid crystal LCD such as its brightness, thickness, fastness and light intensity.

This EL device also need less power, higher contrast, looks better just as bright from all viewing angles, and low fabrication cost to produce than inorganic LCD. Also, in accordance with research, EL device which was composed of biomaterial has more advantage than organic EL device, for example, its efficiency and driving voltage aspect.

In these reasons, bio EL device could be powerful alternative to the next generation display which replaces the organic and inorganic based on EL device. Tajima *et al.* first proposed a biomolecular electroluminescence (EL) device that contained cytochrome c.[85] However, this EL device showed relatively low light intensity, also, it has just emitted light near the red end of the spectrum. Moreover, the bio EL device consisted of cytochrome c could not be applied commercially because of its poor stability in ambient condition.

The bio EL device composed of a biological pigment based heterolayer aligned as the emitting layer, and reported its EL performance. The bio EL performance heterolayer that emits a blue light sustains to be investigated. In our research, the external quantum efficiency (η ext) of the bio EL devices is higher than current organic based EL device. Figure 17 describes a schematic diagram of bio EL device. A bio EL device composed of chlorophyll a was developed by the application of photo-excited properties. The bilayered bio-EL

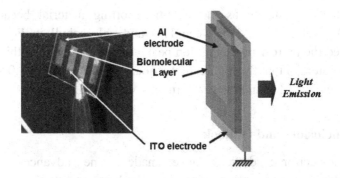

Fig. 17. Schematic configuration of bio electro luminescence (EL) device.

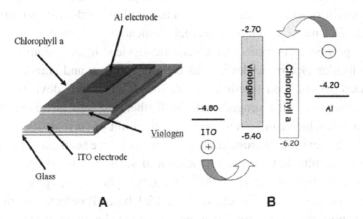

Fig. 18. Schematic diagram (A) and energy diagram (B) of the chlorophyll a/viologen bilayered bio EL device.

device consisted of ITO/viologen/chlorophyll a/Al was developed by introducing an LB technique, and the schematic configuration and energy diagram are described in Fig. 18(A). In this study, it was investigated the EL characteristics which contains chlorophyll a at various thicknesses of LB layers (25 layers and 15 layers). Also, it was investigated an external quantum efficiency of bio EL and the luminescence of the fabricatied bio EL device. The simple fabrication method of a multilayer and the control of molecule orientation are powerful advantages of the LB technique. The fabricated bio EL device, contains viologen as the hole-transporting layer that is valuable as the hole-transporting material because the HOMO level is located between ITO and chlorophyll in Fig. 18(B), displayed the narrow blue emission peak around 455 nm, and this device has the potential that the high external quantum efficiency of $1.0 \times 10^{-3}\%$ in the range of 17 V and the low turn on voltage of 5 V.

4. Conclusions and Outlook

The bioelectronic devices have made some advances towards establishing new strategies for the development of the molecular electronic device. From our research, it is summarized that the proposed nanobioelectronic device can be simply applied in the future as a new candidate for electronic devices, such as biophotodiode, biomemory and bio EL. Our next challenge is the real application and commercialization of the proposed biomolecular electronic device. In this study, a basic biomolecular electronic device has been developed, and should now be extended to the construction of advanced electronic devices. These achievements lay the groundwork for further research on biomolecular electronics, which can help to overcome the limit of current silicon-based electronic devices. Throughout the research, we can acquire more scientific results that are the verification of scientific theory, application of various technological fields (biology, photonics, physics, and chemistry), and establishment of industrial base. Bioelectronic devices are not possible in common use and cannot make an industrial benefit immediately. However, considering present market demand and future prospect of electronics, bioelectronic device's share may grow rapidly in 21th century. If the biodevice could be early developed, it can secure monopolistic technological position in worldwide market. Ultimately, bioelectronic devices can establish new industry field coming from Biotechnology (BT) + Information technology (IT) + Nano technology (NT). Also, it is expected that the bioelectronic devices can contribute to stand the technological advanced country through the creation of high value industry which provide synergy and motive for society. The biodevice typically requires very short time from research to commercialization. Thus, research introduced above should be achieved at early time.

Acknowledgments

This research was supported by the Nano/Bio Science and Technology Program (M10536090001-05N3609-00110) of the Ministry of Education, Science and Technology (MEST), by the Korea Science and Engineering Foundation (KOSEF) grant funded by the Korea government (MEST) (2006-05374), and by the Korea Research Foundation Grant funded by the Korean Goverment (MOEHRD) (KRF-2006-005-J02301), and by Kyungwon University Research Fund in 2009.

References

1. C. M. Lieber and W. Lu, Nat. Mater. 6, 841 (2007).
2. J. R. Heath and M. A. Ratner, Phys. Today. May, 43 (2003).
3. M. C. Petty, in Molecular Electronics from Principles to Practice (Wiley, Chichester, 2007), p. 42.
4. Z. Liu, A. A. Yasseri, J. S. Lindsay and D. F. Bocian, Science 302, 1543 (2003).
5. C. J. Amsinck, N. H. D. Spigna, D. P. Nackashi and P. D. Franzon, Nanotechnology 16, 2251 (2005).
6. R. Held, J. Xu, A. Schmehl, C. W. Schneuder, J. Mannhart and M. R. Beasley, Appl. Phys. Lett. 89, 163509-1 (2006).
7. I. Willner and E. Katz, in Bioelectronics: From theory to applications (Wiley-VCH, Weinheim, 2005), p. 475.
8. K. Keren, R. S. Berman, E. Buchstab, U. Sivan and E. Braun, Science 302, 1380 (2003).
9. N. C. Seeman, Nature 421, 33 (2003).
10. R. J. Tseng, C. Tsai, L. Ma, J. Ouyang, C. S. Okan and Y. Yang, Nat. Nano. 1, 72 (2006).
11. K. Fujibayashi, R. Hariadi, S. H. Park, E. Winfree and S. Murata, Nano. Lett. 8, 1791 (2007).
12. R. J. Kershner, L. D. Bozano, C. M. Micheel, A. M. Hung, A. R. Fornof, J. N. Cha, C. T. Rettner, M. Bersani, J. Frommer, P. W. K. Rothemund and G. M. Wallraf, Nat Nano. 4, 543 (2009).
13. K. Tanaka, G. H. Clever, Y. Takezawa, Y. Yamada, C. Kaul, M. Shionoya and T. Carell, Nat. Nano. 1, 190 (2006).
14. S. M. Iqbal, D. Akin and R. Bashir, Nat. Nano. 2, 243 (2007).
15. P. Yin, H. M. T. Choi, C. R. Calvert and N. A. Pierce, Nature 451, 318 (2008).
16. J. Sharma, R. Chhabra, A. Cheng, J. Brownell, Y. Liu and H. Yan, Science 323, 112 (2009).
17. Y. Ke, S. Lindsay, Y. Chang, Y. Liu and H. Yan, Science 319, 180 (2008).
18. K. Inyama, Photochem. Photobiol. 29, 633 (1979).

19. K. Iriyama, M. Yoshiura and F. Mizutani, Thin Solid Films 68, 47 (1980).
20. N. V. Tkachenko, P. H. Hynninen and H. Lemmetyinen, Chem. Phys. Lett. 261, 234 (1996).
21. H. G. Choi, B. K. Oh, W. H. Lee and J.-W. Choi, Biotechnol. Bioprocess Eng. 6, 183 (2001).
22. Y. K. Kim, M. S. Kwak, W. H. Lee and J.-W. Choi, Biotechnol. Bioprocess Eng. 5, 469 (2000).
23. J.-W. Choi, J. Min, W. H. Lee and S. B. Lee, Biotechnol. Bioprocess Eng. 5, 65 (2000).
24. V. Erokhin, in Protein Architecture. Interfacing Molecular Assemblies and Immobilization Biotechnology, Ed. Y. Lvov and H. Mo"hwald (Marcel Dekker, New York, 1999), p. 99.
25. J. H. Schon, H. Meng and Z. Bao, Science 413, 713 (2001).
26. S. Corni, IEEE Trans. Nanotech. 6, 561 (2007).
27. K. Tomizaki, L. Yu, L. Wei, D. F. Bocian and J. S. Lindsey, J. Org. Chem. 68, 8199 (2003).
28. Z. Liu, A. A. Yasseri, J. S. Lindsey and D. F. Bocian, Science 302, 1543 (2003).
29. A. Alessandrini, M. Salerno, S. Frabboni and P. Facci, Appl Phys Lett. 133902, 86 (2005).
30. J.-W. Choi, Y. S. Nam, S. J. Park, W. H. Lee, D. Kim and M. Fujihara, Biosens. Bioelectron. 16, 819 (2001).
31. J.-W. Choi and M. Fujihira, Appl. Phys. Lett. 84, 2187 (2004).
32. B. Lee, S. Takeda, K. Nakajima, J. Noh, J.-W. Choi, M. Hara and T. Nagamune, Biosens. Bioelectron. 19, 1169 (2004).
33. Y. S. Nam, J.-W. Choi and W. H. Lee, Appl. Phys. Lett. 85, 6275 (2004).
34. J.-W. Choi, B. K. Oh, J. Min and Y. J. Kim, Appl. Phys. Lett. 91, 263902-1 (2007).
35. S.-U. Kim, A. K. Yagati, J. Min and J.-W. Choi, Curr. Appl. Phys. 9, e71 (2009).
36. A. K. Yagati, S.-U. Kim, J. Min and J.-W. Choi, Electrochem. Commun. 11, 854 (2009).
37. A. K. Yagati, S.-U. Kim, J. Min and J.-W. Choi, Biosens. Bioelectron. 24, 1530 (2009).
38. S.-U. Kim, A. K. Yagati, J. Min and J.-W. Choi, Biomaterials 31, 1293 (2010).
39. T. Lee, S.-U. Kim, J. Min and J.-W. Choi, Adv. Matt. 22. 510 (2010).
40. A. Ulman, in An Introduction to Ultrathin Organic Films From Langmuir-Blodgett to Self-Assembly (Academic Press, Inc. , San Diego, 1991).
41. G. G. Roberts, K. P. Pande and Barlow, Phys. Technol. 12 (1981).
42. I. R. Peterson, J. Phys. D. 23, 379 (1990).
43. R. W. Corkery, Langmuir 13, 3591 (1997).
44. Y. Guo, F. Feng and T. Miyashita, Macromolecules 32, 1115 (1999).
45. P. Facci, V. Erokhin, F. Antolini and C. Nicolini, Thin Solid Films 237, 19 (1994).
46. S. P. Spooner and D. G. Whitten, in Photochemistry in Organized and Constrained Media, Ed. V. Ramamurthy (VCH, New York, 1991), p. 692.

47. H. Bourque, T. Taleb and R. M. Leblanc, Chem. Phys. Lett. 302, 187 (1999).
48. Y. S. Nam, J. M. Kim, J.-W. Choi and W. H. Lee, Mater. Sci. Eng. C. 24, 35 (2004).
49. N. Denkov, O. Velev, P. Kralchevski, I. Ivanov, H. Yoshimura and K. Nagayama, Langmuir 8, 3183 (1992).
50. G. M. Whitesides, J. K. Kriebel, J. C. Love, Science Progress 88, 17 (2005).
51. Ozin and Arsenault, in Nanochemistry: a chemical approach to nanomaterials (Cambridge: Royal Society of Chemistry, 2005).
52. G. M. Whitesides and M. Boncheva, PNAS 99, 4769 (2002).
53. J.-M. Lehn, Science 295, 2400 (2002).
54. P. M. Forster and A. K. Cheetham, Angew. Chem. Int. Ed. 41, 457 (2002).
55. S.-U. Kim, Y. J. Kim, C. H. Yea, J. Min and J.-W. Choi, Korean J. Chem. Eng. 25, 1115 (2008).
56. S.-U. Kim, Y. J. Kim, S. G. Choi, C. H. Yea, R. P. Singh, J. Min, B. K. Oh and J.-W. Choi, Ultramicroscopy 108, 1390 (2008).
57. J. Min, S.-U. Kim, Y. J. Kim, C. H. Yea and J.-W. Choi, J. Nanosci. Nanotechnol. 8, 4982 (2008).
58. S.-U. Kim, J. H. Lee, T. Lee, J. Min and J.-W. Choi, Nanoscale film formation of recombinant azurin variants with various cysteine residues on gold substrate for bioelectronic device, J. Nanosci. Nanotechnol. In Press.
59. D. O. Cowan, G. Pasternak and F. Kaufman, Proc. Natl. Acad. Sci. U.S.A. 66, 837 (1970).
60. N. J. Hardy, M. D. Hanwell and T. H. Richardson, J. Mater. Sci.: Mater. Electron. 18, 943 (2007).
61. J. Meyer, M. D. Clay, M. K. Johnson, A. Stubna, E. Munck, C. Higgins and P. Wittung-Stafshede, Biochemistry 41, 3096 (2002).
62. B. La-Rosa, D. Milardo, D. Grasso, R. Guzzi and L. Sprotelli, J. Phys. Chem. 99, 14864 (1995).
63. G. Battistuzzi, M. Borsari, A. Ranieri and M. Sola, J. Am. Chem. Soc. 124, 26 (2002).
64. T. Lee, S.-U. Kim, W. A. El-Said, J. Min and J.-W. Choi, Verification of surfactant CHAPS effect using AFM for making biomemory device consisting of recombinant azurin monolayer, Ultramicroscopy, In press.
65. S.-U. Kim, T. Lee, J. H. Lee, A. K. Yagati, J. Min and J.-W. Choi, Ultramicroscopy 109, 974 (2009).
66. A. Messerschmidt, R. Huber, T. Poulos and K. Wieghardt, in Handbook of Metalloproteins (John Wiley and Sons, NY, 2001), p. 1472.
67. A. Dey, F. E. Jenny, M. Adams, E. Babini, Y. Takahashi, K. Fukuyama, K. O. Hodgson, B. Hedman and E. I. Solomon, Science 318, 1464 (2007).
68. B. J. Lee, S. H. Um, J.-W. Choi and K. K. Koo, Colloids Surf. 30, 307 (2003).
69. J. B. Lee, S. H. Um, J.-W. Choi and K. K. Koo, Langmuir 19, 8744 (2003).
70. J. B. Lee, D. J. Kim, J.-W. Choi and K. K. Koo, Mater. Sci. Eng. C. 24, 79 (2004).

71. G. J. Kavarnos, in Fundamentals of Photoinduced Electron Transfer (VCH. NY, 1993), p. 235.
72. H. Kuhn and F. T. Hong, in Molecular Electronics-Biosensors and Biocomputers (Plenum Press, NY, 1993), p. 3.
73. J. Deisenhofer, O. Epp, K. Miki, R. Huber and H. Michel, Nature 318, 618 (1985).
74. C. Gust and T. A. Moore, Science 244, 35 (1989).
75. Y. Lvov, in Protein Architecture: Interfacing Molecular Assemblies and Immobilization Biotechnology (Marcel Dekker, New York, 1999), p. 125.
76. H. G. Choi, B. K. Oh, W. H. Lee and J.-W. Choi, Biotechnol. Bioprocess Eng. 6, 183 (2001).
77. S. Isoda, S. Nishikawa, S. Ueyama, Y. Hanazato, H. Kawakubo and M. Maeda, Thin Solid Films 210, 290 (1992).
78. M. Sakomura, S. Lin, T. A. Moore, A. L. Moore, D. Gust and M. Fujihira, J. Phys. Chem. A. 2218 (2002).
79. J.-W. Choi, S. W. Chung, S. Y. Oh, W. H. Lee and D. S. Shin, Thin Solid Film 106, 671 (1998).
80. M. Fujihira, K. Nichiyama and H. Yamada, Thin Solid Films 132, 77 (1985).
81. J.-W. Choi, Y. S. Nam, W. H. Lee, D. Kim and M. Fujihira, Appl. Phys. Lett. 79, 1570 (2001).
82. J.-W. Choi, Y. S. Nam, S. J. Park, W. H. Lee, D. Kim and M. Fujihira, Biosens. Bioelectron. 16, 819 (2001).
83. X. D. Cui, A. Primak, X. Zarate, J. Tomfohr, O. F. Sankey, A. L. Moore, T. A. Moore, D. Gust, G. Harris and S. M. Lindsay, Science 294, 571 (2001).
84. G. B. Khomutov, L. V. Belovolova, V. V. Khanin, E. S. Soldatov and A. S. Trifonov, Colloids Surfaces A 198, 745 (2002).
85. H. Tajima, S. Ikeda, M. Matsuda, N. Hanasaki, J. W. Oh and H. Akiyama, Solid State Com. 126, 579 (2003).